STATISTICS FOR GEOGRAPHERS AND SOCIAL SCIENTISTS

By the same Author

- Frontiers in Migration Analysis
- Introduction to Rural Settlement (Revised and Enlarged Edition)
- Introductory Methods in Population Analysis
- Land Utilization: Theory and Practice
- Patterns of Regional Geography: An International Perspective
- Recent Trends and Concepts in Geography
- Systems of Rural Settlements in Developing Countries
- Urban Geography: A Textbook
- Water Resource Management
- Wetlands Management in North Bihar
- नगरीय भूगोल की रूपरेखा

About the Author

R.B. Mandal completed his doctoral and post-doctoral research works from Patna University. A specialist in rural settlement geography, he teaches at R.D. & D.J. College, Monghyr and at the Post Graduate Department of Geography, Bhagalpur University. He has also been working on research projects sponsored by the University Grants Commission, New Delhi. He is the author and editor of several books on geography.

STATISTICS FOR GEOGRAPHERS AND SOCIAL SCIENTISTS

R. B. MANDAL

CONCEPT PUBLISHING COMPANY PVT. LTD.
NEW DELHI-110059

ISBN-13 : 978-93-5125-066-1(P.B)

First Published 1982
Reprinted 2014

Published and Printed by :

Concept Since 1974

Concept Publishing Company Pvt. Ltd.
A/15-16, Commercial Block, Mohan Garden
New Delhi-110059 (India)
T: 25351460, 25351794, *F*: 091-11-25357109
E: publishing@conceptpub.com, W: www.conceptpub.com
Editorial Office: H-13, Bali Nagar, New Delhi-110 015, India.

PREFACE

THE USE of statistics in the social sciences, has become common specially with the recent revolution in scientific thinking. The sphere of influence drawn by the statistical measures are wide whether that may be a pure science, mathematical thinking besides sociological, economic, geographical or any scientific investigation in a logical manner.

The present study is directed towards the compilation of methods developed in different branches of geography. There is an idea, how does statistics fit in geography, the answer may be that there may be statistics for spatial interactions and data available from field observations. Statistics seems to be a means of reproducing scientific ideas through better analysis of data because with the help of statistics it is easy to draw inferences about any observations.

The first part of the book is a compendium of papers contributed by scholars of different fields like demography, sociology, statistics, geography and economics.

The papers contributed are mainly directed towards the analysis of applied economics through review of literature, techniques of sampling, measures of dispersion or central tendency, mathematical modelling, testing polarization of growth in spatial context, techniques of computer programming, nearest neighbour analysis, quantitative analysis of geomorphological studies, population statistics and chi-square test.

In the second section of this book most of the topics have been prepared by the author which are concerned with applied statistics in different branches of geography such as physical, regional, land utilization, rural settlement studies, urban, industrial and transport geography. I do not claim, however, that this study is an exhaustive one which encompasses almost all methods and branches of geography due to certain limitations of the author.

I am grateful to Prof. P. Dayal, Head of the Department of Geography, Patna University, Patna for giving valuable suggestions in the completion of this book. I am also indebted to Dr. L.N. Ram, Reader in Geography, Patna University for his fruitful suggestions.

I am thankful to the contributors of this volume who have not only provided their scholarly papers but given me a lot of encouragement in the final outcome of this volume.

I would welcome constructive criticism and scholarly suggestions regarding the methods delineated in this book so that in the next edition more and more things may be included besides the applied aspects of almost all formulas which are presented here.

RAM BAHADUR MANDAL

Department of Geography
R. D. & D. J. College
MONGHYR-8111201

13 March 1981

CONTENTS

SECTION TWO

APPLIED STATISTICS

by R.B. MANDAL

LIST OF CONTRIBUTORS

Emilio Casetti
Department of Geography, Ohio State University, U.S.A.

Ram Bahadur Mandal
Lecturer in Geography, R.D. & D.J. College, Monghyr, India

S.K. Mishra
Department of Architecture and Regional Planning, I.I.T. Kharagpur, West Bengal, India

Ram Naresh Prasad Sharma
Department of Geography, Magadh University, Bodh Gaya, Bihar, India

Shamsher Singh
Department of Architecture and Regional Planning, I.I.T. Kharagpur, West Bengal, India

S.N. Singh
Department of Statistics, Banaras Hindu University, Varanasi, India

O.P. Singh
Lecturer in Geography, H.R. Degree College, Khalilabad, Basti, Uttar Pradesh, India

Devendra Prasad Singh
Department of Geography, Magadh University, Bodh Gaya, Bihar, India

B.L. Sinha
Lecturer in Geography, Rajendra College, Chapra Bihar, India

Joseph Uyanga
Department of Geography, University of Calabar, Calabar, Nigeria

J.R. Yadava
Lecturer, Department of Geography, D.A.V. College, Kanpur University, Kanpur, India

K.N.S. Yadav
Department of Statistics, Banaras Hindu University, Varanasi, India

LIST OF FIGURES

LIST OF TABLES

GLOSSARY OF SYMBOLS

$a<b$	*a* less than *b*.
$a\leqslant b$	*a* less than or equal to *b*.
$a>b$	*a* greater than *b*.
$a\geqslant b$	*a* greater than or equal to *b*.
$a\neq b$	*a* not equal to *b*.
$a\approx b$	*a* is approximately equal to *b*.
$\infty - \infty$	Plus and minus infinity increasing or decreasing without bounds.
$A \cup B$	the Union of the events A and B.
$A \cap B$	the intersection of the events A and B.
$log\ x$	logarithm of *x* to the base 10.
In x	logarithm of *x* to the base *e*.
d_y/d_x	derivative of *y* with respect to *x*.
d_y	differential of *y*.
$\int$	Integral sign.
Δ	Index of change or dissimilarity.
e	Base of natural logarithm or value of 2.71828 ......
π	Ratio of the circumference of a circle to its diameter of value of 3.14159 ......
$\vert a \vert$	the absolute value of *a*.
θ	it is used as Theta (θ)
α	used as Alpha.
β	Beta
δ	Delta
ϵ	Epsilon
η	Eta
λ	Lambda
μ	Mu
Σ or σ	Sigma
ϕ	Phi
Ω	Omega
χ	Chi

INTRODUCTION

THE PROSPEROUS side of geography in view of research and progressive science seems to rest on the development of quantitative techniques. It is said that "when you cannot measure what you are speaking about, when you cannot express it in numbers, your knowledge is of a meagre and unsatisfactory kind, it may be the beginning of knowledge, but you have scarcely in your thought advanced to the stage of a science, whatever the matter may be". Hence, the recent stress on statistics in American geographical publications is the call of the day.

The general feeling is that geography is a science of areal differentiation of the earth surface and their mapping techniques in relation to man. In such a situation where does statistics fit in. The answer may be given that statistics can also be about maps and areally defined data. At this juncture the interests of geographers and statisticians come together. Recently geography has experienced a reorientation in its research practice[1] and this has come to be known as the "quantitative revolution"[2] in geography. In such a situation whereas in the 1950s an article having correlation coefficient was a rarity, by 1980 there seems to be flood of such literature in geography. This revolution has several implications in geographical thought. This led geography from descriptive regional geography to analytical and theoretical systematic geography. There are several contributions to the analysis of areal data by statistics, i.e., the work of M.G. Kendall (1939)[3] who had used principal component analysis on geographical patterns of crop productivity, the work of G.U. Yule and Kendall (1950)[4] who used modifiable areal units in terms of statistics, while P.A.P. Moran (1940)[5] and R.C. Goery (1954)[6] used spatial auto-correlation along with Hart

(1954)[7] who used spatial mean, mode and point of minimum aggregate travel in geographical research. Hence the new geography is referred to as quantitative geography. This includes mainly the following topics in most of the studies and symposia of the world over : classification and regionalization, interpolation routines, Markov chains, computer simulation, trend surface analysis, time budgets, factor analysis and principal component analysis, network analysis, computer graphics, spatial interaction models, spatial auto-correlation, areal sampling, mathematical and theoretical geography, computer aided instruction and spatial analysis. Most statisticians remain unaware of the recent geographical applications of statistics. For geographers the problems of using statistical methods are of two types: one is the peculiar nature of geographical data, particularly the lack of spatial independence; and the second arises in the geographical application of various statistical procedures which are general to all users of such methods, e.g., normality and computational errors.[8]

In comparison with American and European geographers, Indians are lagging behind so far as the use of statistical techniques are concerned. Statistics and mathematics are scientific tools which must be used by geographers in solving their problems and in formulating theories and principles. Statistics helps researchers in putting their results in unambiguous terms. Any neglect of statistics by geographers has surely put them in a very disadvantageous position vis-a-vis other natural and social sciences. Mathematical and statistical tools are the language in which most of the sciences talk today. Because of this fact, geographers have no escape from the quantitative approach.

The nineteen-seventies have seen the dynamic nature of geographical investigation in view of the changing concepts and methods. In British and American universities this new trend in the field of geography has been exemplified by the increasing attention paid to the statistical techniques in degree courses. In almost all branches of geography they reflect changing approach towards the use of quantitative techniques and data. Climatological investigations have traditionally and necessarily been concerned with numerical data. Economic geography has also long ago utilized quantitative data as a prime source of information, although explanatory studies have tended to rest more on subjective judgements than would in many cases seem desirable. Geomorphology, population studies and various other aspects of human geography in an attempt to render a more accurate and objective assessment of the geography of particular areas have also increasingly turned to more precise numerical data. As geography is engaged in practical fields of planning the need to present conclusion in sound quantitative terms becomes imperative.

The contents of quantitative geography might be mainly grouped under the following sub-heads.[9]

(a) Summarising geographical information;
(b) The distribution of geographical data;
(c) Estimating information;
(d) Spatial patterns;
(e) Contrasts and similarities;
(f) Geographic relationship; and
(g) Reflections.

Meaning of Statistics

Statistics is a science of variability. This also includes scientific investigation through counting and estimation of probability. The testing and verification of all scientific laws or theories, requires collection, classification, presentation and interpretation of quantitative data related to any branch of physical and natural sciences like geography, economics, psychology, sociology, geology, physics, chemistry, pedology, medicine, astronomy, astrology to politics and ethics. The presentation of this numerical data enable the human mind to interpret them properly to draw reasonably accurate conclusions based on duly verified methods.

Bowley[10] has very nicely sketched this view point and his words may be reproduced here as "it is better in fact, to define statistics *a posteriori*. It deals with masses of figures, large number of descriptive groups, series of totals or averages. It is found that special methods become necessary—which depends on particular properties of large numbers methods which are suitable for describing complex groups so that they can be easily comprehended, methods for analysing the accuracy of statements, for measuring the significance of differences, for comparing one statement with another. The estimates to which these methods apply are within the scope of statistics".

Scope of Statistics

The scope of statistics is very wide because it can be applied with advantage to any science worth the name as an instrument to aid scientific enquiry based on the observation of quantitative data relating to other sciences or any branch of them.

The development of statistics through time proves that its methods have been applied for investigation in practically every branch of human knowledge and their application has very often been the basis of the discovery of many universal laws in the sphere of the natural sciences. All scientific experiments have been and are carried out on methods resembling to statistical investigation. As in biology Karl Pearson has given the doctrine of evolution and heredity and its statistical basis. The Pearsonian coefficient of correlation is now a well-known statistical formula to ascertain the correlation between the effects of different variables. The aid of statistical methods is similarly indispensable in ascertaining the averages of temperature, moisture of air and the force of winds, etc., in climatology. Statistics is very closely related with population geography, urban geography, land utilization, geomorphology, rural settlement geography, economic, industrial and transport geography and it is also related in propounding and testing economic theories. Geographers, economists and sociologists have to depend largely on statistical data, as all these studies show the relationship between man and society.

So far as the function of statistics is concerned, statistical evidence helps scientists, politicians, geographers, economists or social reformers to argue with confidence on the good or bad side of the proposition in hand, to enlighten his own vision on the subject. In showing the relative importance of things, statistics renders a very useful service to other sciences. The study of problems like land capability, land classification, law of stream numbers, population concentration and migration, industrial location, per capita income of the people are always to be estimated in relation to total population in order to show their relative position in the society. Statistics provides the basis for all such comparative study. The burning problems of modern times like fluctuations in prices, monetary inflation of a country, their adjustment to the cost of living of the people of a country, the problem of unemployment, the question of mass education with a view to suiting national requirements, the problem of trade policy and industrial development with regard to national and international affairs, the position of means of transport and communication in relation to the development of trade and commerce of the country, the relation of the existing constituents of the money market to matters of trade, industrial development and adjustments of prices are all to be studied and verified with the help of statistical methods. Even if the accuracy of statistical data is not determinable, it is always more reasonable and scientific to judge or estimate questions of such vital social, political, spatial or economic importance against some sort of statistical background even if this is somewhat inadequate or barely accurate.

Application of Statistics

A few decades ago the applications of statistics in geography were mainly concerned with the reduction of large amounts of observed data to the point where general trends became apparent. Statistical methods were admirably suited to the study of aggregate behaviour of individuals and the aggregate data fitting consistently with the concept of a population.

In the next step the statistics rose to meet the need for improved analytical tools in geomorphology, urban geography, population geography, agricultural geography and biological sciences. Simple and better analytical tools were needed to improve the process of interpretation and generalization from sample data. As for example, the farms are faced with the task of maintaining a high level of productivity of field crops. Hence, the agronomist conducts an endless number of experiments to determine differences among yields of various crop varieties, effects of various fertilizers and the best methods of cultivation. With the help of the statistical experiments and accurate results, he makes useful recommendations to the farmer. Thus, statistics as a probabilistic method is of great value to the research workers.

Most of the statistical analysis and probability theory in geography[11] have emphasized the testing of hypotheses concerning the nature of location patterns, the extent to which different phenomena are interrelated and associated over the surface of the earth, the importance and validity of certain regional constructs, and the significance of selected factors in differentiating the regions.

So far as the use of statistics is concerned, it is a phase of development into which geography is now entering and the "frontier" nature of geographic research is reflected in the work currently being undertaken on population migration, cultural diffusion, drainage basin evolution, law of stream numbers, climatic changes, urban systems and market sharing problems in widely separated areas. "At present these geographic models are crudely fashioned but even in this form they are suggestive of fruitful approaches for future geographic research".

First of all in 1866, the father of settlement geography Meitzen recognized the use of modern statistics in geography.[12] It has developed from two different roots.[13] In the seventeenth century the term "political arithmetic" had been devised by Graunt and Petty. This shows that the idea of social problems could be subjected to quantitative analysis and that a causal relationship existed between quantitative parameters.

In the series of German political arithmetic (economics), the location theory was completed, supplementing Von Thunen's work (1826) on agricultural location with the twentieth century masterpieces the works of Weber on industrial location (1909), Christaller (1933) and Lösch (1939) on cities and the tertiary sector. It is from these theories that the first agreement as to significant research questions emerged in quantitative geography.

Later on in 1939, Kendall used the principal component analysis to develop a multivariate index that could portray the geographical distribution of crop productivity in England. Similarly in 1443, Hagood used multiple-factor analysis to define multivariable uniform regions. In 1954, Geary applied his measures of contiguity to evaluate the lack of independence of residuals from regression in studies of spatial association. Auto-correlation is found in the works of B. Matern "spatial variation" a major contribution to the literature of sampling. The ideas of point pattern distribution and spatial diffusion processes came from quantitative plant ecology and epidemiology. Recently Roberto Bachi has revived the idea of centrography with the development of a variety of measures of dispersion and association of spatial distributions. The invention of "social physics"[14] like gravity potential models, mean centres of areal distribution, nearest-neighbour analysis, standard distance method etc., have contributed to the quantitative revolution in geography. The advent of the modern computer, together with the use of factor analysis to reduce many variables to a few factors representing "complexes" of associated characteristics, and numerical taxonomy to derive the optimal classification of observations into regions, has provided a new and objective solution in the field of quantitative revolution in geography.

Limitations and Sources of Error in Statistics

It is necessary to know the limitations of statistical methods and thereby to equip the mind of the investigator with the necessary safeguards in using statistical data as without such an equipment the investigator is likely to fall in confusion and disarray and may very likely prove wrong conclusions having no proper bearing on the problem. It is often said that "anything can be proved by statistics". This may lead to unscrupulous results, by all possible means and reference to figures which are appealing to the layman and not to the general public as well. This is because they cannot have the means to prove and verify the truth of statistical evidence. This process of proving his own wrong or biased conclusions can also sometimes be employed by an investigator who has no clear grasp of the handicaps that are likely to occur in the collection, classification or presentation of statistical data. This is because they may not be aware of the principle that statistics collected with some particular object and the same cannot always be used with equal force in regard to other enquiries, even though similar figures were required there too. Due to change in the circumstances of collection of data the investigator cannot throw the same light on other matters existing under a different set of circumstances, were necessary to draw conclusions in this investigation also. The reader, on the other hand, may be unaware of the extent of inaccuracy of estimates or calculations made by those engaged to collect and classify statistical data and his arguments or inferences may, therefore, become apparently ridiculous or untrustworthy.

Thus we should not hesitate to admit that there are certain limitations of statistical methods, viz., that they can be used without proper reference to context, the degree of inaccuracy of the available data may not be pre-determined which may lead to uncertain conclusions. The data may be used without knowledge of circumstances under which and the purpose with which they were collected and as such, if taken for granted in whatever form they are presented, would unavoidably lead to incorrect and perhaps irrelevant conclusions. The use of incorrect data, may either lead to the interpreter himself being misguided in drawing his own conclusions or to his giving misleading results without letting the public know its defects. In case of dealing with large quantities of comparative data the statistician might himself become biased, without the knowledge of the mental barrier himself, and under this condition the interpretation may be biased and not capable of giving an impartial judgement.

At the end it can be said that limitations of any scientific laws must always be safeguarded with a clear outlook on them but they can never claim to make the science itself useless. Thus statistics is not the science whose function is to prove anything, true or false, good or bad. It is due to a lack of knowledge of the true nature of statistical methods and their limitations, the bias of the interpreter himself and quite often the zeal of the unscrupulous interpreter that such a state of affairs is created.

The Purpose of the Study

In geography the use of statistics rarely if ever serves to master the subject but as a technique to simplify geographical data and problems in order to have some theoretical generalization and principle formulation. The purpose of the present volume is to highlight three aspects of statistics, i.e., firstly, general statistics in the simplest possible manner; secondly, statistics used in different branches of geography such as geomorphology and physical geography, land utilization, rural settlement geography, urban geography, transport geography; population studies along with migration and regional geography; and thirdly, the presentation of scattered nature of edited papers contributed by different scholars both in the field of geography and statistics.

The basic idea behind presenting this book is to review the statistical applications in different branches of geography. This review is undertaken in the hope that the geography student, by familiarizing himself with the strength and weaknesses of these applications, may become more critical in his own use of the techniques and, in so doing, contribute to a more rigorous pursuit of geographic inquiry. The division of the material into different sections reflects the major thrusts of quantitative research in geography but the author does not claim that the list given is exhaustive.

Review of Collected Papers

In the introduction the author traces out the development of quantitative geography through review of literature. Besides this, the meaning and scope, application and limitation of statistics, sources of error and purpose of study have been delineated.

In the second paper on techniques, need, advantage and sources of error and different types of sampling on spatial level have been analyzed besides highlighting a few advantages and disadvantages.

Dispersion is the measure of the scatter in a distribution. The present paper deals with various measures to find out the dispersion of a given distribution. Suitable examples have been given to illustrate different measures like range, mean deviation, standard deviation, Gini coefficient, Lorenz curve, etc.

Special care has been taken to clearly understand the fields of application of these measures in different types of problems, and their limitations and caution in deriving conclusions on the basis of these parameters. With the help of examples derived from actual data relating to the field of spatial, regional and other social sciences, it has been attempted to study the finer differences in the applicability of these measures.

Statistical methodologies like the measures of central tendency reduces a haphazard mass of quantitative data into a single typical figure to enable one to draw a general conclusion about the characteristics of the phenomenon under study. In this chapter different measures of central tendency like averages (arithmetic, geometric, harmonic, progressive and moving), median and mode will be covered highlighting their applications under various situations for the data obtained from geographical and other social surveys. Calculation methodologies for these measures for ungrouped and grouped data will be described with suitable examples.

The purpose of this chapter is not to enter into the complicated mathematics of statistics rather to treat its more elementary and more frequently used phases in these disciplines. It is hoped that this chapter on the measures of central tendency will be of great use for scholars engaged in the field of geography and other allied sciences.

Casetti while delineating "Mathematical Modelling and the Expansion Method" outlined a routine for creating a sequence of clearly identified logical steps, and called it "expansion method".

This method is a technique for generating models. Basically, it consists of the following:

(a) An "initial" model in which some of all the parameters are in non-numerical form is selected.

(b) Some of the later parameters are "expanded" by redefining them as functions of variables that may or may not appear in the initial model.

(c) A "terminal" model is generated by replacing the expanded parameters into the initial model.

Further, it deals with testing a functional relation for stability, agricultural productivity using the expansion method to create custom made functions in order for generating complex mathematical models suited for specific research objectives.

In testing formal polarization of growth in the spatial context Uyanga determined space as a field of forces which in most cases is homogeneous in aggregate. This paper gives an insight on how the existing urban system within a nation can be rationally organized and guided to serve enhanced developmental ends. Polarized regions are those in which connections and flow of goods, services, labour and socio-economic strength are predominantly in one direction. Here sectoral polarization is also tested in the south Australian context in addition to the polarization of the total non-agricultural employment.

The relevance of Computer Programming for analytical studies in the social sciences is increasing very fast. The present paper is an introductory essay on the Techniques of Computer Programming. Computer Programmes may be broadly divided into four classes, namely, statistical techniques, optimization or operations research, mathematical techniques and mapping and other control techniques. Here we will discuss and demonstrate statistical techniques only.

Beginning with the introduction to computer language Fortran IV, the paper provides guidelines on which main programmes for input-output control and the use of sub-routines can be learnt. To enrich the inventory of techniques, some important programmes like those of correlation, regression, central tendency, dispersion, principal component analysis, etc., have been appended.

The paper on nearest-neighbour technique highlights the assessment of sparse, dispersed nucleated settlements with the help of uniform, random and clustered patterns from spatial statistics point of view.

The paper on quantitative methods of geomorphology highlights a comprehensive analysis of quantitative methods like arithmetical, graphical, cartographic and statistical methods of landform analysis.

To the extent that it affects population changes, the pattern of sectorial distribution of workforce is posed as a dimension in regional development dynamics. In south Australia this condition is reflected in the nature of "balance" in regional workforce and industrial distribution. The problem is studied in three contexts: the regional, the urban system and the urban centre. From the threshold point of view, the findings tend to support the hypothesis, in terms of employment, re-investment, population shifts, monetary outflow, and the multiplier effects.

The paper on population statistics reveals that the traditional subjectivity of population geography has been replaced by a more precise and objective method of measurement of data. In this way, the growth of population, vital statistics, fertility (variability, and future estimate of population), physiological density, age and sex structure, economic activities, literacy, rural urban ratio, migration analysis and population resource regions have been analyzed here.

The main application of chi-square (χ^2) technique is in the testing of results for statistical significance, and it may also be used for the analysis of point patterns. We may adopt it in the comparison between areas of equal size and unequal size also. In this method we estimate the probability that the differences we actually observed in the areas are due to chance. If the probability is very low, then we may with confidence seek an explanation. If, however, our estimate indicates a considerable probability, there is no justification in looking for causes. Thus, in this method (χ^2) we compare observed results with these to be expected theoretically on some formulated hypotheses. The difference between two statistics is tested against an 'equal probability hypothesis'—a variety of 'null hypothesis'. It is termed 'null' because it states that two samples from same part of the population, and there is a high probability that the observed differences are due to chance variations. A 'null hypothesis' is more useful than other hypotheses because it is exact. The equation for chi-square (χ^2) is stated as follows: χ^2 in which f_o=frequency of observed or experimentally determined facts; f_e = frequency of expected occurrence on some hypothesis. The difference between observed and expected frequencies are χ squared and divided by the expected number in each case, and the sum of these quotients is χ^2. The more closely observed results approximate to the expected. The smaller the chi-square the greater the probability of a real divergence of experimentally observed from expected results. To evaluate we enter chi-square table with the computed value of chi-square and the appropriate number of degree of freedom $(d_f) = (r-1)\ (c-1)$ in which r is the number of rows and c is the number of columns in which the data are tabulated.

The probability is frequently referred to in a scale from 0 (which never occurs) to 1.0 (that certainly occurs), and 95 per cent confidence is often called the .05 level of significance, meaning that the particular result has only 95 per cent chance of occurring through random variations. Similarly, 99 per cent chance of occurring is referred to as the .01 level of significance. The validity and reliability of this technique may be tested by using the t-test which may be computed as $t = \frac{M_1 + M_2}{2}$ in which M_1 and M_2 = mean sample number 1 and 2 and d is the standard error of difference (say).

In testing hypotheses usually two types of wrong inferences are drawn and they are termed as Type I error and Type II error. Type I errors are made when we reject a null hypothesis by marking a non-significant difference significant, although no true difference exists. If we set a low level of significance P greater than .05 level we increase the likelihood of this error, whereas, if we set up a high level of significance (P less than .05 level), we render such erroneous inferences less likely. Type II errors are made when we accept a null hypothesis by making a significant difference insignificant, although true differences actually exist. The possibility of drawing erroneous inferences is enhanced when we set a very high level of significance.

This method has been tested in the "Synchronization of Rural Periodic Markets in Azamgarh Area" with 158 periodic market centres spread over in an area of 5,744 sq. km. of size subdivision providing socio-economic services to 28,57,484 inhabitants. The three formulated hypotheses are: (a) that the number of market meetings depend upon wide extension of the areas, (b) that the number of of market meetings depend on seven days of a week, and (c) that the number of market meetings depend upon both areas and week days. In this case of first-two hypotheses, an "equal probability hypothesis" a variety of "null hypothesis" have been tested. The first "null hypothesis" has been rejected at .01 level, while the second hypothesis has been accepted at .05 level and the third one "independence hypothesis" a variety of "null hypothesis" have also been accepted. Further the validity and reliability of the chi-square analysis has been tested by the use of *t*-test to find out "significant differences" between six divisions area and in the case of second hypothesis an analysis of insignificant result has also been tested. Thus, consequently it may be asserted that different number of market meetings on different days is only by chance.

In the second section of this book almost all the papers are collected by the author in which discussions have been made regarding the use of statistics in geography which are quite different from general statistics. Owing to the vastness of the subject the author has tried to represent only some of the statistical methods as adopted from the original literature.

REFERENCES

1. P.R. Gould, "Methodological Developments Since the Fifties", *Progress in Geography*, 1969, 1, pp. 3-49.
2. I. Burton, "The Quantitative Revolution and Theoretical Geography", *Canadian Geographer*, 1963, 7, pp. 151-62.
3. M.G. Kendall, "The Geographical Distribution of Crop Productivity in England", *Journal of Royal Statistical Society*, 1939, 102, pp. 21-48.
4. G.U. Yule and M.G. Kendall, *An Introduction to the Theory of Statistics*, 1950.
5. P.A.P. Moran, "The Interpretation of Statistical Maps", *Journal of Royal Statistical Society* (B), 1948, 10, pp. 243-51.
6. R.C. Goery, "The Contiguity Ratio and Statistical Mapping", *The Incorporated Statistician*, 1954, 5, pp. 115-45.
7. J.F. Hart, "Central Tendency in Areal Distributions", *Economic Geography*, Vol. 30, 1954, pp. 48.
8. P.J. Taylor and John Goddard, "Geography and Statistics: An Introduction", *The Statistician*, 1974, Vol. 23, No. 314, pp. 149-55.
9. W.H. Theakstone and C. Harrison, *The Analysis of Geographical Data*, Heinemann Educational Books, London, 1971.
10. A.C. Sinha, *Elements of Statistics*, S. Chand & Co., Delhi, 1952, p. 2.

11. L.J. King, *Statistical Analysis in Geography*, Prentice-Hall, Inc., New Jersey, 1969, p. 5.
12. A. Meitzen, *Siedlung und Agrarwesen der Westgermanen und Ostgermanen*, der Kelten, Romeru, Finnen und Slawen, Berlin, 1895.
13. A. Meitzen, *History, Theory and Technique of Statistics*, Philadelphia, American Academy of Political and Social Science, 1891.
14. B.J.L. Berry and D.F. Marble, *Spatial Analysis: A Reader in Statistical Geography*, Prentice-Hall, Inc., New Jersey, 1968, p. 3.

SECTION ONE

ONE

R.B. Mandal

TECHNIQUES OF SAMPLING

THIS CHAPTER deals with statistical inference and sampling in the spatial context as they have appeared in the geographic literature. The examples are few as the nature of geographers' interests who infrequently deal with samples. Moreover, geographers are mostly concerned with the total set of observations, and inferential statements are not always meaningful in this context. Hence some sample designs have been given here for its proper use.

Sample or sampling is a part of a whole, more or less representative field of observation. It should be drawn cautiously on the plan paper or in the field because a faulty sample design leads to a fallacious conclusion.

There is nothing theoretical about the idea of samples: samples of physical objects, of people, rural settlements, processes, industries or of institutions. To choose anything from any table, map, population, field or industries is not a new thing but it is a common aspect of life. In statistics such picking and choosing requires a scientific knowledge so that our observation may not be biased but it has some logical proof regarding which sampling is suitable for the study of a particular type of distribution. Thus an essential requirement in statistical sampling is the selection of sample units which are representative of the whole set of population. Thus, sampling has two properties: (a) representativeness, and (b) small and adequate selection of items.

Need for Sampling

The investigator takes the help of sampling as it reduces the unmanageable heterogeneous items of observation to a handy sample so that all the types are adequately represented. Thus the sample helps in selecting a few representative items from the aggregate or total field of study. The results of sampling are reliable as it is mathematically proved that the sample mean is the best estimate of the total population mean. It is also useful for measures of location and dispersion.

So far as the beginning of sample survey is concerned in 1886, a poverty survey was done by Charles Booth, but sampling was used in 1912 by Bowleg. Similarly in 1924, an employment survey was carried out, and in 1937-38 a household expenditure survey was carried out on the basis of random sampling design.

ADVANTAGES OF SAMPLING SURVEY OVER CENSUS SURVEY

(i) Saving of time,
(ii) Saving of money,
(iii) Detailed study,
(iv) Accuracy in results,
(v) Administrative convenience, and
(vi) Impracticability of the use of census survey.

Sources of Error

The sample has many advantages over a complete survey or complete enumeration of a finite items of observation. If carefully designed, the sample is not only considerably cheaper but may give results which are more accurate than those of a complete land use survey. The reason is that in survey different ways and efficiency of surveyor is subject to partial error, sometimes considerably larger due to faulty sampling method employed. The smaller sample makes possible the use of more care in the design and execution of each step in the enquiry. In this way, sources of error can be investigated, reduced, eliminated, measured and corrected for the accurate result. Hence, a carefully designed sample may actually be better than a poorly planned and executed land use survey. For highly variable characteristics of results, a sample may have to be so large in order to give the desired results that it is better to use a complete survey or census instead of a sample study. Finally in some problems, such as in crops yield rate measurement, and total requirement of input of fertilizers only a sample can be used.

The sample is only a part of the whole observation. Hence the exact agreement is not possible between sample estimates and actual proportion of land in various uses. The use of any sample in estimates must ensure two properties, i.e., accuracy and precision.

Accuracy relates with correctness in estimation and if there is consistently over and under estimation of the value then the sample is said to be biased. Any sample procedure should produce estimates that differ from true value only by some random disturbance factor.

Precision means the spread of sample estimates of the land use value around the actual value of land. An efficient sample has a relatively small spread of such errors of estimation. Accuracy is the first requirement of any sampling besides accurate procedures and relative efficiency, estimation and rejection of one method. Only in the case of change through time, the biased procedure gives greater efficiency, as in the temporal comparisons the bias may be eliminated.

Disadvantage of Sampling Survey

1. Chances of bias,
2. Difficulties in getting a representative sample,
3. Need for specialized knowledge, and
4. Impossibility of sampling.

Types of Sampling

Broadly, sampling can be classified as:

Subjective or Purposive Selection

It is made purely on the basis of the convenience of the investigator, ranging between a subjective and arbitrary selection of items to "over-emphasize extreme variants rather than major attributes".[1] Here the personal element dominates and predetermined choice of the investigator comes into effect. If the investigator has to select 20 per cent sample settlements from a lot for studying their characteristics and generalising over all the settlements, he may for some reason or the other (nearness, approachability, etc.), have use for some of them and may even try to include them in the samples, whether they represent the group or not.

The problem with the purposive sample is that there is no absolute idea of **what** chance each individual item in the population has of being selected, all one can say is that this chance is not the same for every item. The use of such a biased sample might result, for example, from sampling rural settlements in a given region simply on the basis of being most accessible from a main road **running** through the area. One would accept a smaller variation in spacing than might be obtained from more widespread sampling, but would be unsure of the relationship between these two variances, as also between the mean values of rural settlements spacing with respect to the sample and the region as a whole. It might be expected that the average spacing of sample would be different from that of the total population, but whether it is less on account of the road following the plain land or whether it is greater on account of the road running from marshy low-lying flood plain, one could not be sure.

In reality purposive selection is segmental in the sense that the selection of elements is restricted to a certain segment of the total statistical population.

Simple Random Sampling

Random sampling is drawn haphazardly. Each and every item or individual has an equal chance to be selected. Hence, it is also known as probability sampling. Samples are done by chit or card methods or with the help of random number table. All the items are placed on chits and then required number of chits are taken out. It can be performed with replacement method or without replacement method (Fig. 1.1).

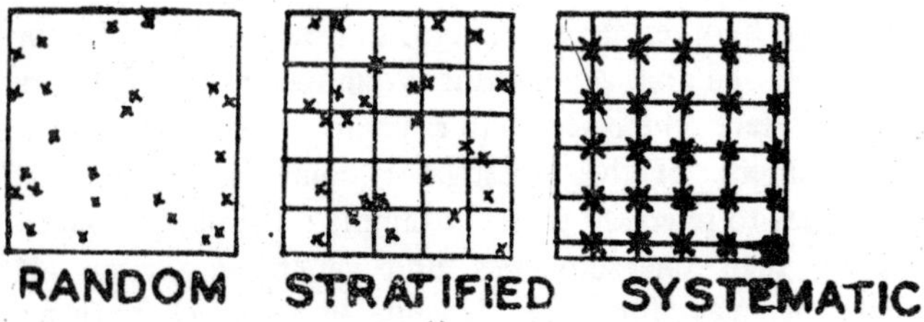

FIG. 1.1: **Stratified Random Sampling**

This is the more reliable method of studying the group characteristics by means of sample to pick up samples from the whole group at random, e.g., selecting every tenth family in a village or every 100th or 75th village from the state. Random sampling, unlike the deliberately chosen example, relates to selecting examples in any pre-determined order leaving it to chance to pick up any instance that come for measurement while proceeding in a pre-determined order. In this selection by chance, the question of personal choice of investigator does not arise and the study of a large number of instances picked up at random from the whole group or class from which the instances have been picked up.

Thus the design of random sampling can be applied to fields, farms or villages which are listed and each is assigned a number. It can also be drawn from random number Table. It is useful in the finite group of discrete units, but the geographer frequently wishes to obtain a sample of a spatially continuous variable, such as height of the land, slope, angle and soil characteristics. In this case a grid is superimposed on a map of the particular phenomenon and a random series of grid intersections selected from the table of random numbers which prevent biased selection.

The random sample method may be inappropriate for the investigation of spatial variations of some phenomena, since it can lead to a biased sample if there is marked spatial clustering of particular categories of features which are being studied. Unfortunately, geographical landscape elements are rarely scattered at random and consequently this method must be used cautiously.

Stratified Random Sampling

The geographical field is rarely homogeneous. Random sampling may lead to unreliable inferences because from one type of field there can occur more units, while lesser from the others, and some classes may remain unrepresented. In order to arrive, at more correct inferences (result), it is advisable to divide the heterogeneous universe into homogeneous classes, called strata, based on certain grounds, and then draw the samples randomly. Thus, the random sampling can be grouped into stratified random sampling.

In geographical investigation it is the stratified random sampling which has proved to be more helpful than the other sampling methods. For example, in an investigation regarding the assessment of per hectare yield of a certain crop, widely grown in a geographical unit of complex nature, it is advisable to take samples from all types of lands within the area under study. The method of stratified random sampling in this particular case will be to divide the area into physiographic units, e.g., uplands, rolling ground plains, lowlands, etc. They will represent the various strata. Within these groups, samples should be taken or drawn from areas representing all possible combinations of the variable factors affecting the yield, e.g., (soil composition and depth), water-balance, etc. This leads to substratification of the area under question (Fig. 1.1).

Random Serial Sampling

This sampling method facilitates the collection of data with some pre-determined plan at arbitrary, but usually equal interval spacing. This method is ideal for linear sampling, i.e., pebble size along a narrow stream or maximum angles of slope along a valley side. Strahler employed this plan in sampling maximum valley side slope angles in the Verdugo Hills, suggesting that the interval spacing should be 5-10 feet in fine-textured badlands and 100 to 150 feet in medium-textured topography. The question of how far apart individual samples must be taken on a continuous surface to be considered as different readings obviously varies with the magnitude of the phenomena measured, and in this instance each sample was a complete orthogonal profile from which the maximum angle was measured.

Systematic Grid Sampling

In this method, the sample is taken at regular intervals, using a map, e.g., at every grid intersection. It is laid in a random manner over the area to be sampled, such that the number of grid squares or intersections is equal to the size of the sample required. It means by listing the features and sampling at regular intervals, e.g., every 10th item in the list.

It is based on the grid method. The soil scientist and geographers have often made use of this method in reconnaissance or grid surveys. The samples are drawn at a systematic interval of 5 or 10 from the grid maps, while selecting the villages for taking soil samples, the area is first divided into small grids with the help of latitudes and longitudes. It is then superimposed over the settlement map of the area and the prints are selected accordingly.

The advantages of this method over that of the random co-ordinates are that the sampling prints are obviously less tedious to locate, and the sampling coverage is more uniform, but there are also a number of disadvantages. In some instances a regular pattern of sampling may not give as representative a sample as a truly random pattern. Another disadvantage is that there is a limit imposed by the grid as to the size of the sample which can be selected. In this way, a completely unrepresentative sample will result if there is a similar regular pattern to the spatial distribution of the phenomena to be measured, which either coincides exactly with the grid pattern or deviates entirely from it. In this method, the size of the grid mesh is largely governed by experience and the magnitude of individual phenomena.

Grids are not always rectangular, however, nor is there spacing always arithmetical. Figure 1.2 illustrates "a radial logarithmic grid

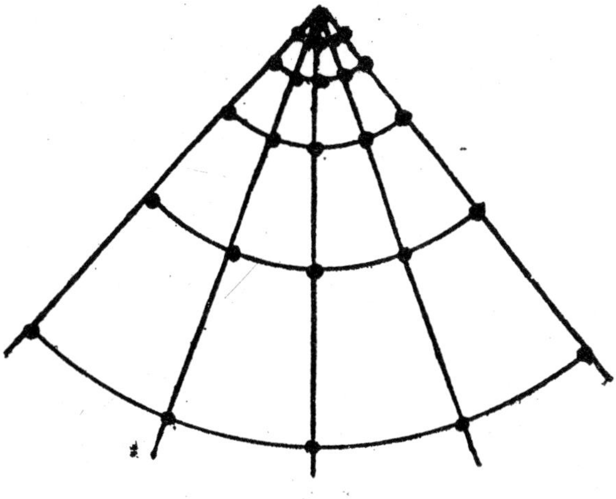

FIG. 1.2: **Radial Sampling**

for sampling the surface slope or pebble size of an alluvial fan where distance from the fan head is important, and pebble characteristics are believed to vary logarithmically with distance".[2]

In geomorphology two-factor and three-factor basic forms of systematic grid methods are useful. The single-factor basis form has been taken under simple stratified sampling. A two-factor form is designed to study the variation of a particular attribute in two directions. Figure 1.2 illustrates such a grid for sampling conditions within a stream channel oriented to give random sampling locations normal and parallel to the direction of flow in order to compare such local variations as those due to the sorting action of cross fluctuations of the currents with regional variations in a down stream direction. Figure 1.2 shows the two-factor form of stratified sampling of the moisture content along and across a beach, the whole being divided into two possible populations born which equally divides the grid.

The three-factor basic form is useful as it combines with time sampling such that a grid may be placed to allow sampling in two significant directions, and such that samples can be taken at different times from the same location. An example of this would be a grid placed over a rapidly-eroding badland valley at the intersections of which height measurements are taken at different times to separate the erosion rates parallel to the valley axis.

Controlled Sampling or Quota Sampling

Since the purposive selection may be subject to considerable bias, an attempt is often made by means of some type of control to reduce this bias. This type of sampling is often referred to as controlled sampling. One example of controlled sampling is the quotes used in opinion poll work. Interviews give the number of persons with various characteristics who must be interviewed in a given locality: so many must be urban dwellers and so many rural and farm dwellers, so many must be male and so many female; so many must have high incomes and so many medium and low incomes, etc. These quotes are calculated before-hand from census and election statistics.

Single-Double or Sequential Sampling

These methods are not only advantageous to take more than one random sample from the same population, but it is absolutely essential in some cases. In sampling inspection taking of one sample is called single sampling and taking of two samples on which to base a decision is called double sampling. If the sample size is not pre-determined but the number of random samples taken is continued until a decision is reached, then this type of sampling is called sequential sampling.

The repeated design of random samples must be taken in order to obtain a valid estimate of experimental error. In an agricultural experiment, for example, every variety to be tested is planted in each of several blocks in order to provide for an uncontrolled random variation which arises from the randomized position of varieties within blocks. In the field of experimental design the use of repeated random samples in the same area at the same time is called replication.

Nested Sampling: This hierarchical or multi-level technique is commonly applied when a large area is to be investigated, within which significant local variations are also thought to occur. To combine those sampling with a large area of coverage this method employed is to divide the total area into major, minor, sub-minor, etc., units, and then select in some random manner first a given number of major upper level units, then a number of minor units from within there, and so on until finally a number if random locations are selected from the smallest areal sampling units. Several of these sub-regions are chosen at random and broken down again to smaller units and this process being continued until the smallest meaningful unit is reached or data cease to be available.[3] Olsen and Potter have applied this nested sampling method in field investigation of sedimentary petrology.[4] It can also be applied to each and every branch of geography with the same effect. This is considered as the best method of sampling in comparison to all previous methods.[5] (Fig. 1.3)

Sources of Error in Sampling

The sample has many advantages over a census or complete enumeration of a finite population. If carefully designed, the sample is not only considerably cheaper but may give results which are just as accurate and sometimes more accurate than those of a census. The reason is that a census is subject to partial error, sometimes considerably larger due to any sampling method employed. The smallness of the sample makes possible the use of more care in the design and execution of each step in the enquiry. In this way, sources of error can be investigated and reduced, eliminated, measured and corrected. Hence a carefully designed sample may actually be better than a poorly planned and executed census. For rare groups and highly variable characteristics, a sample may have to be so large in order to give the desired results that it is better to use a census instead of a sample study. Finally in some problems, such as in destructive tests of the quality of a manufactured article for example ammunition only a sample can be used.

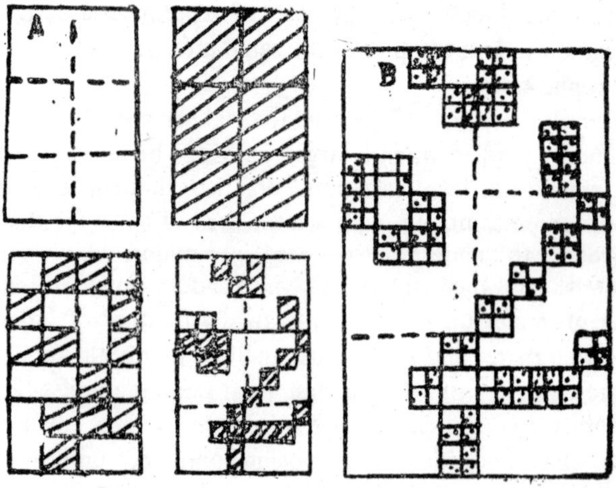

FIG. 1.3: **Five Stage Nested Sampling Procedure for Extracting Scale Components—Tagu Sado Basin, Central Portugal**

The sample is only a part of the whole observation. Exact agreement is not possible between sample estimates and the true population values, proportion of land in various uses, types of rural settlements or industries. The use of any sample estimates must mainly ensure accuracy and precision.

Accuracy relates to correctness in estimation and if there is consistently over and under estimation of the value then the sample is said to be biased. Any sample procedure should produce estimates that differ from the true value only by some random disturbance factor introduced by the fact that a sample is being taken in the first place.

Precision means the spread of sample estimates of the population value around the true value. An efficient sample has a relatively small spread of such errors of estimation. Accuracy is the first requirement of any sampling. Accurate procedures, relative efficiency of estimation errors provide a clue for selection or rejection of one method. In the case of change through time the biased procedure gives greater efficiency; in temporal comparisons the bias may be eliminated.

REFERENCES

1. W.C. Krumbein, "Some Problems in Applying Statistics to Geology", *Applied Statistics*, 1960b, 9, pp. 82-91.
2. G.H. Dury, *Essays in Geomorphology*, Heinemann, London, 1966, pp. 292-305.
3. R.J. Chorley and P. Haggett, *Frontiers in Geographical Teaching*, Methuen & Co. Ltd, London, 1970, pp. 175-77.
4. B.J.L. Berry and A.M. Baker, "Geographic Sampling", *Spatial Analysis: A Reader in Statistical Geography*", Ed. by B.J.L. Berry and D.F. Marble, Prentice-Hall, New Jersey, 1968, pp. 91-100.
5. L.J. King, *Statistical Analysis in Geography*, Prentice-Hall, New Jersey, 1969, pp. 60-68.

TWO

Sham Sher Singh and *S.K. Mishra*

MEASURES OF DISPERSION

MEASURES of central tendency like mean, mode and median provide us with a particular value which concisely characterise a distribution, but this value can be misleading as it might be affected by the extreme values in the distribution. So, unless the measures of central tendency are coupled with description of the shape of the distribution like whether the data is widely scattered or clustered together, they tell very little about the distribution. This scattering or clustering of the values around some measure of central tendency, generally mean, is called the "dispersion" or variation of the distribution. For example, in a class of 15 students, the average marks obtained in a particular subject is 55 but some students are getting more and some are getting less than the average; there is also one minima and one maxima. The simple values of average does not give us the idea as to how the marks are distributed among all the students, or how the marks deviate from the central value, i.e., what is the dispersion or variation in the distribution of marks.

A number of techniques have been developed to know the dispersion in a given distribution. Major ones are the range, the mean deviation, the standard deviation, the Lorenz curve and the Index of qualitative variation, which are very often used by geographers and other social scientists. There are other methods also like Gini coefficient, Garry's contiguity ratio, non-Neumen's ratio and Mahalanobis' 'D' which have not been dealt with here because of two reasons : (i) these are very frequently used in econometrics and less known to other social scientists who are not well equipped to infer from them properly, and (ii) they need understanding of the fundamental methods as pre-requisites.

Therefore, only range, mean deviation and standard deviations and their uses in different forms have been explained.

The Range

Range is the simplest measure of dispersion which is nothing but the difference between the maximum value and the minimum value of a distribution. The highest temperature at Karnal in Haryana was 44.0°C in the month of June and the lowest 0.4°C in the month of January in the year 1975. In this case the annual range of temperature at this particular station is 43.6°C. Likewise, the range can be used in describing the age difference in a group of persons, the difference of the minimum and maximum marks obtained in a class, and the difference in the distribution of income, etc. But the range is a poor measure of dispersion because it tells us only the top, bottom and the breadth of the distribution ignoring how the data are distributed within the range. And, also the range is the measure of absolute dispersion which cannot be compared, e.g., range of minimum and maximum temperature is not comparable with the range between the lowest and the highest value. So, the need for measuring relative dispersion arises which can be obtained by dividing the absolute dispersion by some other value regarded as standard for the purpose. The relative dispersion of a distribution is called the co-efficient or ratio of dispersion. From the range, as a measure of absolute dispersion it is obtained by two ways.

(i) Relative dispersion = Range/mean ...... (1)

(ii) Relative dispersion = Range/sum of extreme values ...... (2)

Interquartile Range

A slightly improved form of the simple range is to divide the distribution into certain segments and then to know what proportion of the cases fall above or below a certain point. The commonly used procedure is to divide the distribution into quarters or segments known as quartiles. The first quartile, Q_1, is the point indicating 25 per cent of the cases below it and 75 per cent above it. The second quartile, Q_2, is the median which is the 50th percentile.* The third quartile, Q_3, is the point with 75 per cent of the cases below it and 25 per cent above it. The 'Interquartile Range' is the distance between the 25th and the 75th percentile. This is generally indicated by the letter Q and is defined by the equation:

$$Q=Q_3-Q_1 \qquad \ldots\ldots(3)$$

Thus, the interquartile range gives us the idea of the range for the middle 50 per cent of the distribution.

TABLE 2.1: **Examination Scores in Rank Order**

Sl. No.	Marks	Sl. No.	Marks
1	75	11	48
2	72	12	47
3	71	13	45
4	68	14	41
5	65	15	39
6	64	16	36
7	59	17	33
8	55	18	30
9	52	19	29
10	50	20	25

The formula for getting the quartiles is

$$Q_1=\frac{(N+1)}{4} \qquad \ldots\ldots(4)$$

where N is the number of cases. For getting Q_2 and Q_3 the value of Q_1 is simply to be multiplied by 2 and 3.

* The term percentile is used here because all the cases in a distribution are taken equal to 100 and then on the basis of the percentages segments are decided.

For the scores in Table 2.1, the $N=20$ and the Q_1 falls between 36 and 39 (the 15th and 16th cases or 5th and 6th cases from the lowest rank in the Table 2.1). Thus, $Q_1=37.50$; Q_2 (the median) = 49.00; and the $Q_3=64.50$. The interquartile range for this distribution is:

$$64.50-37.50=27.00$$

which gives an idea that the middle 50 per cent of the cases have range of 27 marks spread between 37.50 to 64.50. This interquartile range has avoided the effect of extreme scores which fall at the end point of the distribution.

Decile and Other Ranges

In principle the distribution can be divided into segments of any convenient size and the ranges can be constructed. One such range is the 'Decile Range', the distance between the 10th and the 90th percentile. Any percentile can be calculated by dividing $N+1$ by 10 and multiplying it by any number from 1 to 10 as the case may be. For example, in Table 2.1 both percentiles are falling between 2nd and 3rd cases from lower rank order and the value is 29.5. Similarly the value of 90th percentile is 71.5 and the Decile Range is 71.5—29.5=42. Actually lower 10 per cent is the first decile and 90 per cent is the 9th decile. This measure excludes lowest 10 per cent and the highest 10 per cent, considering central 80 per cent only which gives two values.

The major difference between the quartile and decile range is that latter covers more percentage of cases in the distribution. The major shortcoming of both the ranges is that they do not make use of the values of all the cases.

Quartile Deviation

Another better measure of dispersion than the range is "Quartile Deviation" or "Semi-interquartile Range". This is calculated by taking half of the inter-quartile range. Symbolically, it is

$$Q.D. = \frac{Q_3 - Q_1}{2} \qquad \text{......(5)}$$

The use of quartile deviation is made because of the reason that half of the interquartile range will, in a normal distribution, be equal to the difference between the meridian and any quartile.

It means that 50 per cent of cases will lie within the interval defined by the semi-interquartile range $\pm$ median.

In the foregoing example of examination scores the Q.D. is 64.50—37.50/2=13.50, which is the width of the interval in which 50 per cent of the cases, $\pm$ median, fall. For getting the coefficient of quartile deviation, i.e., the relative deviation, the absolute measure is divided by the average size of the two quartiles. Symbolically:

$$\frac{Q_3 - Q_1/2}{Q_3 + Q_1/2} = \text{Coefficient of Quartile Deviation} \qquad \text{......(6)}$$

In case of above mentioned example of examination scores (Table 2.1) this comes out to be

$$\frac{64.50 - 37.50/2}{64.50 + 37.50/2} = .265$$

Quartile Deviation, actually is not a measure of dispersion rather it is a measure of partition as it shows the distance on a scale from the median. But even then, it gives us an idea about the uniformity of the central 50 per cent values in a distribution. If the coefficient of quartile deviation is small it denotes large uniformity. Anyway, except for the fact that it is simple to calculate and understand it does not satisfy any other condition of good measure of dispersion like it is not based on all the observations, and cannot be used for further algebraical treatment.

Mean Deviation

Mean Deviation of any distribution is the sum of the absolute values of deviations from mean divided by total number of cases. Symbolically it is expressed as:

$$\text{Mean Deviation} = \frac{\sum_{i=1}^{n} |X_i - \bar{X}|}{N} \qquad \text{......(7)}$$

where | | means that $\pm$ signs are ignored and absolute values are taken as we know if signs are considered then

$$\sum_{i=1}^{n} X_i - \bar{X} = 0.$$

Calculation of mean deviation is given in the following example (Table 2.2), where consumer price index for working class for selected stations in India for the year 1976-77 (Base year : 1960 = 100) has been taken for illustration.

Here : $\bar{X} = 302.45$

$$\sum_{i=1}^{n} | X_i - \bar{X} | = 119.40$$

$$N = 11$$

So,

$$\text{Mean Deviation} = \frac{119.40}{11} = 10.85$$

TABLE 2.2: **Consumer Price Index for Selected Stations in India for the Year 1976-77 (Base Year : 1960 = 100)**

Sl. No.	Station	Index (X_i)	$X_i - \bar{X}$	$\mid X_i - \bar{X} \mid$
1.	Ahmedabad	281	—21.45	21.45
2.	Ajmer	303	.54	.54
3.	Bombay	298	— 4·45	4.45
4.	Bangalore	321	18.54	18.54
5.	Calcutta	297	— 5.45	5.45
6	Delhi	332	29.54	29.54
7.	Hyderabad	308	5.54	5.54
8.	Jamshedpur	288	—14.45	14.45
9.	Kanpur	294	— 8.45	8.45
10.	Nagpur	297	— 5.45	5.45
11.	Sholapur	308	— 5.54	5.54

Source : RBI Bulletins.

As this measure is the deviation from a central value, it can be calculated taking deviation from median also. In that case the symbolic representation becomes as

$$\text{Mean Deviation from the median} = \frac{\sum_{i=1}^{n} | X_i - \text{med.} |}{n} \quad ...(8)$$

This deviation from median is always less or equal to the mean deviation.

In case of frequency distribution the mean deviation takes the following form:

$$\text{Mean deviation (M.D.)} = \frac{\Sigma f \ | d |}{N} \quad ......(9)$$

where f = frequency

d = absolute deviation of X_i value from mean.

The following illustration (Table 2.3) gives the solution for a frequency distribution, in which size of each (observation) item is 6, 8, 3, 11, 9, 12, 13; and the respective frequencies are 3, 6, 9, 13, 8, 5 and 4.

TABLE 2.3 : **Solution for Frequency Distribution**

Size (X_i)	Frequency (f)	Cumulative frequency	Deviation from mean 8.75 $\mid d \mid$	$f \mid d \mid$
6	3	3	2.75	8.25
8	6	9	0.75	4.50
3	9	18	5.75	51.75
11	13	31	2.25	29.25
9	8	39	0.25	2.00
12	5	44	3.25	16.25
13	4	48	4.25	17.00

Now, here $N = 48$

$$\text{Mean} = \frac{f(x)}{N}$$

$$= \frac{420}{48} = 8.75$$

$$\Sigma f \mid d \mid = 129$$

$$\text{Mean deviation} = \frac{\Sigma f \mid d \mid}{N}$$

$$= \frac{129}{48} = 2.68$$

In case of grouped data (Table 2.4) the mean deviation is calculated in the same way as applied in case of discrete frequency distribution, with the only difference that for each size group the mid-values are taken as representatives of the size of different frequencies. The problem given in Table 2.4 has been solved in Table 2.5 leading to the mean deviation value of the distribution.

TABLE 2.4 : **Number of Students Falling in Different Classes According to the Marks Obtained**

Class (X)	Number of students (f)
0-10	2
10-20	5
20-30	7
30-40	6
40-50	10
50-60	12
60-70	11
70-80	9
80-90	3
90-100	2

TABLE 2.5 : **Solution for the Problem of Table 2.4**

Class	Mid-values	f_i	Cumulative frequency	Deviation from mean $\mid d \mid$	$f \mid d \mid$
0-10	5	2	2	45.97	91.94
10-20	15	5	7	35.97	179.85
20-30	25	7	14	25.97	181.79
30-40	35	6	20	15.97	95.82
40-50	45	10	30	5.97	59.70
50-60	55	12	42	4.03	48.36
60-70	65	11	53	14.03	154.33
70-80	75	9	62	24.03	216.27
80-90	85	3	65	34.03	102.09
90-100	95	2	67	44.03	88.06

$$\text{Mean} = \frac{\Sigma f_i X_i}{N} = 50.97$$

where N = total frequencies (f_i)

X_i = mid-values of size classes

$$\text{Mean deviation} = \frac{\Sigma f_i \mid d \mid}{N} \qquad \text{......(10)}$$

$$= \frac{1218.21}{67} = 18.18$$

The coefficient of mean deviation for relative dispersion is found in the following way:

$$\text{Coefficient of M.D.} = \frac{\text{Mean Deviation}}{\text{Mean}} \qquad \text{......(11)}$$

$$\text{or} \quad \frac{\text{Mean Deviation}}{\text{Median}} \qquad \text{......(12)}$$

The deviation is calculated from the median value. From the previous example the coefficient of Mean Deviation is

$$= \frac{18.18}{50.97} = .356$$

The major drawback of this measure of dispersion is that it cannot be used as basis for further statistical analysis, though it is easy to compute, based on all values and less affected by the extreme values compared to the previous measures.

Standard Deviation

This is the most commonly used measure of dispersion as it has some desirable properties, like elimination of negative deviations as explained in later paragraphs, which render it viable to be treated as the basis for further mathematical calculations. Standard deviation, generally denoted by the Greek letter sigma (σ), is actually the underroot of the variance of a distribution. Variance of a distribution is the sum of the squared deviations from the mean divided by the total number of observations. It is stated in the following formula :

$$\text{Variance } (X) = \frac{\sum_{i=1}^{n} (X_i - \bar{X})^2}{n} \qquad \text{......(13)}$$

$$\text{or} \quad \sigma^2 = \frac{\sum_{i=1}^{n} (X_i - \bar{X})^2}{n}$$

where 'X' denotes the distribution, 'X_i', the individual values varying from 1 to 'n' in the distribution, '$\bar{X}$', the mean of the distribution, and 'n' total number of cases/observations. As the standard deviation is the underroot of the variance it takes the form as follows :

$$\text{Standard Deviation} = \sqrt{\text{Variance } (X)}$$

$$= \sqrt{\frac{\sum_{i=1}^{n} (X_i - \bar{X})^2}{n}}$$

$$\sigma = \sqrt{\frac{\sum_{i=1}^{n} (X_i - \bar{X})^2}{n}} \qquad \ldots\ldots (14)$$

To give an example, the standard deviation of the distribution given in Table 2.2. is

$$\sigma = \sqrt{\frac{2097.53}{11}} = 13.80$$

Standard Deviation can be computed by many ways for different kinds of distribution where results slightly vary. For example, for the ungrouped data like in Table 2.2 the standard deviation can be computed by using formula:

$$\sigma = \sqrt{\frac{X_i^2 - \frac{(X_i)^2}{n}}{n}} \qquad \ldots\ldots (15)$$

$$= \sqrt{\frac{1008365 - \frac{(3327)^2}{11}}{11}}$$

$$= \sqrt{\frac{1008365 - 1006266.2}{11}} = 13.81$$

This method is actually adopted to ease the computational procedure by avoiding the computation of mean. This can be done by other way also. We can assume some value in the distribution as the mean and then use the deviations from the assumed mean for getting the standard deviation. This procedure is called the shifting of the origin as the origin from which the deviations are measured is the mean.

Using this procedure again in Table 2.2, we assume the value 332 as the mean and the deviations we get are:

—51, —29, —34, —11, —35, 0, —24, —44, —38, —35, and —24.

The form of the formula as used above can be used but now

X will be the X' which is the deviation from the assumed mean. The formula becomes as:

$$\sigma = \sqrt{\frac{\Sigma X'^{2}_{i} - \frac{(\Sigma X'_{i})^2}{n}}{n}} \quad \text{......(16)}$$

and the standard deviation comes out as:

$$\sigma = \sqrt{\frac{11701 - \frac{105625}{11}}{11}} = 13.81$$

For a frequency distribution as shown in Table 2.4, following formula is used for getting the standard deviation.

$$\sigma = \sqrt{\frac{\sum_{i=1}^{k} f_i (m_i - \bar{X})^2}{n}} \quad \text{......(17)}$$

where 'm_i' is the mid-point of the class intervals; $\bar{X}$ is the mean of the distribution; f_i is the frequency of each class; n is the total number of frequencies (i.e., $n = \Sigma f_i$); and k is the number of classes. Using this formula, the standard deviation for the distribution shown in Table 2.4 is :

$$= \frac{32211.94}{67} = 480.77 = 21.93$$

But this formula poses some inconvenience as $\bar{X}$ is to be calculated for a frequency distribution and these deviations are to be obtained from $\bar{X}$. To avoid this inconvenience another formula has been evolved where deviations are taken from an assumed mean as was in case of formula 16. The present formula is

$$\sigma = C\sqrt{\frac{\sum_{i=1}^{k} f_i d_i^2}{n} - \left(\frac{\sum_{i=1}^{k} f_i d_i}{n}\right)^2} \quad \text{......(18)}$$

where 'C' is the class interval; 'f_i' is the frequency of the ith class; d_i is the deviation of the ith class from an assumed origin; and 'n' is the total number of observations. This is explained by an example in Table 2.6.

Here assumed origin has been taken at the class interval of 60-70 and the minus class deviations are taken upward and plus class deviations are taken downward. The selection of this class interval as origin is to avoid the cumbersome calculations as the class contains the maximum frequencies. Putting the respective values in the formula we get:

$$\sigma = 10\sqrt{\frac{237}{61} - \left(\frac{-27}{61}\right)^2}$$

$$= 10\sqrt{3.88 - .19} = 10 \times 1.92 = 19.20$$

TABLE 2.6 : **Frequency Deviation Taken Downward**

Class	*f*	*d*	*fd*	d^2	fd^2
20-30	4	−4	−16	16	64
30-40	7	−3	−21	9	63
40-50	8	−2	−16	4	32
50-60	9	−1	− 9	1	9
60-70	12	0	0	0	0
70-80	10	1	10	1	10
80-90	8	2	16	4	32
90-100	3	3	9	9	27

If the grouped data is with unequal class interval this following formula is used which is similar to the formula (18) except that here m_i stands for mid-value of the classes. The formula is

$$\sigma = \sqrt{\frac{\Sigma f_i m_i^2 - \frac{(\Sigma f_i m_i)^2}{n}}{n}} \qquad \text{......(19)}$$

Standard deviation of two or more distribution can be combined by the following formula. This formula is given for two distributions only.

$$\sigma_{12} = \sqrt{\frac{N_1 \sigma_1^2 + N_2 \sigma_2^2 + N_1 (\bar{X}_{12} - \bar{X}_1)^2 + N_2 (\bar{X}_{12} - \bar{X}_2)^2}{N_1 + N_2}} \qquad \text{......(20)}$$

where, N_1 and N_2 are number of items in first and second distribution respectively; σ_1^2 and σ_2^2 are the variances of respective distributions, $\bar{X}_1$ and $\bar{X}_2$ are the means of respective distribution and $\bar{X}_{12}$ is the combined mean of two distributions calculated as

$$\bar{X}_{12} = \frac{N_1\bar{X}_1 + N_1\bar{X}_2}{N_1 + N_2} \qquad \text{......(21)}$$

The relative measure for the standard deviation is the coefficient of variance (V) which is calculated as

$$V = \frac{\sigma}{\bar{X}} \qquad \text{......(22)}$$

Sometimes it is multiplied by 100 to explain it in terms of percentages.

Comments on Standard Deviation

Standard deviation by far is the most important measure of dispersion. Generally this is the only measure which is used in further statistical analyses like correlation, etc. It is also the most frequently used measure of the spread of a series of data. If $\pm\ \sigma$ is measured from the mean of a normal distribution, $\bar{X} \pm 2\ \sigma$ includes about 95 per cent of the items. All these characteristics of standard deviation make it very useful measure of dispersion among all.

Lorenz Curve

It is the graphic method of measuring the deviations from the average, devised by Lorenz for measuring the inequalities in the distribution of wealth. But it can equally be applied for measuring the deviations for any frequency distribution. A major advantage of this is that a comparative graphic representation can be done for many distributions of the same nature in one graph. For example, income distribution of persons of two different cities can be shown in one figure.

TABLE 2.7 : **Number of Operational Holdings and Area Operated by Size Class of Operational Holdings in the District of Kalahandi, Orissa**

Size class (hectares)	No. of Holdings	% of total No. of holdings	Cumulative Percentage	Area under total No. of Holdings	% to total area of operational holdings	Cumulative Percentage
Below 0.5	8002	5.63	5.63	2616.00	.46	.46
0.5— 1.0	16725	11.77	17.40	13011.00	2.30	2.77
1.0— 2.0	40475	28.50	45.90	66024.00	11.70	14.47
2.0— 3.0	12775	8.99	54.89	34690.00	6.15	20.62
3.0— 4.0	14153	9.96	64.85	49121.00	8.70	29.32
4.0— 5.0	7117	5.0	69.86	31608·00	5.60	34.92
5.0—10.0	36114	25.42	95.28	248296.00	43.99	78.91
10.0—20.0	5212	3.67	98.95	68306.00	12.10	91.01
20.0—30.0	933	.66	99.61	21712.00	3.85	94.86
30.0—40.0	230	.16	99.77	7955.00	1.41	96.27
40.0—50.0	109	.08	99.85	4994.00	0.88	97.15
50.0 & above	197	.14	100.00	16085.00	2.85	100.00
Total	142042			564418.00		

Source : Statistical Abstract, Orissa, 1973.

Lorenz curve is, actually, a cumulative percentage curve. Cumulative percentages of one item are plotted against the cumulative percentages of other item. Before drawing the curve various groups of each variable should be converted into percentages and subsequently cumulative percentages should be obtained. These cumulative percentages should then be plotted along the axis of Y and along the axis of X. Along the Y-axis, at the intersection, the scale begins from zero and goes upward upto 100 while it begins from 100 and goes upto zero along the X-axis. When we join the zero of X-axis with 100 of Y-axis, it gives us a line which is possible only if the distribution is equal among all the groups and because of this the line is termed as the line of equal distribution. The actual data is plotted on this graph and a curve is drawn through all the points. The distance of the curve from the line of equal distribution gives an idea of deviation in the distribution. The farther the curve, the greater is the deviation.

For illustration, number of operational holdings (by size class) and the area under them (Table 2.7) have been plotted on Fig. 2.1. The graph indicates that the area under operational holdings is not equally distributed among the number of holdings. It shows quite great deviation.

Index of Qualitative Variation (IQV)

When we come across the measurement of heterogeneity in a qualitative set of data the index of qualitative variation (IQV) proves to be a better tool. This index, actually, explains the percentage variation existing in a set of qualitative data of the maximum possible variation. Thus, it is not the amount of variation, but a ratio and the use of this tool is very often practised when the data is given in the form of categories or groups. For example, there are 350 persons, 75 from one religious group, 100 from another, 90 from third one, and rest (85) from the fourth religious group.

Now, in this set there are four religious groups, capturing their share of persons from the group of 350 persons. Had these persons been equally distributed among four groups there could have been maximum possible differences in the population.

But, this is not so indicating that there is certain value of observed differences. The ratio of these observed differences with the maximum possible, differences is actually the IQV which is represented in percentage terms after multiplying by 100.

Maximum possible differences are found by the following formula:

$$\text{Maximum possible differences} : \frac{K\,(K-1)}{2}\,(N/K)^2 \qquad \ldots\ldots (23)$$

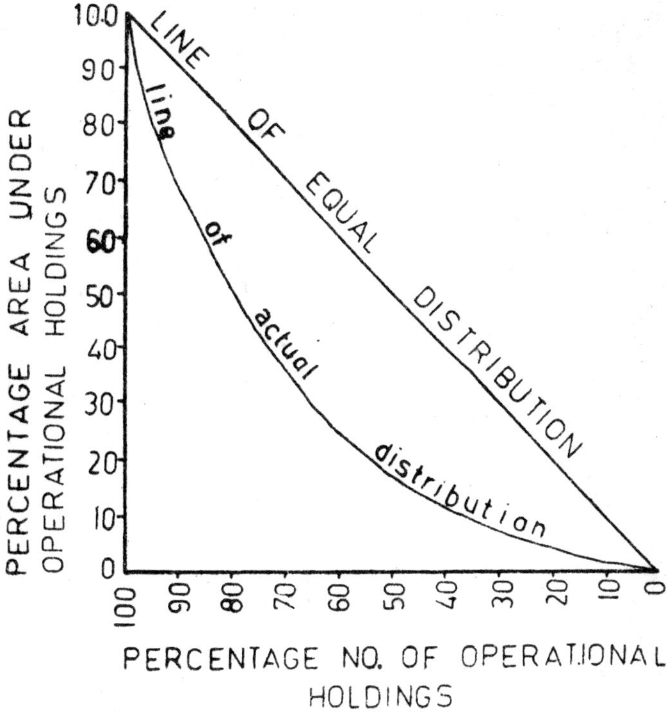

FIG. 2.1 : **Lorenz Curve for Number of Operational Holdings and Area Operated by Size Class of Operational Holdings—District Kalahandi**

where N = Number of cases in all the categories or groups.

K = Number of categories or groups.

In the example, mentioned above 'N' = 350 and 'K' = 4. Therefore, Maximum possible differences are :

$$= \frac{4\,(4-1)}{2}\,(350/4)^2 \qquad = \frac{4\,(3)}{2}\,(87.5)^2$$

$$= \frac{12}{2}\ (7656.25) = 6\ (7656.25) = 45937.5$$

The total observed differences are found by summing the differences between each category and every other category/group.

$$\text{Total observed differences } \sum_{i=1}^{k-1}\left(\sum_{j=i+1}^{k}\left(f_i f_j\right)\right) \quad \text{......} (24)$$

where k = number of categories/group

f_i = frequency of category/group

f_j = frequency of category/group j.

In case of four above mentioned groups the formula becomes :

$$\sum_{i=1}^{k-1}\left(\sum_{j=i+1}^{k}\left(f_i f_j\right)\right) = \sum_{i=1}^{3}\left(\sum_{j=i+1}^{4}\left(f_i f_j\right)\right)$$

$$= f_1 f_2 + f_1 f_3 + f_1 f_4 + f_2 f_3 + f_2 f_4 + f_3 f_4$$
$$= f\,(f_2 + f_3 + f_4) + f_2\,(f_3 + f_4) + f_3\,(f_4)$$
$$= 75\,(100+90+85) + 100\,(90+80) + 90\,(85)$$
$$= 20625 + 17500 + 7650 = 45775$$

Then IQV is

$$\text{IQV} = \frac{45775}{45937.5} = .996 = 99.6\%$$

This means that the present distribution of persons among different religious groups explains 99.6 per cent of the maximum possible differences.

IQV can be calculated from both percentage frequency or absolute number of frequency but the value of IQV does not differ.

IQV must be explained with care as, already mentioned, it is not the amount of variation but the ratio, and the value of the ratio depends upon the number of groups or categories and not on the total number of cases involved in the categories or groups. Higher value of IQV means there is high proportion of variation for the number of categories/groups involved. In other words, higher the IQV, the nearer is the distribution of cases among categories/group to uniformity.

IQV is useful when the heterogeneity for the same variable is measured for two or more sets of data. It can pretty well be used to measure as how varied the population of different cities is in terms of different socio-economic groups.

SUGGESTED READINGS

D.H. Elhance, *Fundamentals of Statistics*, Kitab Mahal, Allahabad.

David Ebdon, *Satistics in Geography—A Practical Approach*, Basil Blackwell, Oxford, 1977.

Gregory, S., *Statistical Methods and the Geographer*, Longman, London, 2nd edition, 1969.

Gupta, C.B., *An Introduction to Statistical Methods*, Vikas Publishing House Pvt. Ltd., New Delhi, 1973.

Handel, J.D., *Introductory Statistics for Sociologists*, Prentice-Hall, Inc., Englewood Cliffs, N.J., 1978.

King, L.J., *Statistical Analysis in Geography*, Prentice, Hall, Inc., Englewood Cliffs, N.J., 1969.

Yamane, Taro, *Statistics—An Introductory Analysis*, Hapur & Row, New York, Evanston & London and John Weatherhill, Inc., Tokyo.

THREE

S.N. Singh and *K.N.S. Yadava*

MEASURES OF CENTRAL TENDENCY AND LOCATION (AVERAGES)

STATISTICAL methodology is a comprehensive term which includes almost all the methods involved in the collection, processing, condensing and analysing of data. The data collected from the field for a number of items vary greatly in their qualitative as well as quantitative nature. Geographical phenomena are much more variable than others; the variation must be appreciably great. For example, the rainfall at a particular region is erratic in nature and shows variation from year to year, month to month and even day to day. The condensation of data in terms of maps. charts, diagrams, etc. is a first and necessary step in rendering a long series of observations comprehensible. But for practical purposes it is not enough, particularly when we want to compare two or more different series of data, e.g., we may wish to compare the distribution of status in two races of man, or the birth rates in India in two successive decades, or the rainfall in two different regions, or the number of wealthy people in two different countries. For such problems, there are certain statistical techniques one of which is a measure of central tendency.

It is found that the observations have a tendency to cluster round a central value, and this characteristic of measurement is called the 'central tendency'. Any statistical measure which gives the point round which the observations have a tendency to cluster is known as a 'measure of central tendency'. The central value of the variable in any series of observations is useful in finding the location of the distribution and so it is also called an average.

Simply, an 'average' of a series of measurements is a single value of the variable which is a satisfactory representative of the distribution.

In this chapter different measures of central tendency will be covered highlighting their applications under various situations for the data obtained from geographical and other social surveys. Calculation, methodologies for these measures for ungrouped and grouped data will be described with suitable examples.

Types of Averages

The measures of central tendency or averages are of different types, but the most common in use are of three types :

1. Mean
2. Median
3. Mode

The mean is further classified as:

(i) Arithmetic mean
(ii) Geometric mean
(iii) Harmonic mean

Since each one of the above measures of central tendency has its own individual characteristics and properties, a decision must always be made as to which would be the most appropriate and useful in view of the nature of the statistical data and the purpose of the inquiry. The qualities desired in a measure should be (a) rigidly defined, (b) easily computed, (c) capable of a simple interpretation, (d) not unduly influenced by one or two extremely large or small values, and (e) likely to fluctuate relatively little from one random sample to another (of the same size and from the same population).

Arithmetic Mean (ungrouped data)

The arithmetic mean of a series of items is obtained by summing up the values of all the measurements and dividing the total by the number of measurements. Let X denotes the variable of measurement and X_1, X_2, X_3..., X_n be n individual values of X, then the mean $\bar{X}$, say is computed as :

$$\bar{X} = \frac{\text{Sum of the observations or values}}{\text{Number of observations}}$$

$$= \frac{X_1 + X_2 + X_3 + \dots + X_n}{n}$$

$$= \frac{\sum_{i=1}^{n} X_i}{n}$$

where Σ, i.e., sigma stands for summation and X_1 is the value of the ith observation.

Example 1

The rainfall record in a month in 10 regions of a State is given below. Compute the average rainfall of the month in the State.

Region:	1	2	3	4	5	6	7	8	9	10
Rainfall: (in mm)	17.6	10.1	11.4	18.5	10.5	14.3	8.9	13.4	10.6	12.5

Solution

$$\text{Mean } (\bar{X}) = \frac{X_1 + X_2 + \dots + X_{10}}{10}$$

$$= \frac{\sum_{i=1}^{10} X_i}{10}$$

$$= \frac{17.6 + 10.1 + 11.4 + \dots + 10.6 + 12.5}{10}$$

$$= \frac{127.8}{10} \qquad = 12.78 \text{ mm}$$

Short Cut Method

This method is applied to avoid lengthy calculations. When the individual readings are large in size, an arbitrary value is selected as a working mean and differences between the working mean and the individual reading are worked out. By summing these differences and dividing by the number of readings we can get the actual mean. Let $X_1, X_2 X_3, \ldots, X_n$ be n individuals readings on the variable and let A be the working mean. Let $d_1, d_2, d_3, \ldots, d_n$ denote the differences between the working mean and individual values $X_1, X_2, X_3, \ldots, X_n$ respectively. Then the mean of X say $\bar{X}$ in term of the mean of differences say $\bar{d}$ is calculated as :

$$\bar{d} = \frac{d_1+d_2+d_3+ \ldots +d_n}{n}$$

$$= \frac{(X_1-A)+(X_2-A)+(X_3-A)+ \ldots +(X_n-A)}{n}$$

$$= \frac{\sum_{i=1}^{n} X_i - nA}{n}$$

$$= \frac{\sum_{i=1}^{n} X_i}{n} - A$$

$$= \bar{X} - A$$

i.e., $$\bar{X} = A + \bar{d} = A + \frac{\sum_{i=1}^{n} d_i}{n}$$

Example 2

The monthly average temperature for a region were given as 17.2, 19.0, 16.4, 21.6, 24.1, 19.7, 27.6, 25.4, 26.1 and 18.4 degree centigrade, respectively from January to December. Find the mean annual temperature.

Solution: Let 21.6 be the working mean. Then from the Table 3.1.

TABLE 3.1 : **Monthly Average Temperature for a Region**

X	$(X—A)$
17.2	—4.4
19.0	—2.6
16.4	—5.2
21.6A	0
24.1	2.5
19.7	—1.9
27.6	6.0
25.4	3.8
26.1	4.5
18.4	—3.2
Total	—.5

$$\text{Mean } (\bar{X}) = A + \frac{\sum_{i=1}^{n} d_i}{n}$$

$$= 21.6 + \frac{(-.5)}{10}$$

$$= 21.6 - .05$$

$$= 21.55^{\circ}\text{ C}$$

Grouped data (Discrete frequency distribution)

In a discrete series, let the individual readings X_1, X_2, . . ., X_n of the variable X occur $f_1, f_2, \ldots, f_n$ times respectively. Then the mean of X is obtained by summing the multiplication of individual readings with corresponding frequencies and dividing the total by the sum of frequency, i.e.

$$\text{Mean } \bar{X} = \frac{X_1 f_1 + X_2 f_2 + \ldots + X_n f_n}{f_1 + f_2 + \ldots + f_n}$$

$$= \frac{\sum_{i=1}^{n} X_i f_i}{\sum_{i=1}^{n} f_i} \quad \text{or} \quad \frac{\sum_{i=1}^{n} X_i f_i}{N}$$

where $N = \sum_{i=1}^{n} f_i$ is the total number of observations.

The various steps involved in the calculation of $\bar{X}$ under this method are:

(i) Multiply the frequency of each row with the concerned variable ($X.f$) and total them. It will be ΣXf.

(ii) Add up all the frequencies (Σf).

(iii) Divide the ΣXf by Σf or N to find out the arithmetic mean ($\bar{X}$).

Short cut method: To find out arithmetic mean in a discrete series by this method the following steps are taken:

(i) Any value of the distribution may be taken as working mean or arbitrary mean, say (A) (preferably it should be near the middle of the frequency distribution).

(ii) Take deviations of the variable X from the working mean ($X—A$) and denote it by dx.

(iii) Multiply each dx by its respective frequency f and denote it by fdx.

(iv) The arithmetic mean ($\bar{X}$) is then calculated with the help of the following formula :

$$\text{Mean } (\bar{X}) = A + \frac{\sum_{i=1}^{n} f_i \, dx_i}{N}$$

Example 3

Table 3.2 gives the number of children born per family in 735 families in a locality. Calculate the average number of children born per family in the locality.

TABLE 3.2 : **Number of Children Born per Family**

Number of children born per family	Number of families	Number of children born per family	Number of families
0	96	7	20
1	108	8	11
2	154	9	6
3	126	10	5
4	95	11	5
5	62	12	1
6	45	13	1

Solution

Computation of the average number of children born per family :

TABLE 3.3 : **Average Number of Children Born per Family**

No. of children born per family (X)	No. of families (f)	$X.f$ (Xf)
0	96	$0 \times 96 = 0$
1	108	$1 \times 108 = 108$
2	154	$2 \times 154 = 308$
3	126	$3 \times 126 = 378$
4	95	$4 \times 95 = 380$
5	62	$5 \times 62 = 310$
6	45	$6 \times 45 = 270$
7	20	$7 \times 20 = 140$
8	11	$8 \times 11 = 88$
9	6	$9 \times 6 = 54$
10	5	$10 \times 5 = 50$
11	5	$11 \times 5 = 55$
12	1	$12 \times 1 = 12$
13	1	$13 \times 1 = 13$
Total	735	2166

Here $N = \Sigma f = 735$; $\Sigma Xf = 2166$

$$\text{Mean } (\bar{X}) = \frac{\Sigma Xf}{\Sigma f}$$

$$= \frac{2166}{735}$$

$= 2.9$ children.

Example 4

The Table 3.4 gives the distribution of units under different heights in a certain region. Compute the mean height of the region:

Heights (in metre) :	200	600	1000	1400	1800	2200
Number of units:	142	265	560	271	89	16

Solution (a) : By direct method

TABLE 3.4 : **Distribution of Units under Different Height**

Height (in metre) X	Number of units (f)	$X.f$ (Xf)
200	142	200 × 142 = 28400
600	265	600 × 265 = 159000
1000	560	1000 × 560 = 560000
1400	271	1400 × 271 = 379400
1800	89	1800 × 89 = 160200
2200	16	2200 × 16 = 35200
Total	1343	1222200

Here $N = \Sigma f = 1343$; $\Sigma Xf = 1322200$

$$\text{Mean height } (\bar{X}) = \frac{\Sigma Xf}{\Sigma f}$$

$$= \frac{1322200}{1343}$$

$$= 984.51 \text{ metres}$$

(*b*) *By short cut method*

Let the working mean $(A) = 1400$

TABLE 3.5 : **Distribution of Units under Different Heights**

Height in metres (X)	$X—A$ (d)	No. of units (f)	$d.f$ (df)
200	—1200	142	—1200 × 142 = —170400
600	— 800	265	— 800 × 265 = —212000
1000	— 400	560	— 400 × 560 = —224000
1400 A	0	271	0 × 271 = 0
1800	+ 400	89	+400 × 89 = 35600
2200	+800	16	+800 × 16 = 12800
Total		1342	—558000

Here $N = \Sigma f = 1343$; $\Sigma fd = -558000$

$$\text{Mean height } (\bar{X}) = A + \frac{\Sigma fd}{N}$$

$$= 1400 + \frac{-558000}{1343}$$

$$= 984.51 \text{ metres.}$$

Grouped Data (Continuous Frequency Distribution)
When the measurements are given in the grouped form, the mean is computed by multiplying the various mid-values with their respective frequencies and dividing the total by the sum of frequencies. If $m_1, m_2, \ldots, m_n$ are the mid-values or class marksc orresponding to frequencies $f_1, f_2, \ldots, f_n$, then the mean is

$$\bar{X} = \frac{m_1 f_1 + m_2 f_2 + \ldots\ldots + m_n f_n}{f_1 + f_2 + \ldots\ldots + f_n}$$

$$= \frac{\sum_{i=1}^{n} m_i f_i}{\sum_{i=1}^{n} f_i} \quad \text{or} \quad \frac{\sum_{i=1}^{n} m_i f_i}{N}$$

where n stands for the number of groups and N denotes the total number of observations.

Short cut method : If the class-groups formed by individual readings are large, the following procedure is adopted for computing arithmetic mean in a continuous series by this method:

(i) Find out mid-values of the class intervals, and assume one of the mid-values (preferably in the middle of the distribution) as working mean (A).

(ii) Calculate deviations of the mid-values from the working mean ($m—A$) denoted by d,

(iii) Multiply each d by its respective frequency to find out fd.

(iv) Calculate arithmetic mean by the following formula:

$$\text{Mean } (\bar{X}) = A + \frac{\sum_{i=1}^{n} f_i d_i}{N}$$

Step-Deviation Method: Further if the deviations are large and intervals among consecutive mid-values are equal, deviations can be made small by dividing them by the class width, i.e., if $d' = \frac{(m-A)}{h}$, the formula for calculating arithmetic mean by this method is

$$\text{Mean } (\bar{X}) = A + \frac{\sum_{i=1}^{n} f_i d_i}{N} \times h$$

where h is the size of the class intervals.

Example 5

The rainfall of 66 districts in a particular year is given below. Compute the average annual rainfall.

Rainfall : (in inches)	0—10	10—20	20—30	30—40	40—50	50—60
No. of Districts :	22	10	8	15	5	6

Solution (a) By direct method

TABLE 3.6 : **Rainfall in Districts in Particular Year**

Rainfall in inches (X)	Mid-Values (m)	No. of districts (f)	$m.f$ (mf)
0—10	$\frac{0+10}{2} = 5$	22	$5 \times 22 = 110$
10—20	$\frac{10+20}{2} = 15$	10	$15 \times 10 = 150$
20—30	$\frac{20+30}{2} = 25$	8	$25 \times 8 = 200$
30—40	$\frac{30+40}{2} = 35$	15	$35 \times 15 = 525$
40—50	$\frac{40+50}{2} = 45$	5	$45 \times 5 = 225$
50—60	$\frac{50+60}{2} = 55$	6	$55 \times 6 = 330$
Total		66	1540

Here $N = \Sigma f = 66$, $\Sigma mf = 1540$

$$\text{Mean } (\bar{X}) = \frac{\Sigma mf}{N}$$

$$= 23.33 \text{ inches}$$

(*b*) *By short cut method*

Let the working mean (A) = 35

TABLE 3.7 : **Rainfall in Inches—Short Cut Method**

Rainfall in inches (X)	Mid-Values (m)	(m—A) (d)	No. of districts (f)	d×f (df)
0—10	5	—30	22	—30 × 22 = — 660
10—20	15	—20	10	—20 × 10 = — 200
20—30	25	—10	8	—10 × 8 = — 80
30—40	35A	0	15	0 × 15 = 0
40—50	45	10	5	10 × 5 = 50
50—60	55	20	6	20 × 6 = 120
Total			66	—770

Here $N = \Sigma f = 66$; $\Sigma\, df = -770$

$$\text{Mean } (\bar{X}) = A + \frac{\Sigma fd}{N}$$

$$= 35 + \frac{-770}{66} = 23.33 \text{ inches}$$

(*c*) *By Step-deviation Method*

From the given data $h = 10$

TABLE 3.8 : **Rainfall in Inches : Step Deviation Method**

Rainfall in inches (X)	Mid-values (m)	$\frac{M-A}{h}$ (d')	No. of Districts	$d'.f$ ($d'f$)
0-10	5	$\frac{5-35}{10} = -3$	22	$-3 \times 22 = -66$
10-20	15	$\frac{15-35}{10} = -2$	10	$-2 \times 10 = -20$
20-30	25	$\frac{25-35}{10} = -1$	8	$-1 \times 8 = -8$
30-40	35A	$\frac{35-35}{10} = 0$	15	$0 \times 15 = 0$
40-50	45	$\frac{45-35}{10} = +1$	5	$1 \times 5 = 5$
50-60	55	$\frac{55-35}{10} = +2$	6	$2 \times 6 = 12$
Total			66	-77

Here $N = \Sigma f = 66$; $\Sigma fd' = -77$

$$\text{Mean } (\bar{X}) = A + \frac{\Sigma fd'}{N} \times h$$

$$= 35 + \frac{(-77)}{66} \times 10 = 23.33 \text{ inches.}$$

Example 6

The daily minimum temperature record of a weather reporting station is given below for 80 days in degrees fahrenheit. Compute the average temperature of the station :

Temperature (in degrees F) :	50-60	60-70	70-80	80-90	90-100
No. of days :	9	13	30	16	12

Solution : Here $h = 10$ and let us assume the working mean $A = 75$

TABLE 3.9 : **Daily Minimum Temperature Record**

Temperature in degrees F (X)	Mid-value (m)	$\frac{M-A}{h}$ (d')	No. of days (f)	$d' \times f$ ($d'f$)
50-60	55	-2	9	$-2 \times 9 = -18$
60-70	65	-1	13	$-1 \times 13 = -13$
70-80	75A	0	30	$0 \times 30 = 0$
80-90	85	1	16	$1 \times 16 = 16$
90-100	95	2	12	$2 \times 12 = 24$
Total			80	9

Here $N = \Sigma f = 80$; $\Sigma fd' = 9$

$$\text{Mean } (\bar{X}) = A + \frac{fd'}{N} \times h$$

$$= 75 + \frac{9}{80} \times 10$$

$$= 76.125 \text{ degrees F.}$$

Evaluation of combined average: If $\bar{X}_1$ and $\bar{X}_2$ be the means of two series of sizes N_1 and N_2 respectively, then the mean of the combined series can be computed as :

$$\bar{X} \text{ (Combined mean)} = \frac{N_1\bar{X}_1 + N_2\bar{X}_2}{N_1 + N_2}$$

Similarly, if $\bar{X}_1$, $\bar{X}_2$, $\bar{X}_3$, ..., $\bar{X}_n$ be the means of n series of sizes N_1, N_2, ..., N_n respectively, then the mean of the combined series is

$$\bar{X} \text{ (Combined mean)} = \frac{N_1\bar{X}_1 + N_2\bar{X}_2 + \ldots + N_n\bar{X}_n}{N_1+N_2 + \ldots + N_n}$$

Example 7

The average rainfall record for first 15 days of a month is 51.6 mm whereas for next 15 days of the month was recorded as 47.2 mm. Find the average rainfall for the given month.

Solution

Here $\bar{X}_1 = 51.6$ mm and $N_1 = 15$

$\bar{X}_2 = 47.2$ mm and $N_2 = 15$

$$\bar{X} \text{ (Combined average)} = \frac{N_1\bar{X}_1 + N_2\bar{X}_2}{N_1+N_2}$$

$$= \frac{15\times51.6+15\times47.2}{15+15}$$

$$= \frac{774+708}{30}$$

$$= 49.4 \text{ mm.}$$

Example 8

The average ages of 250 males and 210 females in a village are 41.6 and 38.5 years respectively. Find the average age combining both males and females together.

Solution

Here $N_1 = 250$, $\bar{\bar{X}}_1 = 41.6$ years

$N_2 = 210$, $\bar{\bar{X}}_2 = 38.5$ years

$$\bar{X} \text{ (Combined average)} = \frac{N_1\bar{\bar{X}}_1+N_2\bar{\bar{X}}_2}{N_1+N_2}$$

$$= \frac{250\times41.6+210\times38.5}{250+210}$$

$$= \frac{10400 + 8085}{460}$$

$$= 40.18 \text{ years.}$$

Geometric Mean

In case of finding the averages of rates and ratios, geometric mean is the more useful measure than others, e.g., in finding the population increase, simple and compound interests, etc.

Case 1: In case of ungrouped data, it is obtained by multiplying together all the values of the variable and extracting the relevant root of the product, i.e., if $X_1, X_2, \ldots X_n$ are n values of a variable under study, then the geometric mean (G.M.) is computed as :

$$\text{G.M.} = (X_1, X_2, X_3 \ldots X_n)^{\frac{1}{n}}$$

To facilitate the computation, one can make the use of logarithms as :

$$\log (\text{G.M.}) = \frac{1}{n} \log (X_1, X_2, X_3, \ldots, X_n)$$

$$= \frac{1}{n}\left[\log X_1 + \log X_2 + \log X_3 + \ldots + \log X_n\right]$$

$$= \frac{1}{n}\left[\sum_{i=1}^{n} \log X_i\right]$$

$$\text{So G.M.} = \text{Antilog } \frac{1}{\text{n}}\left[\sum_{i=1}^{n} \log X_i\right]$$

Thus the logarithm of the geometric mean is equal to the arithmetic mean of the logarithms of individual measurements.

Case 2 : In case of grouped data, if $f_1, f_2, \ldots, f_n$ values be the frequencies corresponding to the individual values $X_1, X_2, \ldots, X_n$, then G.M. is computed as:

$$\text{G.M.} = \left(X_1^{f_1} . X_2^{f_2} . X_3^{f_3} \ldots\ldots X_n^{f_n}\right)^{\frac{1}{\sum_{i=1}^{n} f_i}}$$

$$\text{or } \log \text{G.M.} = \frac{1}{\sum_{i=1}^{n} f_i}\left[f_i \log X_1 + f_2 \log X_2 + \ldots + f_n \log X_n\right]$$

$$= \frac{1}{\sum_{i=1}^{n} f_i}\left[\sum_{i=1}^{n} f_i \log X_i\right]$$

$$\text{So G.M.} = \text{Antilog}\left\{\frac{1}{\sum_{i=1}^{n} f_i}\left[\sum_{i=1}^{n} f_i \log X_i\right]\right\}$$

Example 9

The monthly average temperature figures for a station for five months are given as 16.2, 23.4, 20.6, 33.4 and 16.4 degrees centigrade. Find the mean temperature of the station.

Solution

$$\text{G.M.} = (16.2 \times 23.4 \times 20.6 \times 33.4 \times 16.4)^{\frac{1}{5}}$$

$$\text{or } \log \text{G.M.} = \frac{1}{5}\left[\log 16.2 + \ldots\ldots + \log 16.4\right]$$

$$= \frac{1}{5}\left[1.2095 + 1.3692 + 1.3139 + 1.5238 + 1.2148\right]$$

$$= \frac{1}{5} \times 6.6312$$

$$= 1.3262$$

$$\text{so G.M.} = \text{Antilog } [1.3262]$$

$$= 21.1934 \text{ degrees centigrade.}$$

Solution 10

From the following data calculate the G.M.

Class group :	0-10	10-20	20-30	30-40	40-50
No. of observations :	14	23	27	21	15

TABLE 3.10 : **Data Calculation by the Geometric Mean**

Class group (X)	Mid-values (m)	Log of m	No. of observations (f)	$f \times \log m$
0—10	5	0.69897	14	9.78558
10—20	15	1.17609	23	27.05007
20—30	25	1.39794	27	37.74438
30—40	35	1.54407	21	32.42547
40—50	45	1.65321	15	24.79815
Total			100	131.80365

Solution

Here $\Sigma f = N = 100, \quad \Sigma f \log m = 131.80365$

$$\text{G.M.} = \text{Antilog}\left\{\frac{1}{\Sigma f}\left[\Sigma f \log m\right]\right\}$$

$$= \text{Antilog}\left[\frac{131.80365}{100}\right]$$

$$= \text{Antilog}\left[1.318036\right] = 20.7987$$

Use of Geometric Mean in Computing Rate of Growth

Geometric mean provides a satisfactory measure for computing rate of growth of population phenomena, specially that phenomena which grow at geometric progression:

The formula is :

$$P_t = P_0\,(1+r)^t$$

$$\text{or} \quad r = \left(\frac{P_t}{P_o} - 1\right)^{\frac{1}{t}}$$

where P_t = the value of variable at the end of the period, i.e., at time t.

P_o = the value of the variable at the beginning,

r = the rate of growth per unit of time,

t = number of units of time.

Example 11

Population of India in 1961 and 1971 were 43.9 and 54.8 crores. Find the rate of increase.

Solution

Here P_t = 54.8 ; P_o = 43.9 ; t = 10 years ; r = ?

$$r = \left(\frac{P_t}{P_o} - 1\right)^{\frac{1}{t}} = \left(\frac{54.8}{43.9} - 1\right)^{\frac{1}{10}}$$

$$\text{or } \log r = \frac{1}{10} \log \left(\frac{54.8}{43.9} - 1\right)$$

$$= \frac{1}{10} (.248292)$$

$$r = \text{Antilog } (.0248292) = 1.05884$$

Harmonic Mean

In problems such as work, time and rate, where the amount of work is held constant and an average rate is required, the harmonic mean (H.M.) is utilized. It is defined as the reciprocal of the arithmetic mean of the reciprocals of the given individual readings, i.e., H.M. of X_1, X_2, ..., X_n is defined as :

$$\text{H.M.} = \frac{n}{\frac{1}{X_1} + \frac{1}{X_2} + \ldots + \frac{1}{X_n}}$$

$$= \frac{n}{\sum_{i=1}^{n} \frac{1}{X_i}}$$

where n is the number of observations.
For example the H.M. of 2, 4, 6 is

$$\frac{3}{\frac{1}{2}+\frac{1}{4}+\frac{1}{6}} = \frac{36}{11} = 3.37$$

In case of frequency distribution (grouped data), if $f_1, f_2, \ldots, f_n$ be the frequencies corresponding to $x_1, x_2, \ldots x_n$ then H.M. is computed as:

$$\text{H.M.} = \frac{f_1 + f_2 + f_3 + \ldots + f_n}{\frac{f_1}{X_1} + \frac{f_2}{X_2} + \ldots + \frac{f_n}{X_n}}$$

$$= \frac{\sum_{i=1}^{n} f_i}{\sum_{i=1}^{n} \frac{f_i}{X_i}}$$

Example 12
Compute the H.M. from the data given in example 8 :

TABLE 3.11 : **Class Group Data**

Class-group (X)	Mid-values (m)	No. of observations (f)	$\frac{f}{m}$
0—10	5	14	2.80
10—20	15	23	1.53
20—30	25	27	1.08
30—40	35	21	0.60
40—50	45	15	0.33
Total		100	6.34

Solution

Here $\Sigma f = N = 100$; $\Sigma \frac{f}{m} = 6.34$

$$\text{H.M.} = \frac{\Sigma f}{\Sigma \frac{f}{m}} = \frac{100}{6.34}$$

$$= 15.77$$

Median

The median is another important and useful measure of central tendency. It has the connotation of the middle most or most central value of a set of measurements. It is usually defined as that value which divides a distribution in such a manner that the number of items below it is equal to the number of items above it. The median is thus a positional average. It is better indication of central value when one or two of the peripheral readings are too large or too small because they give the wrong idea of the average when mean is computed. In the analysis of rainfall figures and its nature, this measure of central tendency is frequently used.

Calculation of Median (Ungrouped data)

Case 1: In case of ungrouped data when the number of observations are odd, then median is the middle value after the measurements have been arranged in ascending or descending order of magnitude, i.e., if there are n number of measurements and measurements are arranged in ascending or descending order of magnitude, the median of the measurement is $\left(\frac{n+1}{2}\right)^{th}$ measurement where n is the odd number of observations.

Case 2: If the number of observations are even, median is defined as the mean of the two middle observations when observations are in ascending or descending order of magnitude.

$$\text{i.e., median} = \frac{\left(\frac{n}{2}\right)^{th} \text{value} + \left(\frac{n}{2}+1\right)^{th} \text{value}}{2}$$

Example 13 (a)

The daily minimum temperature record of a weather reporting station is given below for 20 days in degrees F. Find the median temperature value.

68, 62, 75, 82, 68, 71, 68, 71, 62, 68, 74, 59, 74, 68, 60, 71, 59, 73, 73, 58.

Example 13 (b)

The monthly income of 11 families in a small locality is given below. Find the median family income.

(Income in Rupees)

200, 150, 260, 285, 380, 305, 4989, 307, 1280, 233, 403.

Solution 13 (a)

To compute the median, first we arrange the values in ascending order of magnitude as:

58, 59, 59, 60, 62, 62, 68, 68, 68, 68, 68, 71, 71, 71, 73, 73, 74, 74, 75, 82.

The number of observations n is even in this case, i.e., $n = 20$.

So

$$\text{median} = \frac{\left(\frac{n}{2}\right)^{\text{th}} \text{value} + \left(\frac{n}{2}+1\right)^{\text{th}} \text{value}}{2}$$

$$= \frac{68+68}{2} = 68 \text{ degrees F.}$$

Solution 13 (b)

To compute the median, first we arrange the values in ascending order of magnitude as:

150, 200, 233, 260, 285, <u>305,</u> 307, 380, 403, 1280, 4989.

The number of observations n in this case is odd, i.e., n = 11, so the median is the $\left(\frac{n+1}{2}\right)^{\text{th}}$ value i.e. $\left(\frac{11+1}{2}\right)^{\text{th}}$ or 6th value of observation and thus underlined value, i.e., 305 is the median family income of the locality.

Calculation of Median (grouped data)

In case of discrete frequency distribution, median can be obtained with the help of cumulative frequencies as follows:

(i) First find $N/2$ where $N = \Sigma f$.

(ii) Find the cumulative frequency just greater than $N/2$.

(iii) Corresponding value of X (i.e., of variable) is median.

Example 14

Calculate the median height from the data given in example 4.

TABLE 3.12 : **Find the Cumulative Frequency**

Height (in metres) (X)	Number of units (f)	Cumulative frequency (f_c)
200	142	142
600	265	407
1000	560	967
1400	271	1238
1800	89	1327
2200	16	1343
Total	1343	

Solution

$$\text{Here } \Sigma f = N = 1343 \text{ , } \frac{N}{2} = 671.5$$

The cumulative frequency just greater than 671.5 is 967. Corresponding to this cumulative frequency, the value of X is 1000 and thus the median height is 1000 metres.

Example 15

Daily rainfall record of a district for 40 days is given below. Compute the median rainfall.

Rainfall (in inches) :	11	12	13	14	15	16
No. of days :	3	5	10	13	6	3

TABLE 3.13 : **Daily Rainfall Record of a District**

Rainfall (in inches) (X)	No. of days (f)	Cumulative frequency
11	3	3
12	5	8
13	10	18
14	13	31
15	6	37
16	3	40
Total	40	

Solution

Here $\frac{N}{2} = 20$. The cumulative frequency just greater than 20 is 31. The corresponding value of X to this cumulative frequency is 14. Thus the median rainfall is 14 inches.

Continuous Frequency Distribution: The median of the distribution is computed by the following formula:

$$\text{Median} = L + \left(\frac{\frac{N}{2} - f_c}{f} \right) h$$

where L is the lower limit of the median class,

f_c is the cumulative frequency of the class preceding the median class,

h is the class-width of the median class,

f is the frequency of the median class, and

$N = \Sigma f_i$

The main problem in calculation of median from grouped data is to find out the median class. The class corresponding to the cumulative frequency just greater than $N/2$ is called the median class.

Note : In case of a grouped data it is always essential to convert it into continuous form if given in a discrete form.

Example 16

Compute the median temperature for the distribution given in example 6.

TABLE 3.14 : **Median Temperature for the Distribution**

Temperature in °F (X)	No. of days (f)	Cumulative frequency (f_c)
50—60	9	9
60—70	13	22
70—80	30	52
80—90	16	68
90—100	12	80
Total	80	

Solution

Here $\frac{N}{2} = 40$. The cumulative frequency just greater than 40 is 52 and the corresponding class-group is the median class. For this median class we have,

$L = 70$; $f_c = 22$: $f = 30$ and $h = 10$

$$\text{Median} = L + \left(\frac{\frac{N}{2} - f_c}{f} \right) \cdot h$$

$$= 70 + \frac{40 - 22}{30} \times 10 = 76^{0}\text{ F.}$$

Example 17

The following table gives the size of land holding of families in a village. Find out the median holding size.

Area of land: (in acres)	5—9	10—14	15—19	20—24	25—29	30—34
No. of families:	20	35	150	70	44	38

Solution

Since the class groups are given in the discrete form and hence we first have to convert it into continuous form by adding .5 to the upper limits and subtracting .5 from the lower limits as given in column 2 below:

TABLE 3.15 : **Continuous Form and Cumulative Frequency**

Area of land (in acres) (X)	Area of land (in acres) Continuous form	No. of families (f)	Cumulative frequency (f_c)
5— 9	4.5— 9.5	20	20
10—14	9.5—14.5	35	55
15—19	14.5—19.5	150	205
20—24	19.5—24.5	70	275
25—29	24.5—29.5	44	319
30—34	29.5—34.5	38	357
		357	

Here $\frac{N}{2} = 178.5$, the cumulative frequency just greater than 178.5 is 205 and the corresponding class group is the median class. For this median class, we have

$L = 14.5,\ f_c = 55,\ f = 150$ and $h = 5$

$$\text{Median} = 14.5 + \frac{(178.5-55)}{150} \times 5$$

$$= 14.5+4.12$$

$$= 18.62 \text{ acres.}$$

Calculation of Median (Graphic Method)

One of the methods to compute median is the graphic method. In this case take the class intervals (or the individuals readings) on the axis of X and plot the corresponding cumulative frequencies on the axis of Y against the upper limit of the class interval (or against the variate value in case of discrete frequency distribution). The curve obtained by joining the points by means of free hand drawing is the cumulative curve or ogive. For the calculation of median, take a point on the axis of Y that is equivalent to $N/2$ and from this point draw a line parallel to X-axis. This line will cut the curve and from the cutting point, draw a line perpendicular to Y-axis. The distance of the X-axis from origin and the point at which the perpendicular line cuts is the value of median.

Example 18

Find out the median rainfall from the distribution given in example 13 by the graphic method:

Solution

Figure 3.1 shows the cumulative frequency curve formed between the upper limits of classes and corresponding cumulative frequencies. The point $N/2$ is shown on Y-axis and the dotted line parallel to X-axis cuts the cumulative curve at C. Perpendicular line cuts the X-axis at M. The distance OM is the median value.

Mode

The mode or modal value of the distribution is that value which occurs most frequently. For example, at any station the average number of occurrences for thunder storms or days with snowfall, wind direction etc. are most realistically presented by modal value. In case of frequency distribution the mode is that value which has maximum frequency. If two or more observations occur the same number of times then there is more than one mode and the distribution is multi-modal as against uni-modal.

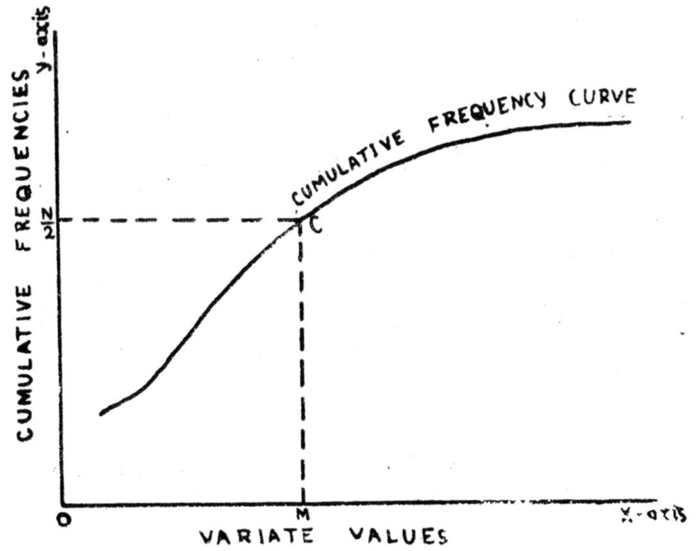

FIG. 3.1 : **Cumulative Frequency Curve**

Calculation of Mode

The mode in a series of individual measurements can be located either of two ways :

(i) The data should first be placed in an array so that items having the same value can be identified and quickly counted, the value of that item which occurs most of the times is the modal value.

(ii) The data should be converted into a discrete series.

Example 19

Find the modal temperature value from the values given in example 11 (a).

Solution

(i) Array: 58, 59, 59, 60, 62, 62, 68, 68, 68, 68, 68, 71, 71, 71, 73, 73, 74, 74, 75, 82.

Here mode = 68° F.

(ii) Discrete series :

X:	58,	59,	60,	62,	68,	71,	73,	74,	75,	82
f:	1	2	1	2	5	3	2	2	1	1

Mid-Value

Here the value 68 occurs the maximum number of times, hence it is mode.

Discrete Series (grouped data)

In case of discrete frequency distribution, mode can be located by inspection of the distribution alone. The size having the maximum frequency will be reckoned as mode.

Example 20

Compute the modal size of children born per family in the locality from the data given in example 3.

Solution

The highest size of frequency in the given distribution is 154 and corresponding to this frequency the number of children born per family is 2. Hence the modal size of children born per family in the locality is 2.

Continuous Series (grouped data)

The steps in the computation of mode from a continuous frequency distribution are :

(i) Determine the modal class interval. It is the class interval with the maximum number of frequencies in it. This can be found out by just observing the series.

(ii) Determine the value of mode by applying the following formula :

$$\text{Mode} = L + \left(\frac{f - f_p}{2f - f_p - f_s} \right) . h$$

where,

L is the Lower limit of the modal class

f is the frequency of the modal class

f_p is the frequency of the class preceding the modal class

f_s is the frequency of the class succeeding the modal class and h is the class width of the modal class.

Example 21

Compute the modal agricultural holding of the village from the data given in example 15,

Solution

The maximum frequency 'f' in the distribution is 150 which corresponds to class group 15-19, i.e., 14.5—19.5 in continuous case (see column 2, example 15). Hence modal class is 14.5—19.5. Now mode is computed as:

$$\text{Mode} = L + \left(\frac{f-f_p}{2f-f_p-f_s}\right) \times h$$

Here $L = 14.5, f = 150, f_p = 35, f_s = 70$ and $h = 5$.

$$\text{Mode} = 14.5 + \frac{150-35}{150\times2-35-70} \times 5$$

$$= 14.5 + \frac{115}{195} \times 5$$

$$= 17.45 \text{ acres.}$$

Example 22

Compute the modal temperature for the distribution given in example 6.

Solution

The maximum frequency 'f' in the distribution is 30 which corresponds to class-group 70-80 and hence the modal class is 70-80. Corresponding to this modal class we have,

$$L = 70 \,;\, f = 30 \,;\, f_p = 13; f_s = 16 \text{ and } h = 10$$

$$\text{Mode} = L + \frac{f-f_p}{2f-f_p-f_s} \times h$$

$$= 70 + \frac{30-13}{2\times30-13-16} \times 10$$

$$= 70 + \frac{17}{31} \times 10$$

$= 75.48° F.$

Sometimes mode is also computed with the help of mean and median. For a symmetrical distribution mean, median and mode coincide and if the distribution is moderately asymmetrical, the mean, median and mode are approximately related by the formula:

$$\text{Mode} = 3 \text{ Median} - 2 \text{ Mean.}$$

Example 23

If the mean and median of a moderately asymmetrical series are 12.9 and 12.1 respectively, what would be its most probable mode?

Solution

Mean = 12.9, Median = 12.1, Mode ?

$$\text{Mode} = 3 \text{ Median} - 2 \text{ Mean}$$
$$= 3 \times 12.1 - 2 \times 12.9 = 36.3 - 25.8 = 10.5.$$

ACKNOWLEDGEMENT

The authors are grateful to Mr. L. Sharma for numerical calculations.

FOUR

Emilio Casetti

MATHEMATICAL MODELLING AND THE EXPANSION METHOD

Background

A VAST and growing segment of contemporary scientific research in the social sciences make use of mathematical models relating to constants and/or variables and/or random variables.

The progress of this research involves a recursive process in which two logically distinct phases can be identified. One phase is concerned with formulating or revising disciplinary conceptual frameworks that define (a) the issues and phenomena to be studied, (b) what constitutes pertinent information and data, (c) the procedures through which information and data are obtained, and finally (d) the 'important' relations among phenomena that are singled out for investigation. The second phase of this process is concerned with modelling. It involves constructing mathematical models, extracting their quantitative and qualitative implications, estimating the models' parameters, and testing hypothesis.

The conceptual and modelling phases of the research process interact, and in fact research unfolds *via* recursive interactions between the two. The results of the modelling phase of the process, in the form of implications of models or of tests of hypotheses, produce modifications of the previous conceptual framework. This in turn generates new tasks to be carried out during a subsequent modelling phase, and so on.

Most of the attention of the scholars working with models tends to be focussed on parameter estimation, or on extracting the implications of propositions formalized into specific mathematical models, or on producing models required by the pursuit of specific research leads within a disciplinary context. The logical processes by which models are arrived at tend to remain in the background and generally are not viewed at as a distinct object of enquiry. And yet an awareness of the mental and logical routines by which models are generated has the potential to render modelling simpler, more efficient, and more likely to produce better results.

The Expansion Method

Casetti (1972) outlined a routine for creating or modifying models, made of a sequence of clearly identified logical steps, and called it "expansion method". The expansion method is a technique for generating models. Basically, it consists in the following:

(a) an "initial" model in which some or all of the parameters are in non-numerical form is selected;

(b) some or all of the letter parameters are "expanded" by redefining them as functions of variables that may or may not appear in the initial model;

(c) a "terminal" model is generated by replacing the expanded parameters into the initial model.

To demonstrate the method at the formal level, let us assume an initial model of the logistic type

(1) Y = K/(1+exp(A+B*T))

where Y and T are variables and K, A, and B are "letter" parameters. Suppose we wish to expand K into a linear function of T, and A into a quadratic function of a variable S that does not appear in the initial model. Then we write

(2) K = K0 + K1*T

(3) A = A0 + A1*S + A2*(S**2)

K0, K1, A0, A1, and A2 are expansion parameters.

The terminal model is obtained by replacing (2) and (3) into (1):

(4) Y = (K0+K1*T)/(1+exp (A0+A1*S+A2*(S**2))

The expanded model (4) is more complex than the initial one, and in it Y is a function of the two variables T and S.

Most mathematical models can be logically related to simpler ones, by thinking them as the result of some expansion of the simpler models' parameters. For instance, quadratic polynomials can be regarded as expansions of linear polynomials. To show it, take as initial model the polynomial Y=M+N*X and expand N into a linear function of X. Namely, N=N0+N1*X. Then, the expanded model Y=M+N0*X+N1*(X**2) is a quadratic.

As a technique to relate models to one another, the expansion method is quite general, since any model can be expanded, and in turn just about any model can be thought of as the result of the expansion of a simpler model. The primitive within this frame of reference is a model consisting of a variable set equal to a constant value, because no expansion can generate it. On the other hand, any primitive Y=A can be converted by repeated expansions into a wide variety of different functional relations.

The expansion method's capability to routinize the creation of mathematical models renders it suitable for a number of practical applications. The two applications that will be discussed and demonstrated in this paper are concerned respectively with testing relations for "stability", and with creating custom made complex functions for specific research purposes. The two types of applications will be considered in sequence.

Testing a Functional Relation for Stability

Assume that we have a set of observations on the three variables Y, X1, and X2, and that the functional relation of primary interest to us is

(5) Y = A + B*X1

However we suspect that A and B in (5) might be different for data with different X2 values. In order to clarify the matter we could divide the data into subsets homogeneous with respect to X2, estimate the coefficients of equation (5) for each of these subsets and then investigate the estimated A and B to detect whether they change in trendlike fashion with increasing X2. An alternative is to redefine the two parameters into functions of X2. This would expand the initial model (5) into a terminal model that allows the direct testing of hypotheses concerning the stability of A and B with respect to X2. Specifically, suppose we choose to expand A and B into linear functions of X2. In order to do so we write

(6) A = A0 + A1*X2

(7) B = B0 + B1*X2

and then replace (6) and (7) into (5) to obtain the expanded equation

(8) Y = A0 + A1*X2 + B0*X1 + B1*X1*X2

The parameters of (8) can be estimated using the entire data set. If the estimated A1 and B1 are not statistically significant, the relation Y=A+B*X1 is "stable", or at least does not have the type of instability that can be detected by linear expansions. It could still have, though, an instability that could be brought to the surface by some other expansion of the two coefficients. If instead either A1 or B1 or both are significant, this indicates that the relation Y=A+B*X1 changes with X2 in the manner that can be made explicit by replacing the estimated A1 and B1 into equations (6) and (7).

In the sections that follow the technique outlined in the preceding sections will be demonstrated by applying it to an analysis of the determinants of change in agricultural labour productivity.

Agricultural Productivity Example

Casetti (1979) has shown that the rates of growth of agricultural labour productivity is positively related to indices of population pressure, and tends to be larger when agricultural production expands rapidly (Verdoorn Law). It could be questioned whether the estimated functional relations on which these conclusions are based are stable with respect to development levels. In other words, do population pressures and Verdoorn effects affect the rates of growth of agricultural labour productivity to the same extent and in the same direction in countries at different levels of economic development ?

In the sections that follow, the pertinent background issues will be outlined. Then the stability of the relations between agricultural productivity growth, population pressures, and rates of growth of agricultural product will be investigated using the expansion method.

In recent years, the issue whether population pressures have a favourable impact on economic development has been the object of heated controversies with direct implications on labour productivity (Robinson and Horlacher 1971; Simon 1977; Clark 1978). Population pressure is sometimes operationalized in terms of high population density, taken to imply a high ratio of population to resources, and sometimes as high rates of population growth. The specific implications of each of these have not been clearly separated, and at any rate the population optimists have tended to view both in positive terms.

It has been argued that population pressures stimulate innovation and technical progress and, in general, are conducive to get individuals and nations to work harder and to innovate more. Also, it has been claimed that denser and faster growing populations benefit from scale economies. On the opposite side of these issues, negative effects of high population pressures have been associated with diseconomies of scale and external diseconomies due to congestion, population, and resource depletion. Also it has been claimed that high levels of population growth are associated with lower savings and at the same time tend to channel a greater portion of available savings to "demographic" investments as opposed to the investments with a stronger impact on economic progress (Coale and Hoover, 1958). The empirical analyses aimed at identifying whether faster population growth is associated with higher rates of development remained inconclusive (Easterlin, 1967; Chesnais and Sauvy, 1973; Simon, 1977). Instead, recent investigations of the effects of population density on development seem to be consistent with the optimists' views (Szimanski 1974; Stryker 1977).

Theoretical propositions regarding the effects of population pressures can be translated into statements concerning the determinants of labour productivity in agriculture. However, in the agricultural sector there is a complicating factor that needs to be taken into account. This factor is the movement of population out of agriculture that has long since been recognized as a concomitant of economic development. Since it can be presumed that in the balance, the people who move out of agriculture are the ones who provide the least productive labour inputs to it, and since the exodus of people from the countryside is associated with increases in farm size that facilitate technological progress, it ought to be presumed that in agriculture higher growth rates of labour force are associated with lower rates of growth of productivity (Kumar, 1973, p. 182 ff.; Hayami, 1971). But the rate of growth of the agricultural labour force is related to the rate of growth of population. Which in turn implies that the gross effect of the rates of growth of population on productivity can be negative even if the corresponding net effect is positive. This net effect is the effect of population growth on productivity growth when the rate of change of the agricultural labour force is controlled for.

Verdoorn (1949) suggested that there is a positive linear relation between rate of growth of productivity and rate of growth of output. His statement was based on data for manufacturing as a whole, and for individual industrial sectors, in different countries. Also most of the subsequent attempts to verify Verdoorn Law have been based on manufacturing data (Salter, 1960; Kennedy, 1971; Cripps and Tarling, 1973). However a variant of Verdoorn Law sometimes referred to as Kaldor Law has been tested for other economic sectors as well. Kaldor Law relates growth rates of labour productivity to growth rates of employment rather than of output (Kaldor, 1966; Cripps and Tarling 1973; Rowthorn, 1975; Cornwall, 1976).

The empirical tendency for expanding economic sectors to experience higher rates of productivity growth can be explained by some combination of the following: (a) expanding sectors may benefit from external economies and from economies of scale, (b) economic growth is associated with an inflow of capital with embodied technological progress, (c) the most productive workers who are carriers of modernity values and are trained in the latest technologies find greater opportunities for employment in expanding economic sectors, and finally (d) modern technologies and "learning by doing" are more likely to start within the expanding segments of an economic sector but eventually diffuse throughout the entire sector. These factors of productivity growth may differ in intensity and manner of operation in various circumstances, but there seems to be no ground for presuming that they are restricted to manufacturing.

The hypotheses that percentage rates of change of labour productivity in agriculture are significantly and positively related to population density (D), to percentage rates of change of population (P) when the rate of change of agricultural labour force (LA) is controlled for, and to percentage rates of agricultural product (YA) were tested by Casetti (1979).

The data used in that investigation, and that will be reemployed in the stability analysis described here, pertain to the following variables:

(1) Percentage rate of change of population to be denoted by P.

(2) Percentage rate of change of labour force in agriculture, denoted by LA.

(3) Percentage rate of change of GDP originating in agriculture, denoted by YA.

(4) Population density in people per square kilometre in 1970, denoted by D.

(5) GNP per capita in the year 1965, in US 1965 dollars, denoted by Y.

(6) Percentage rate of change of labour productivity in agriculture, denoted by AP.

All the percentage rates of change in the data set are "per year" and refer to the decade 1960-70. P and YA were obtained by averaging the rates for the year 1950-65 and 1965-70, published in World Bank (1976, p. 392 ff). LA was derived from data on percentage labour force in agriculture and total labour force for the years 1960 and 1970 (World Bank 1976, p. 514 ff). Y is from Taylor and Hudson (1972, p. 314 ff.). Finally, A was calculated from YA and LA using the definition AP=YA—LA

The initial data set included all countries with a 1973 population of 2 million people or more. Then all observations with missing values on one or more variables were removed. Also removed were Saudi Arabia because of its high ratio of oil revenues to population, and Singapore, because of its exceptionally high population density. The analysis in Casetti (1979) was carried out by estimating the coefficients of the equations.

(9) AP = A + B*D

(10) AP = A + B*P + C*LA

(11) AP = A + B*YA

The estimated regression equations are reproduced in line 1 of Tables 4.1 to 4.3. The estimated regression coefficients in these equations are all positive and significant at the 5 per cent significance level or better on the basis of t values.

TABLE 4.1 : **Agricultural Productivity Example—Regression Tests of Population Density Effects**

Eq/No	Constant	D	D*Y	D*(Y**2)	R**2
1	2.8738	8.618E-3	—	—	.080
	(7.29)	(2.45)			
2	3.0103	3.351E-4	7.495E-3	—	.135
	(7.71)	(.06)	(2.09)		
3	3.2210	—.0137	.0542	—.0220	.256
	(8.69)	(—2.11)	(3.71)	(—3.29)	

In this and in subsequent tables *t*-values are in parentheses under the respective regression coefficients.

In order to test whether the population pressure effects and the Verdoorn effects on agricultural productivity are different for countries at different development levels, the B coefficients in equations (9), (10) and (11) were expanded into linear and into quadratic functions of Y, where Y stands for GNP per capita and is taken as an index of development. The expansion was carried out only on the B coefficients of the original equations since they are associated with the three "effects" tested for. On the other hand, both linear and quadratic expansions were considered in order to allow for the possibility that countries at intermediate development levels would experience effects different from those experienced by countries closer to the end points of the development continuum. The expanded equations are

(12) AP = A + B0*D + B1*Y*D

(13) AP = A + B0*P + B1*P*Y + C*LA

(14) AP = A + B0*YA + B1*Y*YA

(15) AP = A + B0*D + B1*Y*D +B2*(Y**2)*D

(16) AP = A + B0*P + B1*Y*P + B2*(Y**2)*P + C*LA

(17) AP = A + B0*YA + B1*Y*YA + B2*(Y**2)*YA

The regression estimates for these expansions are shown in lines 2 and 3 of Tables 4.1 through 4.3 Table 4.1 contains the results of the analysis of the population density effects. Clearly the best result is represented by equation (3) that implies that the population density effect is indeed a function of development, and tends to be stronger for countries at intermediate development levels. Instead, the effect of population growth on the growth of agricultural productivity shown in Table 4.2 do not appear to be significantly different at different levels of development.

The results of the tests of Verdoorn effects are given in Table 4.3. In regression 2 and 3 the variable YA is included. However, since the regression coefficients of YA in these regressions are not statistically significant the two equations were re-estimated without YA terms (equations 4 and 5, Table 4.3). In all cases the expansion terms are strongly significant and show that the effect of the growth rate of agricultural product on agricultural productivity is strongest at intermediate development levels. An extended discussion of these results is not appropriate here. It will suffice to point out that factors favourable to agricultural productivity appear to operate more strongly in the countries at intermediate development levels. In any case this analysis shows

TABLE 4.2 : **Agricultural Productivity Example—Regression Tests of Population Growth Effects**

Eq/No	Constant	LA	P	P*Y	P*(Y**2)	R**2
1	1.0543 (2.08)	—1.0618 (—13.11)	1.0363 (4.73)	—	—	.754
2	1.0851 (2.13)	—1.0846 (—12.3)	1.0697 (4.44)	—.1041 (—.67)	—	.755
3	1.3079 (2.53)	—1.0298 (—11.18)	.8242 (3.15)	.5661 (1.39)	—.2711 (—1.78)	.767

TABLE 4.3 : **Agricultural Productivity Example—Regression Tests of Verdoorn Effects**

Eq/No	Constant	YA	YA*Y	YA*(Y**2)	R**2
1	2.1330 (3.07)	.4188 (2.19)	—	—	.065
2	1.3039 (2.11)	.3020 (1.82)	.7012 (5.08)	—	.322
3	1.6090 (2.67)	.0346 (.18)	1.5705 (4.53)	—.4219 (—2.51)	.389
4	2.2347 (6.30)	—	.7361 (5.30)	—	.289
5	1.6968 (4.60)	—	1.6048 (5.51)	—.4371 (—3.33)	.389

the expansion method to be a flexible, easily implementable tool to investigate whether relations among variables are stable or instead whether they change gradually when some other variable changes.

The application of the expansion method to generate complex functions suited for the analysis of spatial temporal trends is demonstrated in the sections that follow.

Using the Expansion Method to Create Custom-Made Functions

The expansion method is especially well suited for generating complex custom made models suited for specific research objectives. In the sections that follow this type of application will be discussed and demonstrated by generating functions suitable to describe and estimate the spatial temporal dynamics of urban population densities (Casetti, 1973).

A large body of research has established that urban population densities tend either to decline with distance from city centres, or to increase at first and then to decline (Berry *et al.* 1963; Casetti, 1969; Newling, 1964, 1966, 1969). Both types of trends can be expressed and estimated using a quadratic exponential function of the form.

(18) D (S) = exp (A + B*S + C* (S**2))

where D stands here for population density in people per unit of area, S is distance from the city centre, and A, B, and C are parameters. The parameter A is logarithm of population density at the city centre. If $C = 0$ and $B<0$ or $B = 0$ and $C<0$ or $B<0$ and $C<0$ equation (18) identifies a trend characterized by a decline of population density with distance from the city centre. Specifically, if $C = 0$ and $B<0$ equation (18) specializes to the well known negative exponential function introduced by Clark (1951) and that has been used to fit a great many sets of density distance data. When $B = 0$ and $C<0$ equation (18) specializes to the Gaussian function used in population density work by Tanner (1961).

If $B>0$ and $C<0$ equation (18) will represent a density "crater" characterized by population densities that at first increase and then decline as the distance from the city centre is increased. In the density crater case the maximum population density is attained at distance S = —B/(2*C) from the city centre, rather than at $S = 0$ as it is the case when density "cones" occur.

At last four patterns of change of population density surfaces over time have been discussed in the literature (Berry *et al.* 1963; Blumenfeld, 1954; Newling, 1964, 1966, 1969). The first of them is characterized by an increase in population density throughout an urban area, that does not alter the gradient of a cone-shaped density surface. This trend has been observed for cities in the less developed areas and for western cities prior to the 20th century. The second time pattern is characteristic of cities in developed countries in recent times, and consists in a flattening of cone-shaped density surfaces. The third pattern involves the transition from a density "cone" to a density "crater" and differs from the preceding one because of the more pronounced decline of central densities, that in fact turns the central density peak into a depression. This trend seems to be associated with major increases in city size, especially in developed countries in recent times. The fourth pattern has been observed in major cities characterized by a density crater and consists in a widening of the crater.

Clearly these spatial temporal trends correspond to changes over time in the parameters of equation (18) that can be easily specified. Namely :

(a) A trend of the first kind takes place if either B = 0 and C<0, or B<0 and C = 0, or B<0 and C<0, and if A increases over time while B and C remain unchanged.

(b) A trend of the second kind occurs if either B = 0 and C<0, or B<0 and C = 0, or B<0 and C<0, and either B or C or both decline over time.

(c) A trend of the third kind can be produced by various changes of A, B and C over time, and takes place if S = —B/2C changes over time from a negative or zero value to a positive one.

(d) Finally a spatial temporal trend of the fourth kind is produced by changes of B and/or C over time such that S= —B/(2*C) is greater than zero and increases over time.

Suppose that a body of urban population density data covering successive time periods is available, and that these data are to be tested for consistency with anyone of the trends described above. Applying the expansion method, the coefficients A, B, and C of equation (18) would be redefined into some function of time, namely

(19) A = A(T) = A(A0, A1, T)

(20) B = B(T) = B(B0, B1, T)

(21) C = C(T) = C(C0, C1, T)

Two-parameter functions are assumed here, but no additional conceptual difficulty is involved if functions with more parameters are used. Then A, B, and C in equation (18) are replaced respectively with A(Y), B(T), and C(T) so that a new function D(S, T)

(22) D(S, T) = exp (A(T) + B(T) *S + C(T) * (S**2))

is obtained that is defined with respect to S and T, and has A0, A1, B0, B1, C0, and C1 as its parameters. The parameters of

D(S, T) would then be estimated from the data available. These estimates can then be used in order to check whether the conditions for the occurrence of one of the trends under consideration are met. Parameter estimates that are not significantly different from zero should be taken to be zero.

Any one of a large number of functions of time with not necessarily vanishing first derivatives could be used for expanding the coefficients of Newling's equation. For instance, the following functions could be used:

(23) A(T) = A0*exp (A1*T)

(24) B(T) = B0*exp (B1*T)

(25) C(T) = C0*exp (C1*T)

However, upon substitution into equation (18) these functions would yield an equation

(26) D(S, T) = A0*exp (A1*T) + B0*exp (B1*T) + C0*exp (C1*T)

That is intrinsically not linear and therefore would require a cumbersome iterative estimation procedure. Unless valid considerations to the contrary exist, it is preferable to expand A, B, and C so as to obtain an expanded function linear in the parameters, capable of estimation using an ordinary linear regression computer program. In this sense any expansions of A, B, and C into polynomials in T are acceptable since they would yield a D(S, T) linear in the parameters. The question remains of how many terms these polynomial expansions should contain. In general, it would seem desirable not to exceed the minimum number of terms that allows the identification of the spatial temporal trends investigated. Here linear expansions are adequate for the purpose intended.

(27) A = A0 + A1*T

(28) A = B0 + B1*T

(29) C = C0 + C1*T

By substituting (27), (28) and (29) in (18) the following equation is obtained

(30) D(S, T) = exp (A0 +A1*T + B0*S + B1*S*T + C0* (S**2) + C1* (S**2)*T)

The coefficients of (30) can be easily estimated by ordinary multiple regression.

An application of the technique discussed in the prceeding sections was carried out by fitting equation (30) to multitemporal density distance data for Montreal and Philadelphia. The data used are gross population densities for concentric rings centred at the cities's CBD's, published in Blumenfeld (1954) and Montreal (1965). The reader interested in the mechanism of the calculation of the data is referred to the sources. The distance of each ring from the respective CBD's for use in the regression was approximated by using the mean of the distances of the boundaries of the ring.

Equation (30) was fit to the data using a stepwise regression computer program. Regression steps with the highest R squared and in which all regression coefficients are T-significant at the 5 per cent level were selected. The regression results are given in Table 4.4.

TABLE 4.4 : **Urban Population Densities Example—Estimates of Spatial Temporal Trends**

Eq/No	Constant	S	S**2	T*(S**2)	R**2
1	3.8649	—	—.0791	.0017	.890
	(28.42)		(—10.88)	(3.69)	
2	4.4966	.2137	—.0611	.0009	.962
	(38.24)	(3.12)	(7.15)	(11.12)	

Equations 1 and 2 are based respectively on the Montreal and Philadelphia data.

In order to analyze the regression results shown in Table 4.4 let us view them as estimates of the functions A(T), B(T), and C(T) defined when equation (18) was expanded. For Montreal we have

(31) $A(T) = 3.8649$

(32) $B(T) = 0$

(33) $C(T) = -.0791 + .0017*T$

These show that over the 20 years time span considered C(T) is negative and increasing and B(T) equals zero. Which indicates a spatial temporal trend of the second kind, corresponding to a flattening of a density cone.

For Philadelphia we have

(34) A(T) = 4.4966

(35) B(T) = —.2137

(36) C(T) = —.0611 + .0009*T

that also identify a spatial temporal trend of the second kind over the 50 years time horizon covered by the data set used.

Conclusion

The expansion method is a technique for generating or modifying mathematical models. It consists in taking an initial model that has at least some "letter" parameters, and in redefining some or all of them into functions of relevant variables. This paper discusses the application of the expansion method for testing the "stability" of functional relations, and for generating complex mathematical models suited for specific research objectives.

SUGGESTED READINGS

Berry, B.J.L., Simmons, J.M., and Tennant, R.J., "Urban Population Densities: Structure and Change", *Geographical Review*, 55 (1963) 389-405.

Blumenfeld, H., "The Tidal Wave of Metropolitan Expansion", *Journal of the American Institute of Planners*, 20 (1954) 3-14.

Casetti, E., "Alternate Urban Population Density Models. An Analytical Comparison of their Validity Range", in A.J. Scott (ed.) *Studies in Regional Science*, London: Pion Limited, 1969, pp. 105-16.

Casetti, E., "Generating Models by Expansion Method: Applications to Geographical Research", *Geographical Analysis*, 4 (1972) 81-91.

Casetti, E., "Testing for Spatial Temporal Trends: An Application to Urban Density Trends Using the Expansion Method", *Canadian Geographer*, 16 (1973) 27-37.

Casetti, E., "Correlates and Determinants of Agricultural Productivity Growth: A Test of Some Hypotheses",[(1979) forthcoming on *Agricultural Geography*, Noor Mohammad (ed.)

Chesnais, J.C., Sauvy A., "Progress Economiques et Accroissement de la Population: une Experience Commentee", *Population*, 28 (1973) 843-57.

Clark, C., Urban Population Densities", *Journal of the Royal Statistical Society*, Series A, 114 (1951) 490-96.

Clark, C., "Population Growth and Productivity", *Research in Population Economics*, 1 (1978) 143-54.

Coale, A.J.; Hoover E.M., *Population Growth and Economic Development in Low Income Countries*, Princeton University Press, Princeton, 1958.

Cornwall, J., "Diffusion Convergence and Kaldor's Laws", *Economic Journal*, 86 (1976) 307-14.

Cripps, T.F., Tarling R.J., *Growth in Advanced Capitalist Economies 1950-1970*, Cambridge University Press, Cambridge, 1973.

Easterlin, R.A., "Effects of Population Growth in the Economic Development of Developing Countries", *The Annals of The American Academy of Political & Social Science*, 369 (1967) 98-108.

Hayami, Y.; Ruttan V.W., *Agricultural Development: An International Perspective*, The John Hopkins Press, Baltimore, Maryland, 1971.

Kaldor, N., *Causes of the Slow Rate of Growth in the United Kingdom*, Cambridge University Press, Cambridge, 1966.

Kennedy, K.A., *Productivity and Industrial Growth: The Irish Experience*, Clardon Press, Oxford, 1971.

Kumar, J., "Population and Land in World Agriculture—Recent Trends and Relationships", *Population Monograph Series No. 12*, Institute of International Studies, University of California, Berkeley, California, 1973.

Montreal, "City Planning Department, the Wave of Metropolitan Expansion", *Technical Bulletin* 1, Montreal 1965.

Newling B.E., Urban Population Densities and Intraurban Growth", *Geographical Review*, 54 (1964) 440-42.

Newling, B.E., "Urban Growth and Spatial Structure: Mathematical Models and Empirical Evidence", *Geographical Review* 56 (1966) 213-25.

Newling, B.E., "The Spatial Variation of Urban Population Densities", *Geographical Review*, 59 (1959) 242-52.

Robinson, W.C.; Horlacher, D.E., "Population Growth and Economic Welfare", *Reports on Population/Family Planning*, 6 (1971)1-40.

Rowthorn, R.E., "What Remains of Kaldor's Law?", *The Economic Journal*, 85 (1975) 10-19.

Salter, W.E.G., *Productivity and Technical Change*, Cambridge University Press, Cambridge, 1960.

Simon, J.L., *The Economics of Population Growth*, Princeton University Press, Princeton, N.J., 1977.

Stryker, J.D., "Optimum Population in Rural Areas: Empirical Evidence from the Franc Zone", *The Quarterly Journal of Economics*, 91 (1977) 177-93.

Szymanski A., "Economic Development and Population", *Studies in Comparative International Development*, 9 (1974) 53-69.

Taylor, C.L., Hudson M.C., *World Handbook of Political and Social Indicators*, Yale University Press, New Haven, Connecticut, 1972.

Tanner, J.C., "Factors Affecting the amount of Travel", *Road Research Technical Paper 51*, London: Department of Scientific and Industrial Research, 1961.

Verdoorn, P.J., "Fattori che Regolano lo Sviluppo della Produttivita del Lavoro", *L'Industria*, 1 (1949).

World Bank, *World Tables 1976*, The Johns Hopkins Press, Baltimore, Maryland, 1976.

FIVE

Joseph Uyanga

TESTING FORMAL POLARISATION OF GROWTH IN SPATIAL CONTEXT

IN ABSTRACT terms "a field of forces" defines a space, which can also be seen in terms of concentrations or homogeneous aggregates. Conceived as "centres, poles or focii from which centrifugal forces emanate and to which centripetal forces are attracted; Peroux (1950), the growth pole refers to the growth conditions with this set of characteristics. Each pole being a "Centre of attraction and repulsion, has its proper field which is set in the field of all other Centres" (p. 27). Darwent (1969) discusses and unravels some of the confusions surrounding these concepts and evaluates the usefulness of the concepts in spatial analysis. The geonomical aspects of these ideas have been developed by Friedmann (1967), Hansen (1968) and Boudeville (1968). Only a few tests of the spatio-temporal aspects have so far been carried out, the most interesting of which is Casetti *et al.* (1971).

An important normative question about the growth pole/centre notions is that which concerns the regional location of investments over time and space. The growth pole/centre concept can be given a defined direction if it is analytically clarified, for instance the findings may provide some insights on how the existing urban system within a nation can be rationally organised and guided to serve enhanced developmental ends.

This paper defines polarised regions as those collections of geographic spaces in which connections and flows of goods, services, labour, and socio-economic strengths are *predominantly* in one direction. The boundary of a polarised region is therefore that line beyond which flows and connections are predominantly in some other directions that is towards some other poles. This definition has many features of John Friedmann's centre-periphery model. The polarised region can exist at any scale. It is thus not only compatible with the central place structure of a hierarchy of cities, but also is considered to involve a continuous space-time process.

This paper adopts the formulation by Casetti and others already referred to, but tests the model this time not in the U.S.A. but in the South Australian context. To further their work, sectorial polarization is also tested, in addition to the polarisation of the total non-agricultural employment.

The Model

(1) Consider a growth situation (*G*) "distributed" over a two dimensional space and let $G(x, y, t)$ be a measure of the intensity of growth at some point with coordinates x and y at time t.

(2) Let *G* be defined over a region *S*, that for convenience will be assumed to be reasonably compact, and over a time interval *T*. Assume that *G* is differentiable at least twice almost everywhere with respect to x, y, and everywhere with respect to t belonging to *T*, *G* corresponds to a surface, the elevation of which identifies the spatial distribution of the intensity of growth. Changes of *G* over time involve a warping and stretching of the surface because of the differentiability assumption. Positive change in growth intensity occurs if:

$$\frac{dG}{dt} > 0, x, y \in S, t \in T \qquad \text{......(1)}$$

If growth is polarised with respect to one pole the implication will be that there is one region such that at any other region the rate of growth of the phenomenon is a decreasing function of distance from the polarised regions. It is specified that $\overline{x}$, $\overline{y}$ are co-ordinates of the pole and $S(x, y)$ the Euclidean distance of a region's (x, y) from $(\overline{x}, \overline{y})$.

A condition of polarised growth is implied by the following two conditions:

$$\frac{dG}{dt} > 0 \text{ for all } t \text{ and } S \qquad \text{.....(2)}$$

$$\frac{d^2G}{ds\,dt} < 0 \text{ for all } t \text{ and } S \qquad \text{......(3)}$$

A condition of declining growth can be obtained by modifying (3) to become $\frac{dG}{dt} < 0$ for $S = S^*$ where S^* is the threshold of negative growth.

Specification

If a set of observation is given concerning variables of growth (or phenomena indicating growth surrogate at a given time) and distances from an hypothesised growth centre each of these observations will consist of three sets of variables, level or intensity (G_i); time, (t_i) and distance (S_i). Let the function, $G = G(S, t)$ be defined as possessing non-zero derivatives with respect to t, and to s and t. Let the estimated function be represented as $\hat{G}$. Following equations (2) and (3); the necessary and sufficient conditions of polarised growth could be given by

$$\frac{d\hat{G}}{dt} > 0 \quad \text{......(4)}$$

and
$$\frac{d^2\hat{G}}{dt\,ds} < 0 \quad \text{......(5)}$$

and if any function

$$G = a_0 + a_1 t + a_2 S + a_3 St \quad \text{......(6)}$$

is fitted to the data, polarised growth requires that

$$\frac{d\hat{G}}{dt} = \hat{a}_1 + \hat{a}_3 S > 0 \quad \text{......(7)}$$

for all distances (S) in the sample, and that

$$\frac{d^2\hat{G}}{dt\,ds} = \hat{a}_3 < 0 \quad \text{......(8)}$$

An Empirical Example

For convenience equation (8) was fitted to the data for 25 regions in South Australia. (Table 5.1) in order to assess the pattern of polarisation relative to Adelaide, the largest city in that State. Statistical regions constitute the units of data collection. The model was fitted for different dependent variables using the conventional least squares regression method.

TABLE 5.1: **Statistical Divisions : Data Base (Employments by Sectors)**

	Distance from Adelaide (m/s)	Employment in Mfg. (Gm)			Employment in Commerce, Finance and Property (Gc)			Total non-agricultural employment		
		1961	1966	1971	1961	1966	1971	1961	1966	1971
City of Adelaide	0	2803	2069	1402	1880	1515	1651	11601	9359	7886
Inner Western Suburb	2	7250	4856	3951	2339	2140	2433	19180	11282	9821
Inner Eastern ,,	2	12114	11281	8611	11169	9767	11691	4311	42034	40470
North East ,,	6	12350	12530	13201	6768	8690	11093	3211	37359	44677
Eastern Suburban	5	5624	7708	6801	5481	8496	11299	34481	32775	35083
South Eastern Hills	9	5204	5920	5156	5584	6467	8696	20688	24821	27185
South West Coastal Suburb	15	8876	12593	14242	8418	11101	15829	31586	42787	51352
Western Suburb	4	5532	8867	8463	3352	8709	11536	18760	31443	34510
North Western Suburb	5	23853	24109	21321	11682	11381	13920	58012	58775	56851
Para	15	5929	11945	11745	2346	4894	7617	14533	28270	34450
Central Plains	80	734	663	483	1433	1339	1467	8372	8033	7689
York Peninsula	150	228	328	317	102	572	719	754	3940	1418
Kangaroo Island	100	47	48	22	158	180	193	1251	1324	1294
Northern Ranges	75	1368	1735	1872	1034	1284	1621	7319	8434	8647

Southern Ranges	40	1969	1805	1694	1748	1673	1973	13945	12711	12168
Upper Murray	149	1441	1976	1195	1523	1797	2840	12101	13052	12904
Lower Murray	48	891	1169	1192	918	1121	1312	6094	7104	7672
Mallee	121	72	73	29	332	389	356	2866	3235	2744
Upper South East	152	367	412	278	1259	1463	1708	7615	8548	8526
Upper Eyre	165	2992	3659	3749	2215	2638	3196	12718	14822	14941
Lower Eyre	181	84	128	67	437	506	625	3921	3578	3790
Spencer	293	640	882	592	1133	1474	1771	7405	8681	8207
Frome	184	6426	3424	9947	2816	3288	4536	20383	24355	27662
Lower South East	139	462	369	316	520	725	764	5197	4965	4515
Far North	306	1295	1496	1143	246	316	461	6839	6360	6839

Source : Australian Bureau of Statistics, South Australia.

t = Time period 1961 through 1971
s = Distance in miles from the City of Adelaide.

It is hypothesised that if the two conditions, specified above, hold, employments increased and that the increases are smaller, the greater will be the distance from the city of Adelaide. This model is repeated for the manufacturing and tertiary sectors of the economy in order to discover which sector is the most polarised of the lot.

Summary of Equations

$$Gm = 4862.05 + 9.301t + 0.6715s - 0.552\,st \quad R = .88$$

$$Gc = 2795.12 + 7.546t + 0.622s - 0.181\,st \quad R = .74$$

$$Gn = 7448.91 + 8.682t + 0.707s - 0.592\,st \quad R = .91$$

$$\frac{dGm}{dt} = 9.301 - 0.552s$$

$$\frac{d^2Gm}{ds\,dt} = -0.552$$

$$\frac{dGC}{dt} = 7.446 - 0.181s$$

$$\frac{d^2Gc}{ds\,dt} = -0.181s$$

$$\frac{dGn}{dt} = 9.682 - 0.592s$$

$$\frac{d^2Gn}{ds\,dt} = -0.592$$

In all the parameters, the first derivations with respect to time are positive, and negative with respect to distance. They hypothesis that increases in total and non-agricultural sectoral employment are polarised in Adelaide appears to hold. Total non-agricultural employment tends to be smaller for regions which are far from Adelaide.

The spatial pattern of this polarisation is however not well portrayed by the above discussion. This is affected by the trend surfaces based on a rather indirect way of evaluating polarisation of development. Twenty-eight socio-economic variables were submitted for the principal component analysis, Uyanga (1976). The standardised areal scores (Z) on the first principal component which accounted for 46.3 per cent of the variations "explained" by three major components, was correlated on distance from Adelaide. The equation $Z = 21.7 - 0.22d$ (where d is distance from Adelaide) was obtained. The coefficient of correlation was —0.701. This shows an inverse relationship between the index of regional development, as summarised by this component, and distance from Adelaide.

A grid system was superimposed on the study area. Northings and eastings, the independent variables affecting areal scores, were read from 48 locations selected from regular triangular intervals. Trend surfaces were constructed from these readings. The cubic surface is shown on Fig. 5.1. This portrays the regional develop-

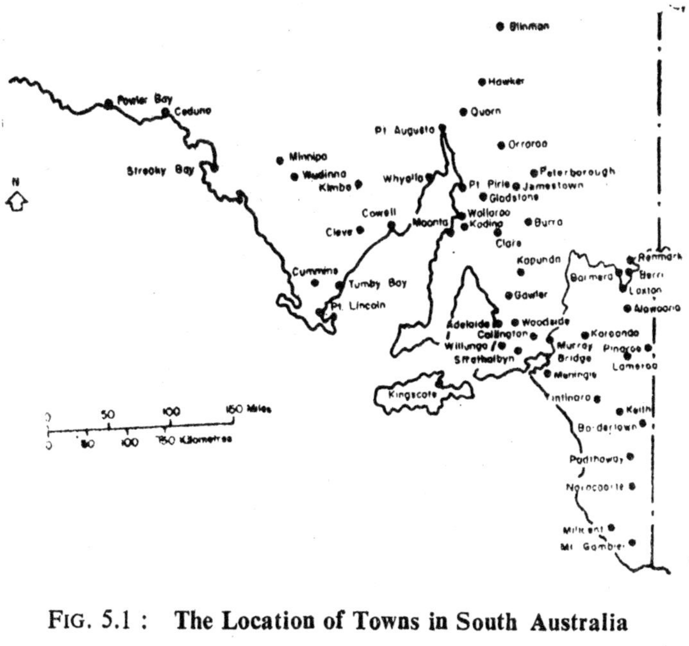

FIG. 5.1 : **The Location of Towns in South Australia**

ment surfaces representing the areal pattern of progressive declining influence of Adelaide. The positive residuals (Fig. 5.2) from the cubic surface (Fig. 5.3) indicate that the effect of Adelaide as a polarising force is underestimated in the medium and large country towns of Whyalla, Port Lincoln, Mount Gambier, and Loxton. These provide smaller scale countervailing local poles of growth. The dominance of Adelaide in the economic landscape clearly gives weight to the conclusion that there is a marked spatial polarisation of economic growth in the State of South Australia.

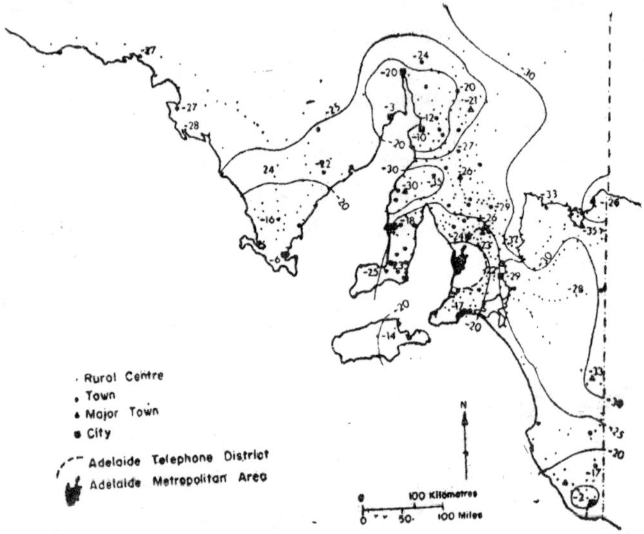

FIG. 5.2 : **The Cubic Trend Surfaces of Growth in South Australia**

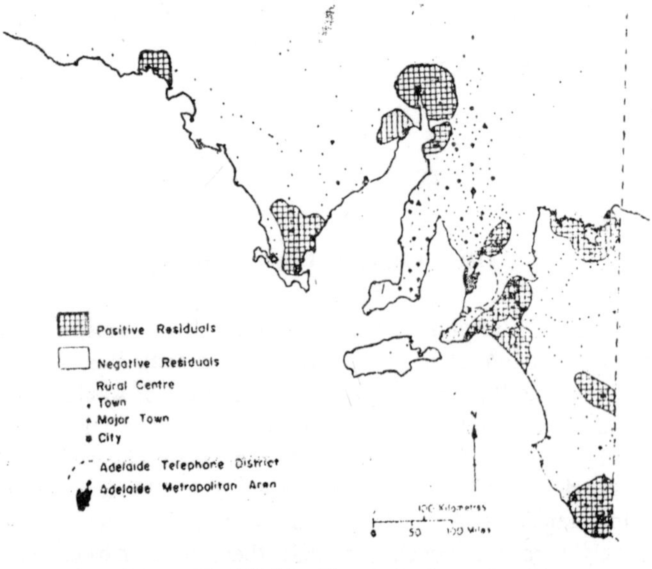

FIG. 5.3 : **The Cubic Trend Surfaces: Residuals**

It might be quite enlightening to conclude by observing the effects of this spatial process through the negative residuals. The negative residuals coincide with Adelaide's "trade shadow" where between a half and two-thirds of secondary goods and services are obtained from Adelaide, Smailes (1969), (Fig. 5.4). These troughs in the development space are the areas where a large number of

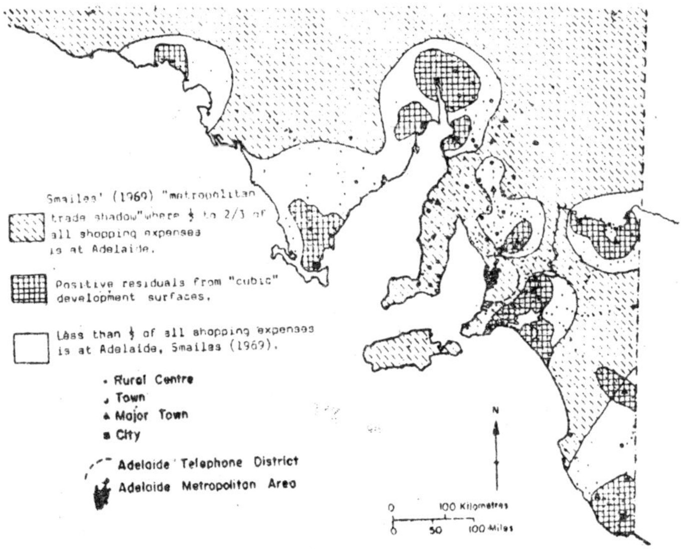

FIG. 5.4 : **The Cubic Trend Surfaces Residuals : Their Relation with Smaile's (1969) Adelaide Trade Shadow**

people by-pass their local centres entirely for many comparison goods in Adelaide (Williams, 1965). To that extent the net effect of polarisation and growth of Adelaide on its region is a sort of backwash whereby not a parallel growth, through higher and increased local demand, capital investment and employment is generated but a counterposed decline especially in the small centres in favour of polarised growth in Adelaide.

SUGGESTED READINGS

Boudeville, J. R. "Les notions d'espace et dintegration", *L'Espace et les poles de Croissance*" Press Universitaire de France, Paris, 1968.

Casetti, E.; King, L.J. and Odland J., "The Formalization and Testing of Concepts of Growth Poles in a Spatial Context", *Environment and Planning*, Vol. 3, 1971, pp. 377-382.

Darment, D.F., "Growth Poles and Growth Centres in Regional Planning—A Review", *Environment and Planning*, Vol. 1, 1969, pp. 5-32.

Friedman, J. R., "A General Theory of Polarised Development, *mimeo* Santiago, Chile, 1967.

Hansen, N.M., *French Regional Planning*, Indian University Press, Bloomington, Indiana, 1968.

Perroux, F., "Economic Space, Theory and Applications", *Quarterly Journal of Economics*, Vol. 64; 1950, p. 54.

Smailes, P. J., "Metropolitan Trade Shadow", *Tijds voor Econ. en. Soc. Geographic*, Vol. 60, 1969, pp. 329-345.

Uyanga, J., "The Regional Development Pattern of South Australia", *Australian Geographical Studies*, Vol. 14, 1976, pp. 15-32.

Williams, M., "The Influence of Adelaide on Rural Shopping Habits in Counties Frome, Dalhousie and Victoria", *Australian Geographer*, Vol. 9, No. 5, 1965, pp. 312-315.

SIX

S.K. Mishra

TECHNIQUES OF COMPUTER PROGRAMMING WITH FORTRAN IV

INTRODUCTION

THE USE OF computers for sophisticated analysis in planning and social sciences is gaining momentum day by day. As a matter of fact, the use of a computer for numerical analysis (statistical or mathematical) of a sizeable complex of data is inevitable. It not only helps an analyst in drawing conclusions from the otherwise highly time and labour consuming and equally tiresome job of manual computation, but it also widens the scope of building and testing a variety of hypotheses and models exclusively necessary for arriving at more realistic and useful inferences and decisions.

However, in India social scientists face twofold difficulties in this regard. Firstly, computer services are not available in most of the research institutions. Secondly, most of the social scientists are not properly trained in using computers even if available at an approachable end. Readymade programs are usually not publicly available and it is not often easy to take the help of a programmer in a computer centre if the user is himself not easily able to express his needs to a programmer and make him understand what exactly the user wants to do.

Fortunately, more and more academic institutions are installing computers and it may be hoped that by the next decade computer facilities would be easily available to most academic and planning institutions. Nonetheless, the second type of problem discussed above would linger unless useful programs are published abundantly, and curricula in academic institutions would include courses in computer programming.

The objective of this paper is to introduce to the readers the techniques of computer programming. It has been assumed that the reader is a stranger to this field. Details have been avoided, though some useful references and bibliography have been appended for those who want to go keenly for further studies. For the use of the readers some important programmes have also been presented. These programmes have been run successfully and ensure results if used properly.

An Introduction to Computer

The computer is an electronic machine. It works on the binary logic (0, 1) mechanically expressed in terms of presence (1) or absence (0) of current and magnetic charge. Flow of current and, thus, existence of magnetic charge (or otherwise) is used for process determination and control of the system. The functioning of the computer is controlled through programs. A program is a set of logically consistent series of directives synchronised according to the plan of a work or job through which the flow of current is controlled at stages finally yielding the desired results. A computer is an automatic machine in which various electronic circuits work at stages as per the directions provided by the programs.

Let us go in for simulation for a while. A computer is a miniature of a human brain. It has a memory. It has decision power. It can control various limbs of its body and order them to do certain activities. Now, for some illustrations.

Suppose, one is asked "Write my name on this piece of paper". The first requirement for a successful result is that the person so asked must understand the meaning of the words and the whole sentence together If the person does not know English, he will not do the job. Next, he must know the name of the person asking so. If every thing is alright, the person must know writing. Then, he will pick up a pen, or pencil, etc., and would do the required job. Now, starting from listening, understanding, recalling the name from memory, picking up a suitable writing device, up to the final work—there is a long sequence of mnemonic actions—controls, decisions etc.

The whole process may be divided into four broad categories of sub-processes. They are (1) Input—pen and paper and source program i.e. "Write my name on this piece of paper". (2) Compilation – or understanding and preparation of the complete sequence of actions for doing the job and establishing relations with memory and the subroutine of the mode of operation. (3) Execution or using inputs according to the program prepared by the brain. (4) Output. And then stop.

Computers also work in the same fashion. Computers know certain languages which must be used for getting work done. As human beings may know many languages, so do computers. As we have languages like Hindi, English, Bengali, German, Greek, etc. so do computers have FORTRAN, COBOL, ALGOL, PL/1 etc. All languages have their grammar, so have computer languages. Computer programs are written in any of these languages. In this paper we would deal with FORTRAN only. For other languages, see any appropriate book given in the suggested readings.

Mode of Input and Output

Input—Data or source program – may be supplied to the machine by various means. They are (1) Punched cards, (2) Punched paper tapes, (3) Magnetic tape etc. Similarly, output may be obtained in (1) Punched cards, (2) Punched tapes, (3) Print, (4) Magnetic tape, etc. (We avoid auto-visual modes here). The most popular mode of input is 'punched card' and that of output is Print, or punched card.

The Language : Fortran (IV)

We start with an example :—

Example 1

```
1. READ 2, A, B, C
2. FORMAT (3F8.1)
3. D=(A + B)/C
4. PRINT 5, D
5. FORMAT (F 12.4)
```

The first statement (No. 1) orders the computer to read (input) values for A, B and C (Let A = Income from land, B = Income from salary and C = Total no. of members in the family) whose layout in the mode of input is specified by the reference 2 to the FORMAT

Numbered 2. Let us for the time being not go for understanding the meaning of 3 F 5.1, we would discuss it later. Next, statement No. 3 orders the computer to compute the value of D which equals sum of A and B divided by C. Then PRINT (output) D according to the layout—referred to by No. 5—specified in format statement no. 5 FORMAT (F 10.4): Again, we will avoid explaining F 10.4 here.

Now one more example :—

Example 2

```
120 READ 21, A, B, C
21 FORMAT (3 F 8.1)
D=(A + B)/C
PRINT 54, D
54 FORMAT (F 12.4)
GO TO 120
```

Compare this programme with the earlier one. In this the number of the first statement has been changed and also reference to the format is now 21. The 3rd statement is not numbered at all. The 4th statement also goes unnumbered. However, the 4th statement refers to Format No. 54 which was numbered 5 earlier. Lastly, we order the computer to go to statement no. 120.

All these mean that in computer programming statement numbers need not be in sequence. However, when we refer to some number the statement designated by that number is taken into consideration. All statements may not be numbered, nevertheless, if a statement is referred to, it must bear the relevant number.

Now, suppose, our data of A, B and C are as follows (Say, for five families).

These data would be punched on 5 cards, one row each. The computer would first read the 1st card containing A, B and C values for the 1st family. It would compute value of D and print. Then it would go back to statement No. 120 that is—READ 21, A, B, C and thus, next, it would read the 2nd card containing data on A, B and C for the 2nd family. Then it would compute D, print the value of D for second family and would again go back to statement No. 120 for reading 3rd card and thus it would continue.

TABLE 6.1 : **Family Income and Number of Persons in Family**

Family No.	Income		No. of persons in family
	A	B	C
1	110.0	360.0	4
2	330.0	1210.0	5
3	1680.0	1810.0	4
4	2358.0	3598.0	7
5	3521.0	321.0	4

Now, look at the program below.

Example 3

```
    130 READ 210, B, D, C
210 FORMAT (3 F 8.1)
    A = (B + D)/C
    PRINT 53, A
    53 FORMAT (F 12.4)
    GO TO 130
```

If data cards are punched as in example 2 the 1st column will be read as B, second column as D and the 3rd column as C. A will be computed and printed. It is interesting to see that from the given set of input in an order results are not different even if they are read under different names. However, computationally the operations must be identical. In example 3 we would get wrong results if we compute D = (A + B)/C.

Example 4

```
130 READ 210, B, D, C
210 FORMAT (3 F 8.1)
    Q = B
    P = D
    R = C
    A = (Q + P)/R
    Z = A
```

```
      PRINT 53, Z
53 FORMAT (F 12.4)
      GO TO 130
```

This program (example 4 above) also would yield the same result as in example 2 or 3, because Q, P and R have been defined to be identical to B, D and C respectively and Z would represent A of the example 3.

Now, suppose we want to print values of D in example 2 only if they exceed Rs. 500 per person. This can be done as follows :

Example 5

```
120 READ 21, A, B, C
21 FORMAT (3F 8.1)
      D = (A + B)/C
      IF (D . GT . 500.0) PRINT 54, D
54 FORMAT (F 12.4)
      GO TO 120
```

Note that IF (D . GT . 500) PRINT 54, D means that if value of D is greater than 500 then print the value of D according to the layout format specified by number 54—otherwise do not print and proceed. It may be learnt here that

.GT. means greater than
.LT. means less than
.GE. means greater than equal to
.LE. means less than equal to
.NE. means not equal to
.EQ. means equal to.

Also note that .GT. etc are between two dots.

Dimension and Subscripted Variables

Recall example 2 in the section 4.1. We have five families and corresponding to each one we have data on A, B and C and we require D to be computed. We have, thus, five values for each A, B and C. If we write A(1), B(1) and C(1) (whose numerical values are 110.0. 360.0 and 4 respectively) for A, B and C of 1st family, A(2), B(2) and C(2) (whose numerical values are 320.0, 1210.0 and 5 respectively) for 2nd family and so on up to 5th family, this type of expression of variables A, B and C is called subscripted and may be expressed as

A(I), B(I), C(I), I = 1, 5, 1

that is value of I or index varies from 1 to 5 with difference of one.

If a subscripted variable is used in a program it must be declared so in the 1st statement of the program. Usually, this is accomplished by DIMENSION statement. The program in example 2 would be written in terms of subscripted variables as follows :

Example 6

```
      DIMENSION A(5), B(5), C(5), D(5)
      READ 21, ( (A(I), B(I), C(I)), I = 1, 5, 1)
21 FORMAT (3 F 8.1)
      I = 0
20 I = I + 1
      D(I) = (A(I) + B(I))/C(I)
      PRINT 54, D(I)
54 FORMAT (F 12.4)
      IF (I . LT .5) GO TO 20
      STOP
      END
```

In the example 6 above we start with index set equal to zero. In statement no. 20 the old index is replaced by new value with an increment of one (1) over the old value. Further, subsequent operations are accomplished with A(1), B(1) and C(1) yielding D(1) since index value I is one (1), and the output D(1) is printed. Next, when the statement *IF (I.LT.5) GO TO 20* is in operation, the index I is tested against 5. Since upto now I is not equal to 5 (as a matter of fact I is 1), hence the control is transferred to statement no. 20 where I is replaced by I + 1, that is to say now I becomes 2 and further operations continue with A(2), B(2) and C(2) yielding D(2). However, this process continues with changing values of I uptill I remains less than 5. As soon as I equals 5 control is not transferred to 20— rather it comes down to the subsequent (*STOP*) and the operation stops.

The same job could be done by using DO statement as shown in the following example

Example 7

```
      DIMENSION A(5), B(5), C(5), D(5)
      READ 21, ((A(I), B(I), C(I)), I = 1, 5)
21 FORMAT (3 F 8.1)
      DO 25 J = 1, 5, 1
      D(J) = (A(J) + B(J))/C(J)
      PRINT 54, D(J)
25 Continue
54 FORMAT (F 12.4)
      STOP
      END
```

In example 7 we introduce DO Loop. This loop starts with the statement *DO 25 J = 1, 5, 1* and terminates on *25* CONTINUE. It means "Continue performing all operations that come inside this loop for all values of J, J taking values from 1 to 5 with an increment of 1 i.e., J = 1, 2, 3, 4, 5". Thus, within this loop D(1), D(2), D(3), D(4) and D(5) will be computed and printed. Do statement is a very useful statement and is often applied for repetitive operations changing with index values. CONTINUE is a dummy statement used as a termination point of Do Loop.

Format Statement

We have used FORMAT statements in all the preceding examples although they have never been explained. Now, we will explain FORMAT statement.

FORMAT statement contains the information on the layout of data. There may be four possible specifications of the layout of data. They are

1. How many data are put in one card
2. What is the nature of data—(a) Integer, (b) real, (c) exponential, (d) alphabetical or textual, (e) blank.
3. What is the width of a datum—or total number of characters contained by a datum.
4. If data contain decimal, how many characters after decimal are there.

An integer datum is expressed by I format; a real datum (containing integer part, decimal and some digits after decimal) is expressed by F format; an exponential datum is expressed by E format; an alphabetic or textual datum is expressed by A format ; and blank is expressed by X format.

Recall Example 7, statement No. 21. The format *FORMAT (3 F 8.1)*. The layout of data this format represents is characterised by the following specifications :

1. There are three data in one card (recall, they are values of (A(I), B(I), C(I)), I=1, 5). Since I=1, 5, and one card contains only three data, total no. of cards required for input would be five.
2. The nature of data is real (in other words, floating point) and hence, the format specifications is F.
3. The total width of a datum is eight
4. Total number of digits after decimal is one.

Let us now take an example of a set of integer (or fixed point) data. Suppose, we have to read the following table. The data on income is in whole Rs. and no fraction of a rupee is considered.

TABLE 6.2 : **Data on Income**

Family No.	Income (Monthly) Rs.
1	422
2	318
3	718
4	1120
5	913
6	319

Then, the relevant statement would be

Example 8

```
DIMENSION INCOME (6)
READ 35, (INCOME (K), K = 1, 5)
35 FORMAT (6 I 5).
```

The format *35 FORMAT (6 I 5)* means here that all the six data on income of six families are put in one card. The nature of data is integer and each datum has a width of 5. It may be noted here that all the figures do not have the same width. **However,** for homogeneity's sake, all the data have been brought to the same width by putting blanks or zeros in the left side of the figures.

At this juncture it may be noted that in Fortran IV two types of variables are considered. They are (1) Integer variables—strictly starting with the letters I or J or K or L or M or N. (2) Real variables—starting with any letter *other* than I, J, K, L, M and N. In both cases, a variable name should never exceed six letters or six alphanumeric characters in all. However, these restrictions are relaxed by "type statements". Any variable, whatever letter it may start with, shall be considered real or integer if it has been declared so at the very outset of the program. Thus, if we declare the variable INCOME as real (although it is integer—since it starts with I), the type statement would be

REAL INCOME (6)

and it will be replacing *DIMENSION INCOME (6)* in the example 8 above. It is to be noted that these type statements also work as DIMENSION statements.

For other formats the readers are requested to refer to the books on computer programming listed in the bibliography.

Algebraical expressions and their FORTRAN version

Following are a few frequently used algebraical notations and their FORTRAN versions.

Algebraical Notation	*Fortran Version*
1. $A + B$	A + B
2. $A - B$	A — B
3. $A \times B$ (or A.B)	A * B
4. $\frac{A}{B}$ (or A/B)	A/B
5. A^2	A**2
6. A^3	A**3
7. $\sqrt{A}$	SQRT (A)
8. $\sqrt[5]{A}$	A** (1/5)
9. A_{ij}	A (I, J)

10. {(A + B) . (C — D)} . (K + P) ((A+B)*(C—D))*(K+P)
(Note that there are no different parentheses and brackets etc. in the FORTRAN Language, nor there is any lower case letter).

11. $\frac{a}{b} + \frac{c}{d}$ (A/B) + (C/D)

12. log (a) A L O G 10 (A)

13. In (a) or ea LOG (A)

14. antilog (a) (10.0)**A

15. exp (a) (2.71828)**A

16. cos a COS (A)

17. tan a TAN (A)

18. sin a SIN (A)

19. $Q = \sum_{i=1}^{n} bi$

```
  Q = 0.0
  DO 1 I=1, N
  Q = Q + B(I)
1 CONTINUE
```

20. $G = \prod_{i=i}^{M} pi$

```
  G = 1.0
  DO 1 I = 1, M
  G = G* P (I)
1 CONTINUE
```

21. $Ri = \sum_{j=1}^{N} Pij \quad i=1, m$

```
  DO 2 I = 1, M
  R(I) = 0.0
  DO 1 J = 1, N
1 R (I) = R (I) + P (I, J)
2 CONTINUE
```

22. $Qi = \sum_{J=1}^{N} Bij \quad i=l, m \quad i=j$

```
  DO 2 I = 1, M
  Q (I) = 0.0
  DO 1 J = 1, N
  IF (I.EQ.J) GO TO 1
  Q (I) = Q (I) + B (I, J)
1 CONTINUE
2 CONTINUE
```

MAIN PROGRAM AND SUBPROGRAMS

Main program is a set of statements by which the following five functions are performed.

1. DIMENSION OF relevant variables.
2. Input of data by READ and other relevant statements.
3. Output of data/results by print and other relevant statements.
4. Calling of subroutine subprograms or use of Function subprograms.
5. Performing certain logical or arithmetic or both types of operations.

Subprograms are subsidiary, supplementary or servile programs performing certain specified logical or arithmetic or both types of operations for the use of and at the call of the main program. A subprogram may use one or more subprograms subsidiary to it, properly nested in sequence. A subprogram cannot use itself as its subprogram. Subprograms are mainly of two types— (a) Function subprograms, (b) Subroutine subprograms. Function subprograms are used simply by their name with relevant argument. For example, SQRT (A) is the use of function subprogram SQRT (A). The example of the use of subroutine subprograms performed by CALL will be illustrated in due course.

Examples

The table below presents rainfall (in an area) during twelve weeks of rainy season. Compute average rainfall.

TABLE 6.3 : **Weekly Rainfall Receipt for 12 Weeks**

Rainfall in inches	1	2	3	4	5	6	7	8	9	10	11	12
	10.1	15.2	18.0	17.3	15.8	20.1	25.3	20.2	18.3	15.2	10.8	6.5

Denoting rainfall as a variable named X and using formula $\bar{X} = \sum_{i=1}^{n} x\,i/n$ where n is the total number of observations (12 here), the main program will be as follows.

Example 9

```
      DIMENSION X (12)
      N=12
      READ 1, (X (I), I = 1, N)
 1    FORMAT (12 F 5.1)
      X BAR = 0.0
      DO 2 I = 1, N
      X BAR = X BAR + X (I)
 2    CONTINUE
      X BAR = X BAR/N
      PRINT 3, X BAR
 3    FORMAT (F 10.3)
      STOP
      END
```

It is to be noted here that the complete operation has been performed in the main program only. However, this specified work of computing average of X BAR may be done by introducing subroutine also.

Example 10

```
C     MAIN PROGRAM
      DIMENSION X (12)
      N = 12
      READ 1, (X (I), I = 1, N)
 1    FORMAT (12 F 5.1)
      CALL MEAN (X, N, X BAR)
      STOP
      END
C     SUBROUTINE MEAN FOR COMPUTING X BAR
      SUBROUTINE MEAN (X, N, X BAR)
      X BAR = 0.0
      DO 2 I = 1, N
      X BAR = X BAR + X (I)
 2    CONTINUE
      PRINT 3, X BAR
 3    FORMAT (F 10.3)
      RETURN
      END
```

It is interesting to note that in the example set above, the main program is exactly used for giving dimension of the variable X, reading X and then calling the subroutine MEAN. The computation is performed in the subroutine and results also are printed therein.

Given any other subroutine, say, for standard deviation, the main program could obtain the results by simply calling the subroutine. Below, we give the subroutine for standard deviation. In this subroutine X BAR, X and N are the inputs and standard deviation (SD) is the output. This subroutine can be called only after the computation of X BAR is done.

Example 11

```
C   Subroutine SDEN for computing Standard Deviation.
    SUBROUTINE SDEN (X, X BAR, N, SD)
    DIMENSION X (12)
    SD = 0.0
    DO 1 I = 1, N
    SD = SD + (X (I) — X BAR)**2
1   CONTINUE
    SD = SQRT (SD/N)
    PRINT 2, SD
2   FORMAT (F 12.5)
    RETURN
    END
```

If we insert a new statement

```
    CALL SDEV (X, X BAR, N, SD)
```

in the main program (in example 10) in between CALL MEAN (X, N, X BAR) and STOP, the main program will be able to compute standard deviation of the rainfall tabulated earlier.

For better conception we set one more example. Below we present the table listing marks obtained by ten candidates in mathematics and physics. We require computation of the Averages, standard deviations and coefficient of correlation of marks obtained in the two subjects.

TABLE 6.4 : **Marks in Mathematics and Physics**

Student Roll No.	1	2	3	4	5	6	7	8	9	10
Marks in Maths.	60	65	70	72	72	68	72	60	50	56
Marks in Physics	65	70	70	68	76	80	50	60	40	40

Designating SUB 1 for maths and SUB 2 for Physics, and reading data on marks as real variables (assuming 60.0 for 60, 65.0 for 65 etc), we may write the main program as follows.

Example 12

```
C
C    MAIN PROGRAM FOR COMPUTING CORRELATION
     COEFFICIENT ETC.
     DIMENSION SUB 1(10), SUB 2(10) N = 10
     READ 1, (SUB 1(I), I = 1, N)
     READ 1, (SUB 2(I), I = 1, N)
   1 FORMAT (10 F 5.1)
     CALL MEAN (SUB 1, N, X BAR 1)
     CALL MEAN (SUB 2, N, X BAR 2)
     CALL SDEV (SUB 1, X BAR 1, N, SD1)
     CALL SDEV (SUB 2, X BAR 2, N, SD2)
     CALL CORR (SUB1, SUB2, X BAR 1, X BAR 2, SD1, SD2,
                                              N, R)
     STOP
     END

C    SUBROUTINE CORR FOR CORRELATION
     SUBROUTINE CORR (SUB 1, SUB 2, X BAR 1, X BAR 2,
                                     SD1, SD2, N, R)
     DIMENSION SUB1 (10), SUB2 (10)
     R = 0.0
     DO 1 I = 1, N
     R = R + (SUB1(I) —X BAR 1)* (SUB2(I) —X BAR 2)
   1 CONTINUE
     R = (R/N)/(SD1 * SD2)
     PRINT 2, R
```

```
2 FORMAT (F 10.6)
  RETURN
  END
```

Some Useful Subroutines

Assuming that by now the reader would be able to write a main program for proding appropriate dimension to the variables of his interest and reading input data, some very important subroutine subprograms have been appended with the paper. They may be used by CALL statement according to the need of the user. Considering the scope and limitations of this paper, it is not possible to give listing of all subroutines required for performing statistical analysis for geographers and social scientists. However, following are the subroutines which have been appended.

1. SUBROUTINE GINIC for computing Gini coefficient
2. SUBROUTINE CORRLN for computing inter correlation matrix
3. SUBROUTINE RANK for computing rank correlation
4. SUBROUTINE REGNL for regionalisation—or for computing synthetic Index by Principal Components method.
5. SUBROUTINE MREGR for multiple regression with provision of treatment of multicollinearity by Ridge regression method and Principal Component based orthogonal transformation method. This program may also be used for fitting bivariate regression.
6. SUBROUTINE MOMENT for computing moments up to desired order.
7. SUBROUTINE DISTBN for finding distribution through frequency counting.

SUGGESTED READINGS

Carl, F. Kossack and Clandia I., Hen Schke, *Introduction to Statistics and Computer Programming*, Holden-Day, San Francisco.

Chinal J., *Design Methods for Digital System*, Springer Verlag Berlin Heidelberg, New York.

Das Gupta, S., *Computer Programming Fundamental* (*Fortran IV*), Khanna Publishers, 2-B, Nath Market, Nai Sarak, Delhi-110006.

Donalt H. Sanders, *Computers in Society*, McGraw Hill Book Company.

Earnshaw, J.D.; Blockford, W.A.R.; *Computer Programming in Algol*, Pitman Publishing, London.

Forsythe, G. E.; Malcolm M.A.; and Moler, C.B.; *Computer Methods for Mathematical Computation*, Prentice Hall, Inc., Englewood Cliffs, N.J.

Joseph Fass, C., *Fortran IV: Programming and Applications*, Holden-Day, Inc., San Francisco.

Kiviat, P.J., Villan R. Lueva and Markowitz, H.M.; *The Script II: Programming Language*, Prentice Hall Series in Automatic Computation.

Robert, V. Jamison *Introduction to Computer Science Mathematics*, McGraw Hill Book Co.

Rice, John; Rice, John R.; *Introduction to Computer Science*; *Problems, Algorithms, Languages, Information and Computers*, with Algol and Fortran Appendices by Edward Desautels, Halt, Rinehart and Winston Inc., New York, Chicago, San Francisco, Atlantia, Dallas, Montreal, Toronto, London and Tokyo.

Rajaraman, V.; and Radhakrishna, T.; *Introduction to Digital Computer Design*, Prentice Hall, New Delhi, India.

Rajaraman, V., *Principles of Computer Programming*, Prentice Hall, New Delhi, India.

Sherman, *Techniques in Computer Programming*, Prentice Hall, Inc., Englewood Cliffs, N.J.

Stoutemyer, Soute, *PL/I Programming for Engineering and Science*, Prentice Hall, Inc., Englewood Cliffs, N.J.

APPENDIX

** ** ** **

SOME IMPORTANT SUB-ROUTINES

Subroutine No. 1

```
      SUBROUTINE GINIC (X, N, GINI, XX)
C     COMPUTES GINI COEFFICIENT
C     **          **          **          **
C
C     N=NO. OF OBSERVATIONS, X=INPUT SERIES,
      XX=ARRANGED SERIES
C     DEVELOPED BY S. K. MISRA DATED 10-9-78
C     COMPUTATION OF GINI COEFFICIENT
C     ON RETURN XX IS NEW X SERIES ARRANGED
      IN DESCENDING ORDER
C     **          **          **          **
C
      DIMENSION X (60), XX (60)
C
C     PRESERVE X FOR FURTHER USE IN MAIN
      PROGRAM
      DO 101 I=1, N
  101 XX(I) = X(I)
      AN=N
      N1=N—1
C     ORDERING OF THE IN DESCENDING ORDER
      KS=0
    1 KS=KS+1
    2 K=0
    3 K=K+KS
      KK=K
    4 KK=KK+1
      IF (KK—N) 45, 45, 6
   45 IF (X(K)-X(KK)) 5, 4, 4
    5 Y=X(K)
      Z=X(KK)
      X(K)=Z
```

```
      X(KK)=Y
      IF (KK—N) 2, 2, 6
    6 IF(KS—N1) 1, 1, 7
    7 PRINT 133, (X (K), K=1, N)
  133 FORMAT (10F 8. 2)
C     COMPUTATION OF GINI COEFFICIENT
      AV = 0.0
      DO 8 I=1, N
    8 AV=AV+X(I)
      AV=AV/AN
      SUM=0.0
      DO 9 I=1, N
      T=I
    9 SUM=SUM+X(I)*T
      P=2.0/( (AN**2)*AV)
      R=1.0/AN
      GINI=1.0+R—(P*SUM)
      PRINT 10, GINI
   10 FORMAT (2X, 26HGINI COEFFICIENTS VALUE= ,
     •F14.4)
      DO 102 J=1, N
      CHNG=XX (J)
      XX (J)=X (J)
      X (J)=CHNG
  102 CONTINUE
      RETURN
      END
```

Subroutine No. 2

```
      SUBROUTINE CORNL (X, M, N, AV, SD, MISRA)
      DIMENSION X (13, 21), AV (21), SD (21), C (21, 21)
C     M=NO OF OBSERVATIONS N=NO OF VARIABLES
      PRINT 15, ( (X (J, I), I=1, N), J=1, M)
   15 FORMAT (10F 10.4)
      AN=M
      DO 1 I=1, N
      AV (I)=0.0
      DO 2 J=1, M
      AV (I)=AV (I) + X(J, I)
    2 CONTINUE
```

```
 1 AV (I)=AV (I)/AN
   PRINT 5
 5 FORMAT (' AVERAGES   ** — ** — ** — ')
   PRINT 6, ((AV (I), I=1, N)
   DO 4 I=1, N
   DO 4 II=1, N
   A=0.0
   B=0.0
   AB=0.0
   DO 3 J=1, M
   DA=X (J, I)—AV (I)
   DB=X (J, II)—AV (II)
   A=A+DA**2
   B=B+DB**2
   AB=AB+DA*DB
 3 CONTINUE
   SD (I)=SQRT (A/AN)
   C (I, II)=AB/(SQRT (A*B))
 4 CONTINUE
 6 FORMAT (10F 10.3)
   PRINT 7
 7 FORMAT (' STANDARD DEVIATIONS     **—**—')
   PRINT 8, (SD (I), I=1, N)
 8 FORMAT (10F 10.4)
   PRINT 9
 9 FORMAT (' CORRELATION  MATRIX ***——      ')
   PRINT 10, ( (C (I, II), II=1, N), I=1, N)
10 FORMAT (10F 10.5)
   IF (MISRA.NE.0) RETURN
   DO 11 I=1, M
   DO 11 J=1, N
11 X (I, J)=(X (I, J)—AV (J))/SD (J)
   PRINT 12
12 FORMAT (' NORMALISED VARIABLES ')
   PRINT 13, ( (X (I, J), J=1, N), I=1, M)
13 FORMAT (10F 10.4)
   RETURN
   END
```

Subroutine No. 3

```
      SUBROUTINE RANK (X, Y, N, NO, R, RC)
C
C     COMPUTES RANK CORRELATION AND SERIAL
      CORRELATION
C
C     IF NO IS NEG. GET RANK CORRELATION, IF NO
      IS ZERO GET BOTH RANK
C     CORRELATION AND COEFFICIENT OF CONCUR-
      RENT DEVIATION, IF NO IS
C     POSITIVE GET COEFF. OF CONC. DEVIATION
      ONLY
C     N=NO. OF OBSERVATIONS. X, Y ARE SERIES TO
      BE CORRELATED
C     PROGRAM BY S. K. MISRA 18-1-1979
C
      DIMENSION X (N), Y (N)
C
      AN=N
      R=0.0
      RC=0.0
      IF (NO . GT . 0) GO TO 3
      DO 1 I=1, N
    1 R=R+(X(I)—Y(I))**2
      R=R*6.0
      R=1.0—(R/(N**3—N))
      PRINT 2, R
    2 FORMAT (42H VALUE OF RANK CORRELATION
      COEFFICIENT IS=, F 10.6)
      IF (NO) 6, 3, 3
    3 N1=N—1
      NC=0
      DO 4 I=1, N1
      J=I+1
      IF (X(J).GT.X(I)) L=1
      IF (X(J).EQ.X(I)) L=0
      IF (X(J).LT.X(I)) L= —1
      IF (Y(J).GT.Y(I)) M=1
      IF (Y(J).EQ.Y(I)) M=0
```

```
      IF (Y(J).LT.Y(I)) M= —1
      AK=L+M
      IF (ABS(AK).EQ.2.0) NC=NC+1
      IF (AK.EQ.1.0) NC=NC+1
    4 CONTINUE
      RC=(NC*2)/N
      RC=RC—1.0
      PRINT 5, RC
    5 FORMAT (49H VALUE OF COEFFICIENT OF CON-
      CURRENT DEVIATION IS=, F10.6)
    6 RETURN
      END
      SUBROUTINE PAL (X, N, AV, SD, C)
C     COMPUTS COMPOSITE INDEX BY PAL'S METHOD
      DIMENSION X(60, 3), AV(3), SD(3), C(3, 3), Y(60)
      M=2
      MISRA=0
      CALL    CONR (X, M, N, C, AV, SD, MISRA)
      DO 1 I=1, N
       X(I, 3)=X(I, 2)*SD(1)+X(I,1)*SD(2)
      X(I, 3)=X(I, 3)/(SD(1)*AV(2)+SD(2)*AV(1))
       PRINT 2, I, (X(I, J), J=1, 3), I
    2 FORMAT (13, 2X, 3F15.6, 13)
    1 CONTINUE
      M=3
      CALL    CONR(X, M, N, C, AV, SD, MISRA)
      RETURN
      END
```

Subroutine No. 4

```
      SUBROUTINE  REGNL  (X2, J1, N1, K1, XD, CX,
      BIG, MISRA, CD1, AV, SD, B1)
C
C     **            **            **            **
C     •
C     SUB-ROUTINE FOR REGIONALISATION  BY KAN-
      DALL'S METHOD
C     **            **            **            **
C
      DIMENSION XX(13, 15)
```

```
      DIMENSION    CD1(20, 20), AV(20), SD(20), CX(20),
                                        XD(500), S1(20)
      AN=J1
C     J1=NO. OF ROWS, N1=NO. OF COLUMNS,
      X2=INPUT DATA
C     NO. OF ROWS ARE OBSERVATIONS, NO. OF COL.
      ARE NO. OF VARIABLES
      DO 1 I=1, J1
      PRINT 2, (X2(I, J), J=1, N1)
    2 FORMAT (1OF 12.4)
    1 CONTINUE
      DO 4 J=1, N1
      AV(J)=0.0
      DO 3 I=1, J1
      AV(J)=AV(J)+X2(I,J)
    3 CONTINUE
    4 AV(J)=AV(J)/AN
      DO 6 J=1, N1
      DO 6 JJ=1, N1
      AS=0.
      ASS=0.
      ASR=0.
      DO 5 I=1, J1
      AD=X2(I, J)—AV(J)
      ADDD=X2(I, JJ)—AV(JJ)
      AS=AS+AD**2
      ASS=ASS+ADDD**2
    5 ASR=ASR+AD*ADDD
      SD(J)=SQRT(AS/AN)
      CD1(J, JJ)=ASR/(SQRT(AS*ASS))
    6 CONTINUE
      PRINT 7
    7 FORMAT(' CORRELATION  MATRIX ')
    8 FORMAT (12F10.6)
      DO 10 J=1, N1
      PRINT 8, (CD1(J, JJ), JJ=1, N1)
   10 CONTINUE
      PRINT 11
   11 FORMAT (' STD DEVIATIONS ')
      PRINT 12, (SD(I), I=1, N1)
```

```
   12 FORMAT (12F10.5)
      IF (MISRA) 2041, 13, 2041
   13 DO 14 J=1, N1
      DO14I=1, J1
      XNRM=(X2(I, J)—AV(J))/SD(J)
      X2(I, J)=XNRM
   14 CONTINUE
 2041 DO303I=1, N1
      B1(I)=0.0
      DO 302 J=1, N1
      B1(I)=B1(I)+CD1(I,J)
  302 CONTINUE
  303 CONTINUE
      KK=0
  330 KK=KK+1
      IF(KK--KI) 331, 331, 311
  331 BIG=B1(1)
      K=2
  304 DO 305I=K, N1
      IF(BIG.LT.B1(I)) GO TO 306
  305 CONTINUE
      GO TO 307
  306 BIG=B1(I)
      K=(I)
      GO TO 304
  307 DO 308 I=1, N1
      J=I
      CX(J)=B1(I)/BIG
  308 CONTINUE
      DO 310I=1, N1
      B1(I)=0.0
      DO 309J=1, N1
  309 B1(I)=B1(I)+CD1(I,J)*CX(J)
  310 CONTINUE
      GO TO 330
  311 PRINT 312
  312 FORMAT (1X, 13H EIGEN VECTOR)
      PRINT 313, (CX(J), J=1,  N1)
  313 FORMAT (8F8.5)
      PRINT 314, BIG
```

```
  314 FORMAT (F14.5, 20H EIGEN VALUE LARGEST)
      SUMB=0.0
      DO 3090I=1, N1
 3090 SUMB=SUMB+CX(I)**2
      IF(MISRA) 1111, 1112, 1111
 1111 ALPHA=BIG/SUMB
      DO 3091 I=1, N1
      CX(I)=CX(I)*ALPHA
      SD(I)=SD(I)*BIG
 3091 CX(I)=CX(I)*AV(I)
      TL=0.0
      DO 3092 I=1, N1
 3092 TL=TL+(CX(I)/SD(I))
      DO 3093 I=1, N1
      SD(I)=SD(I)*TL
 3093 CX(I)=CX(I)/SD(I)
      GO TO 1113
 1112 SUMB=SQRT(ABS(SUMB/BIG))
      DO 1114 I=1, N1
      CX(I)=CX(I)/SUMB
 1114 CONTINUE
 1113 DO 206 I=1, J1
      XD(I)=0.0
      DO 207 J=1, N1
  207 XD(I)=XD(I)+X2(I, J)*CX(J)
  208 CONTINUE
      PRINT 209
  209 FORMAT (2X, 'SERIAL NO', 2X, 'INDEX OF DEVE-
                                          LOPMENT')
      DO 211 I=1, J1
      PRINT 210, I, XD(I), I
  210 FORMAT (5X, I3, 6X, F14.8, 8X, 13)
  211 CONTINUE
      PRINT 17273
17273 FORMAT (' WEIGHTAGES ... ... ... +       ')
      PRINT 17274, (CX(J), J=1, N1)
17274 FORMAT(12F10.5)
      RETURN
      END
```

Subroutine No. 5

```
C       PROGRAM FOR MULTIPLE REGRESSION
        SUBROUTINE MREGR(X, XX, J1, N, I2, I1, C, CV,
        AV, SD, BETA, A, L, M, ALPHA, B, 1Y, TTEST, TA,
        G, RHO, ARSD, AGG, RIDGE, NRIDGE, DELTA,
        ORTH, ACOR, ORTHX, ZORTH, 2, BORTH, CORTH,
        WORTH, GORTH, HORTH, XINDX, SPACE)
        DIMENSIONX (484, 12), C(12, 12), CV(12, 12), AV(12),
        SD(12), BETA(12, 12), 1A(6, 6), L(6), M(6), ALPHA(6),
        Y (484, 12),     TTEST(10, 10), TA(10),
        2G(6), BB(6)   , B(6)   , XX(12,11,44)   , RIDGE(6, 6)
        DIMENSION ACOR(6, 6), ORTHX(484, 6), ZORTH(6),
        BORTH(6), CORTH(6)
        DIMENSION WORTH(6), CORTH (6, 6), HORTH
        (6,6), XINDX (484, 6), SPACE(484, 6)
C
C       **              **              **              **
C
C       DEVELOPED BY S.K. MISRA DATED 18.8.1978
C       J1=TOTAL NO. OF OBSERVATIONS N=NO. OF
        INDEPENDENT VARIABLES
C       12=TOTAL NO. OF DEPENDENT VARIABLES,
        11=TOTAL NO. OF DEPENDENT
C       PLUS INDEPENDENT VARIABLES
C       IF NRIDGE SET NONZERO/R ORTH SET NONZERO
C       RITY BY RIDGE REGRESSION METHOD AND
        PRINCIPAL COMPONENTS METHOD
C       THIS PROGRAM HAS A BUILT IN FACILITY OF
        DEALING WITH MULTICOLLINEA
C
C       **              **              **              **
C       *               *               *               *
C
        IB=I2+1
        O=10.0
        O=0**30
        O=1.0/0
        IF(ORTH.EQ.0.0) GO TO 50001
        DO 50002 J=IB, I1
```

```
      JORTH=J—I2
      DO 50002 I=1, J1
      ORTHX(I, JORTH)=X(I, J)
50002 CONTINUE
      DO 50004 J=1, N
      AV(J)=0.0
      DO 50003 I=1, J1
50003 AV(J)=AV(J)+ORTHX(I, J)
50004 AV(J)=AV(J)/J1
      DO 50006 J=1, N
      DO 50006JJ=1, N
      ORTH1=0.0
      ORTH2=0.0
      ORTH3=0.0
      DO 50005 I=J1,
      ORTHDI=ORTHX(I, J)—AV(J)
      ORTHD2=ORTHX(I, JJ)—AV(JJ)
      ORTH1=ORTH1+ORTHD1**2
      ORTH2=ORTH2+ORTHD2**2
50005 ORTH3=ORTH3+ORTHD1*ORTHD2
      SD(J)=SQRT(ORTH1/J1)
      ACOR(J, JJ)=ORTH3/(SQRT(ORTHI*ORTH2))
50006 CONTINUE
      CALL BROOT (ACOR, N, ZORTH, BORTH, CORTH,
      WORTH, GORTH, HORTH)
      DO 50007 J=1, N
      DO 50007 I—1, J1
      ORTHX (I, J)=(ORTHX (I, J)—AV(J))/SD(J)
50007 CONTINUE
      DO 50009 K=1, N
      DO 50009  I=1, J1
      XINDX (I, K)=0.0
      DO 50008 J=1, N
50008 XINDX (I, K)=XINDX(I, K)+ORTHX(I, J)*HORTH
      (K, J)
50009 CONTINUE
      DO 50010 J=1, N
      JORTH=J+I2
      DO 50010 I=1, J1
      CHAC=XINDX (I, J)
```

```
      XINDX (I, J)=X(I, JORTH)
      X(I, JORTH)=CHAC
50010 CONTINUE
C     GO FOR IDENTIFICATION OF VARIABLES
      NTWO=N*2
      DO 50011 J=1, N
      JORTH=J+I2
      DO 50012 I=1, J1
50012 SPACE (I, J)=X(I, JORTH)
50011 CONTINUE
      NAUG=N+1
      DO 50013 J=NAUG, NTWO
      JORTH=J—N
      DO 50013 J=1, J1
50013 SPACE (I, J)=XINDX (I, JORTH)
      CALL CORRLN (SPACE, J1, NTWO, AV, SD, C)
50001 AZXC=0.0
      DO 6 I=1, I1
      AV(I)=0.
      DO 5 J=1, J1
    5 AV (I)=AV(I)+ X(J, I)
    6 AV (I)=AV(I)/J1
      PRINT 7, (AV(I), I=1, I1)
      DO 9 I=1, I1
      DO 9 II=1, I1
      SD1=0.
      SD2=0.
      CVR=0.
      DO 8 J=1, J1
      S=X(J, I)—AV(I)
      SS=X(J, II)—AV(II)
      IF(S.LE.O)GO TO 8
      IF(SS.LE.O) GO TO 8
      SD1=SD1+S**2
      SD2=SD2+SS**2
      CVR=CVR+S*SS
    8 CONTINUE
      SD(I)=SQRT(SD1/J1)
      CV(I, II)=CVR/J1
    9 C(I, II)=CVR/(SQRT(SD1*SD2) )
```

```
      PRINT 12, (SD (I), I=1, I1)
      PRINT 11, ( (C (I, II), I=1, I1), II=1, I1)
      DO 10 I=1, N
      DO 10 J=1, N
      IN=I+I2
      JM=J+I2
   10 A(I, J)=CV(IN, JM)
      DO 4001 I=1, N
      DO 4001 J=1, N
 4001 RIDGE (I, J)=A(I, J)
30001 FORMAT (4F 16.5)
      PRINT 30001, ( (A (I, J), J=1, N), I=1, N)
      DO 199 K=1, I2
      DO 4003 I=1, N
 4003 B (I)=0.0
      DO 14 K1=IB, I1
      K2=K1-I
   14 G (K2)=CV (K, K1)
      DELTA=1.0
      IF (NRIDGE.EQ.0) GO TO 4011
 4005 DELTA=DELTA+0.001
 4011 DO 4002 I=1, N
      DO 4002 J=1, N
      IF (I.NE.J) A (I, J)=RIDGE (I, J)
      IF (I.EQ.J) A (I, J)=RIDGE (I, J)*DELTA
 4002 CONTINUE
      CALL INVERT (A, N, BB, L)
      DO 16 IP=1, N
      BETA (IP, K)=0.
      DO 16 J=1, N
   16 BETA (IP, K)=BETA (IP, K)+A(IP, J)*G(J)
      PRINT 166, (BETA (IP, K), IP=1, N)
      IF (NRIDGE.EQ.0) GO TO 4012
      DO 4004 I=1, N
      IF (ABS(B(I)-BETA (I, K) ). GT.0.001) GO TO 4005
 4004 CONTINUE
 4012 ALP=0.
      DO 17 JP=IB, I1
      ID=JP-I2
   17 ALP=ALP+BETA (ID, K)*AV (JP)
```

```
      ALPHA (K)=AV (K)—ALP
      PRINT 177, ALPHA (K)
      DO 18 J=1, J1
      Y (J, K)=0.
      DO 1800 IL=1, N
      KT=IL+I2
 1800 Y (J, K)=Y(J, K)+BETA (IL, K)*X(J, KT)
   18 Y (J, K)=Y(J, K)+ALPHA (K)
      PRINT 12234, (Y(J, K), J=1, J1)
12234 FORMAT (' EXPY ', 12E9.3)
      AGG=0.0
      AGH=0.0
      AGHG=0.0
      ARSD=0.0
      DO 2001 J=1, J1
      DAGG=Y(J, K)—AV (K)
      DAGH=X(J, K)—AV(K)
      ARSD=ARSD+(X(J,K)—Y(J,K) )**2
      AGG=AGG+DAGG**2
      AGH=AGH+DAGH**2
 2001 AGHG=AGHG+DAGG*DAGH
      ERSD=0.0
      JKC=J1—1
      DO 2003 J=1, JKC
      ERSD=ERSD+( (X(J+1,K)—Y (J+1, K) )  —(X(J,K)
      —Y (J,K) ) )**2
 2003 CONTINUE
      DURBIN=ERSD/ARSD
      PRINT 2004, DURBIN
 2004 FORMAT (' VALUE OF DURBIN  WATSON STATIS-
      TIC IS=' , E15.3, ' IT IS EXPECTED AUTOREGRES-
      SION OR NOT ')
      AGG=AGG/J1
      AGHG=AGHG/J1
      AGH=AGH/J1
      RHO=AGHG/SQRT (AGG*AGH)
      PRINT 2002, ARSD, AGG, RHO
 2002 FORMAT (' SUM OF SQUARES OF RESIDUALS=' ,
      E15.3,' VAR OF  EXPECT  VALUE  1=' , F15.4, 'R
      VALUE=' ,F6.3)
```

```
      ARSD=ARSD/J1
      TEST=0.
      DO 191 IL=IB, I1
      IP=IB—I2
  191 TEST=TEST+CV(K, IL)*BETA (IP, K)
      TEST=ARSD*J1/(J1—N—1)
      TEST=ABS (TEST)
      TEST=SQRT (TEST)
      DO 192 IL=1.
  192 TTEST  (IL,K)=BETA  (IL,K)/TEST/(SQRT  (ABS(A
      (IL,IL) ) ) )
      STPTLL=J1—(N+1)
      DO 194 IL=1,N
  194 TA(K)=ALPHA (K)*(SQRT (STPTLL) )/TEST
C
C     ————
C
    7 FORMAT (13E8.3, '=AVERAGE')
   12 FORMAT (13E8.3, '=SDEV')
   11 FORMAT (12F9.5, 'CORRELTN')
  166 FORMAT ('OBETA VALUE=' , 6F 14.6)
  177 FORMAT ('AALPHA=' , F14.6)
   19 FORMAT (11F10.4, 'EXPDVALU')
      PRINT 193, (TTEST (IL,K), IL=1,N)
  193 FORMAT ('TTTEST', 6F14.6)
      PRINT 195, TA(K)
  195 FORMAT ('AALPHA TEST=' , F14.6)
      PRINT 188,K
  188 FORMAT (20('KK', I2) )
C
C     ————
C
  199 CONTINUE
      PRINT 30001, ( (A (I,J),J=1,N), I=1,N)
      RETURN
      END
```

Subroutine No. 6

```
      SUBROUTINE MOMENT (X,N,M,S,L)
C
C
C     COMPUTES MOMENTS UP TO M ORDER
C     IF L IS ZERO RAW MOMENTS AROUND ZERO IS
      OBTAINED
C     IF L IS NONZERO MOMENTS AROUND ARTH.
      MEAN IS OBTAINED WHERE MOMENT
C     OF 2,3,4 ORDERS REPRESENT VARIANCE,
      SKEWNESS, KURTOSIS RESPECTIVELY
C
C     -----
C
      DIMENSION X(100), S (100)
      AN=N
      AV=0.
      IF (L) 1, 3, 1
    1 DO 2 I=1, N
    2 AV=AV+X(I)
      AV=AV/AN
    3 DO 5 J=1,M
      S(J)=0.
      DO 4 I=1,N
      D=X(I)-AV
    4 S(J)=S(J)+D**J
    5 S(J)=S(J)/AN
      DO 8 I=1,M
    6 PRINT 7,I,AV, S(I)
    7 FORMAT (' MOMENT OF THE', I2, 'ORDER
      AROUND', E8.3, ' IS', E12.3)
    8 CONTINUE
      RETURN
      END
```

Subroutine No. 7

```
      SUBROUTINE DSTBN (X,N,ALOW, AUP, NRQ, M,
      NO, MF)
C
C
C   **              **              **              **
C
C     X,N,M,NO ARE INPUT FROM MAIN PROGRAM
C     IF NO=0 LOWER AND UPPER LIMITS ARE
      GIVEN FROM MAIN PROGRAM
C     OTHERWISE CONSTRUCTED IN THIS SUBROUTINE
C     X IS THE ARRAY IN N DIMENSIONS=OBSERVED
      SERIES
C     ALOW=LOWER LIMIT, AUP=UPPER LIMIT
      (VECTORS IN M DIMENSION)
C     NRQ=CODE SERIAL OF MEMBERS IN A CLASS
C     MF IS FREQUENCY IN M DIMENSIONS
C
C   **              **              **              **
C
      DIMENSION X(N), ALOW(M), AUP(M), NRQ (M,N),
      AN=N MF(M)
      IF(NO) 1,7,1
C     COMPUTE CLASS LIMITS WITH MEAN AND
      STANDARD DEVIATION
    1 AV=0.0
      DO 2 I=1,N
    2 AV=AV+X(I)
      AV=AV/AN
      SD=0.
      DO 3 I=1, N
      D=X(I)—AV
    3 SD=SD+D**2
      SD=SQRT(SD/AN)
      AM=M/2
      MM=M/2
      NM=AM
      IF (MM.EQ.NM) GO TO 5
      MH=(M—1)/2
```

```
      HM=MH
      DO 4 I=1, M
      ALOW (I)=AV+(AI-(HM+1.5) )*SD
      AUP (I)=ALOW(I)+SD
    4 CONTINUE
      GO TO 7
    5 DO 6 J=1, M
      AJ=J
      ALOW (J)=AV+(AJ—(AM+1.0) )*SD
      AUP (J)—ALOW (J)+SD
    6 CONTINUE
C     GO FOR TALLY MARKING PROCESS TO FIND
                                        FREQUENCIES
    7 DO 9 J=1, M
      MX=0
      MF (J)=0
      DO 8 I=1, N
      IF (X(I).GE.AUP (J) ) GO TO 8
      IF (X(I).LT.ALOW (J) ) GO TO 8
      MF (J)=MF (J) + 1
      MX=MX+1
      NRQ (J, MX)=I
    8 CONTINUE
    9 CONTINUE
C     PRINT OUTPUT          **              **
      PRINT 10
   10 FORMAT (' SL  NO.  LOWER LIMIT UPPER LIMIT
      FREQUENCY ')
      DO 12 J=1, M
   11 FORMAT (15, F14.5,' ', F14.5, ' ', I5)
   12 CONTINUE
      PRINT 13
   13 FORMAT ( 'CLASS NO.        **        **        **
      1 MEMBERS OF THE CLASS             **        ** ')
      DO 15 J—1, M
      PRINT 14, J, (NRQ( J, MX), MX=1, N)
   14 FORMAT ( '      ', 15, '      ', 2714)
   15 CONTINUE
      RETURN
      END
```

SEVEN

B.L. Sinha

NEAREST NEIGHBOUR TECHNIQUE AS A MEANS OF SPATIAL DISTRIBUTION

Analysis in Two-Dimensional Situations

THE LOCATIONAL pattern of settlements and the evolution of their pattern and spacing in any area is governed by a number of forces (such as the nature of topographical conditions, transport intensity, resources of the region, the historical background, agricultural prosperity, trends of population growth, migration, etc.) which vary from one region to another. The result is that we observe varying patterns in the spatial distribution of settlements. One important question arises here as to whether there is any identifiable pattern in the spatial distribution of settlements. The importance of spatial relations among settlements is obvious, even though the different patterns have not been fully explained. Though various models and explanations of these patterns have been forwarded,[1] there has been insufficient work in establishing the patterns for different countries in comparable terms. Nevertheless, the conventional visual assessment of the settlements patterns as sparse, dispersed, nucleated, agglomerated or dense has been substituted by more reasonable scientific judgements of the precise character of such distributions with the help of nearest neighbour index.

Nearest Neighbour Index

This spacing index of map analysis technique contributes a high degree of objectivity and enables geographers to rank patterns of settlements along a scale extending from complete clustering to complete uniformity, thus distinguishing three kinds of basic distributions of points (i.e. settlements on an area) viz., (i) uniform, (ii) Random and (iii) Clustered. This is a straight line measurement of the distance separating any phenomenon and its nearest neighbour in space and was first initiated and successfully used, almost 27 years back, by statistical plant ecologists such as Dice[2], Clark and Evans,[3] Moore[4] and Thompson,[5] to describe and analyse distributions of phyto-sociological populations. But over the years it has been adopted by a number of other disciplines and in geography the method was first initiated by Dacey[6] and applied principally to urban and rural settlement patterns by other workers[7] particularly in the United States. Recently, attention has been paid to the application of nearest neighbour analysis in other branches of geography, but published works of this kind are few and far between. It must also be confessed that despite the promises and justifications of the technique as an illustrative process, it has yet to command widespread popularity. One obvious reason for the relatively slow progress of this important analytical technique is that, the subject is technical and based on mathematics. In the articles dealing with the method in much more details, the authors have not expressed the nearest-neighbour formula in its simplest forms and the beginner in the subject gets confused. This paper is an attempt to present the formula in a most elemental, simple and non-mathematical manner so that beginners in the subject with only a slight mathematical background may have no difficulty in understanding this analytical technique. This will enable even the beginners to use the technique in the course of their map work and familiarity with the approach will encourage them to think spatially, making them arrive at logical conclusions regarding location pattern assessment.

Steps in the Computation of the Nearest-Neighbour Statistic

In visual assessment of the patterns, the interpretations offered by individuals are highly likely to differ from one another. Nearest-Neighbour analysis provides an objective yardstick which reduces the subjective impressions to a minimum and enables one to predict the average distance that would separate points from their nearest neighbour if the points were located at random throughout an area. This predicted value is usually referred to as $\gamma \bar{E}$.[8] Once it is calculated, it can be compared with the observed average distance between nearest-neighbours which is known as $\bar{D}O$. The nearest-neighbour statistic or index of spacing (R) in the real world is concerned with assessing the extent of contrasts between actual patterns and their theoretical random counterparts and is expressed as :

$$R=\bar{D}O/\bar{\gamma}E$$

Derivation of $\bar{\gamma}E$

Clark and Evans[9] used the following construction to predict the average distance between randomly distributed nearest-neighbours :

$$\bar{\gamma}E=1/2\sqrt{\rho}, \ldots \qquad \ldots\ldots(1)$$

Where ρ denotes the number of points in the pattern divided by the area under study (i.e. n/a). King[10] transformed this formula to read :

$$\bar{\gamma}E = 1/(2\lambda^{\frac{1}{2}}) \qquad \ldots\ldots(2)$$

Where λ like ρ in the Clark's form, represented an expression of the density of points. This was subsequently simplified by Pinder[11] to express :

$$\bar{\gamma}E = 0.5\sqrt{(a/n)} \qquad \ldots\ldots(3)$$

Where a=the area of the region and n=the number of points or settlements.

Pinder's formula is the most simple one and the result is that a predicted average distance can be computed simply by a division, a square root and a multiplication removing the reciprocal from the calculation. The table below which is based on the investigations of King who worked the spacing of towns in 20 sample area of the United States[12] is cited as an evidence of the Pinder's simplification where his values for $\bar{\gamma}E$ have been compared with the results derived from King's version and further compared to the erroneous formula noted by Haggett.[13]

TABLE 7.1 : **The Spacing of Towns in Four Sample Areas of the United States**

Sample area	No. of towns	Area (Km.^{2})	$\bar{\gamma}E$ (King's formula)	$\bar{\gamma}E$ (Pinder's formula)	$\bar{\gamma}E$ (Haggett's formula)
1	96	10236.0	5.16	5.16	0.05
2	69	8291.2	5.69	5.69	0.04
3	132	9771.7	4.30	4.30	0.06
4	82	6920.6	4.59	4.59	0.05

Derivation of $\bar{D}O$

The actual average distance is obtained by measuring the distance between each settlement on the map and its nearest-neighbour. The total of these distances between nearest-neighbours, divided by the number of settlements gives $\bar{D}O$:

$$\bar{D}O = \frac{\Sigma d_1 \ldots\ldots d_n}{n} \qquad \ldots\ldots(4)$$

Where d=distance between the settlement and its neighbour, n=the number of settlements and

Σ = the sum of.

Calculation of R

The final stage is the analysis and computation of the nearest-neighbour statistic or index of spacing (R) which is the ratio of the actual mean distance between nearest-neighbour settlements in a given area ($\bar{D}O$) to the mean expected distance of random

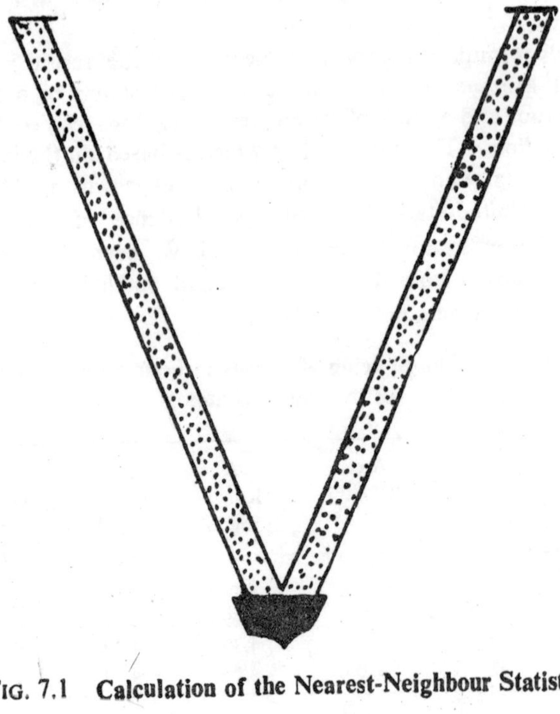

FIG. 7.1 Calculation of the Nearest-Neighbour Statistic

distribution of the same number of settlements (or points) in the same area, i.e. $\bar{\gamma E}$ (fig. 7.1) and is thus expressed:

$$R=\frac{\bar{D}O}{\bar{\gamma E}}=\frac{\bar{D}O}{0.5\sqrt{\frac{a}{n}}} \qquad \ldots(5)$$

and it is used to describe the nature of the pattern under examination.

Step 1

Average expected distance between nearest-neighbours in a random pattern $= \bar{\gamma E}$

$$= 0.5\sqrt{\frac{a}{n}}$$

a = area of the place
n = the number of settlements (or points).

Step 2

Average observed distance between nearest-neighbours

$$= \bar{D}O$$

$$= \frac{\Sigma d_1 \ldots \ldots d_2}{n}$$

d = distance between the settlement and its nearest-neighbour
n = the number of settlements (or points)
Σ = the sum of

Step 3

$$\text{Nearest-neighbour statistic (R)} = \frac{\bar{D}O}{0.5\sqrt{\frac{a}{n}}}$$

The Concept of Randomness

The Central Concept of nearest-neighbour analysis is randomness. Where there is total absence of a rhythmic or systematic pattern of points (settlement locations that may be sufficely mapped by point representation) in a particular region, the distribution of points is called random. Randomness implies that there is no strong pattern forming forces deciding how settlements are arranged. In random pattern the location of each component point is totally uninfluenced by the location of the remaining points and a set of points over a region have the same chance of occurring on any sub-area as any other point.[14] To produce random pattern two lists of 'random numbers' are required. The list contains figures in which each digit is in no way determined by any of the other digits and published random number tables are available.[15] By selecting any two lists from the table and placing them side by side so that the values can be read in pairs—the first value of the pair provides the easting and the second the northing for the location of a point—any number of points can be plotted but all the time the pattern is being produced by forces operating at random.[16] A pattern that is not random is either more clustered than random or more uniform than random. Nearest-neighbour method is used to determine the degree of deviation from random of any point distribution. In short, complete randomness is the midpoint of a continuum of spatial patterns extending from complete clustering to perfect uniformity (Fig. 7.2)

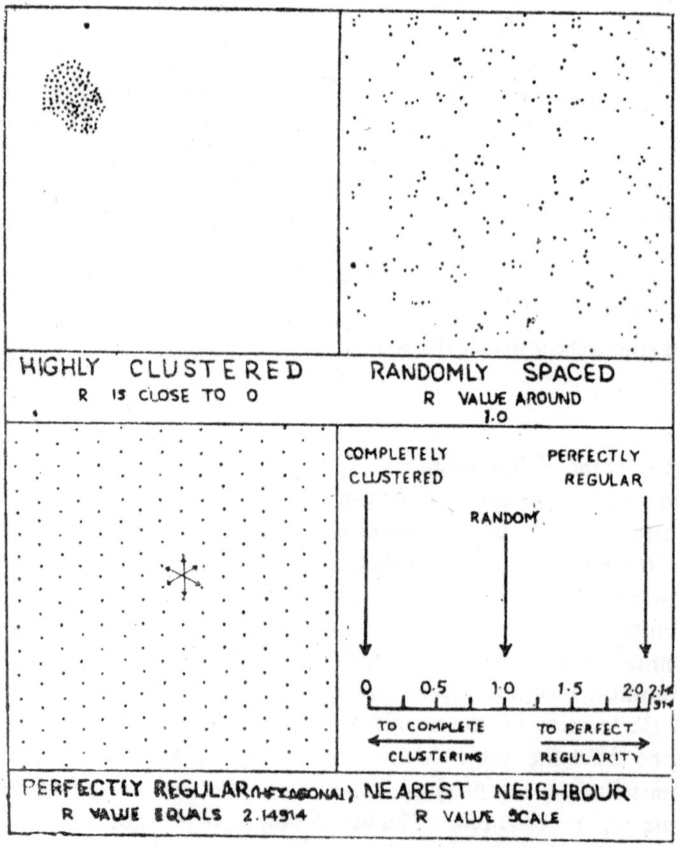

FIG. 7.2 : **Point Patterns and Index of Randomness**

The Explanation of R

The nearest-neighbour statistic (R) values lie along a continuous scale extending from a theoretical low of zero when all the settlements are concentrated at a single point through 1.00 (random) to a maximum value of 2.1491 when the pattern of settlement is perfectly regular, i.e. triangular or hexagonal. By placing one or more investigations on this scale, appropriate interpretation can be made. The findings of many areas can be inserted on this scale and comparisons can be observed between location tendencies in those areas. Many scholars regard this possibility of objectives comparisons as the principal advantage of nearest-neighbour analysis. When R value is equal to 1.0, the $\overline{\gamma E}$ and $\overline{DO}$ are exactly identical and it can be deduced that the location pattern might be the result of forces working at random, though recognition of this point is very critical and is only a possibility. If the value of R falls between 0—1, the observed average distance is also small relative to the expected value. In this case the points are closer together than in a random situation and the pattern is known to be clustered. There may be, of course, various degrees of clustering, the value of 0.0 denoting complete clustering Conversely, if R falls between 1—2.1491, the observed average distance is also greater than expected. Here the points tend to be regularly spaced and at the extreme end is the value 2.1491 which denotes arrangement of points along lattice of equilateral triangles when each point is equidistant from six other points. (Fig. 7.2)

The Test of Significance $\overline{\gamma E}$

Up to this point the explanation of R values seems to be straightforward. Once the region and the total number of settlements (or points) are determined, it is possible to predict the average distance between randomly distributed nearest-neighbours on the formula noted above, which, in spite of showing clear-cut results, may, at times, give a false impression of precision.

Assuming 62 settlements as randomly distributed over an area of 100 Km^2, the $\overline{\gamma E}$, on the basis of the formula noted above, would be 0.635. But using the same region and the same number of points, if another 40 random patterns were generated, the $\overline{\gamma E}$ for these patterns should also be the same. In real world situation, however, the mean distances between nearest-neighbours when actually computed for all these random patterns, only a few would coincide with the prediction. As a matter of fact, the predicted value would nearly approximate to the average of all the observed values. This leads us to think that a value of $\overline{\gamma E}$ should never be considered as representing an unquestionable definition of randomness, rather it should be regarded as the centre of a zone, the width of which denotes the range of average distances likely to be produced by many number of random patterns based on the same number of settlements in the same region. It is, therefore, essential in certain circumstances to devise a technique for assessing the probable extent of the range of randomness, i.e. it is necessary to test the significance of $\overline{\gamma E}$ on the basis of 'standard error of $\overline{\gamma E}$ ($S.E.\ \overline{\gamma}\ E$) which is formulated by Clark and Evans[17] as under:

$$S.E.\ \overline{\gamma E} = 0.26136/(n\lambda)^{\frac{1}{2}} \qquad \ldots(6)$$

but according to Pinder[18] it can be written as

$$S.E.\ \bar{\gamma}E = \frac{0.26136}{\sqrt{(n \times n/a)}} \qquad \ldots(7)$$

As already noted n and a respectively represent the number of settlements and the area studied. After calculating the standard error, it is easier to determine the likely range associated with a particular expected average distance. The standard error is doubled and the product added to, and subtracted from γE. The resulting figures denote the upper and lower limits of the range (at the 95% confidence limits).[19] In the example noted above, the range would be determined as under :

$$\begin{aligned} \text{area} &= 100 \text{ Km}^2 \\ n &= 62 \text{ points} \\ \gamma\bar{E} &= 0.5\sqrt{100/62} \\ &= 0.635 \\ S.E\cdot \bar{\gamma}E &= \frac{0.26136}{\sqrt{62 \times 62/100}} \\ &= 0.042 \\ 2 \times S.E.\ \overline{\gamma}E &= 0.084 \end{aligned}$$

Therefore the likely range of randomness is

$$0.635 \pm 0.084$$
$$= 0.551 \text{ to } 0.715$$

In the above example, the upper limit of the range indicates that $\bar{D}\ O$ must exceed 0.719 before it is unlikely to be paralleled by a random situation. This, in its turn, implies that unless R exceeds 1.13 (i.e. 0.719/ $\bar{\gamma}$E), the presumption that the observed pattern embraces a significant element of regularity cannot be defended. Similarly, towards the other end of the range, $\bar{DO}$ should be less than 0.551 before it can be said that the pattern is prominently clustered. In the above instance, R would not confirm clustering if it surpassed 0.88 (i.e. 0.551/ $\bar{\gamma} E$). And the concept of a range of randomness can further be extended to give a 'range of random matching' on the nearest-neighbour scale where the range on the 'R' scale would depend on the size of both the number of points and the area.

Another index considered for the test is the measure of standard variate of the normal curve (C)[20] which is used to test the significance of the difference between $\bar{DO}$ and $\bar{\gamma} E$ and is calculated as follows :

$$C = (\bar{DO} - \bar{\gamma} E)/ \sigma \bar{\gamma} E \qquad ...(8)$$

Some Points to Remember

1. Nearest-neighbour analysis does not, of course, distinguish between a single and a multiclustered pattern. For instance, if pairs of settlements are situated on either side of a stream, each pair at some distance from the other, the R value will approach very close to zero, denoting thereby that the settlements are clustered around a point.

2. It averages out sub-patterns which may exist within the region. If so warranted, each sub-pattern should be investigated separately.

3. Delimitation of the study area is a difficult problem as it affects the R value. If the analysis is to yield meaningful result, the shape of the area should not be elongated or linear. Usually, the nearest-neighbour analysis is used during the investigations into the nature of two-dimensional point patterns [However, in unidimensional (Linear) situation, a revised formula is often used to predict the average distance between randomly distributed neighbours].[21] The spatial context selected for the analysis must be logically justifiable and the decision must lie with the researcher and can only be made with reference to the nature of the distribution and the specific factors that may affect it.

4. No point outside the boundary of the chosen area is included in the analysis. But if the nearest-neighbours of any point

in the given region is outside of its boundary, this distance may be included in the calculation.

5. If the two points are the nearest-neighbours of each other the same distance is measured twice.

6. And in the end, a study of spacing of point pattern through nearest-neighbour analysis is not only significant as an academic exercise, but also act as a base for decision making in administrative, industrial, commercial and social affairs.

Example : It is essential now to work out some example cases to illustrate the whole procedures of the nearest-neighbour analysis. In an attempt to study the settlement pattern, the map of the Mewat region was divided into several grids of 10×10 sq. miles (Aziz 1974 cf. Mahmood, Aslam).[22] Two such grids are represented in Figs. 7.3 and 7.4 showing the location of rural settlements

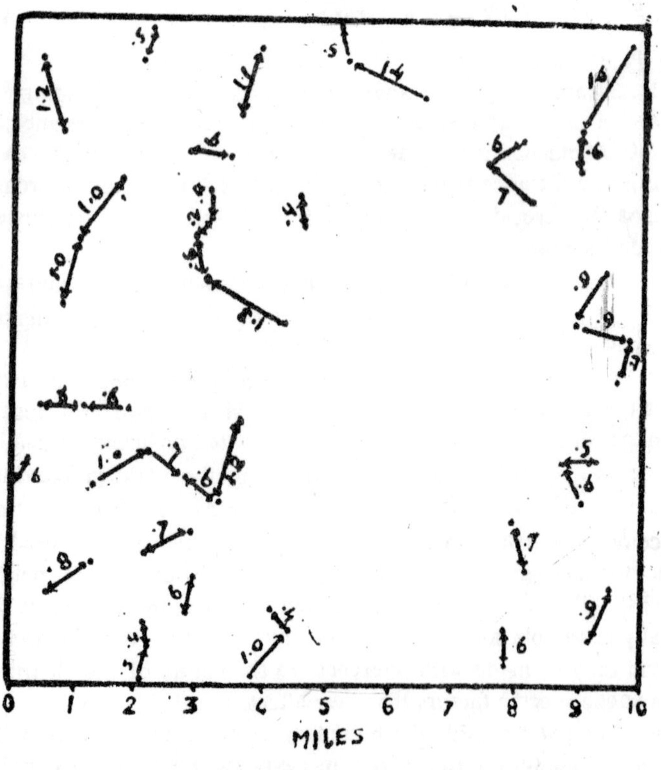

FIG. 7.3 : **Distance between each pair of Nearest Point**

in the region. In first grid (Fig. 7.3) the distance between each pair of nearest point is added together. It equals 44 miles and is divided by 62 (the number of such pairs) to give $\overline{D}O$ i.e.

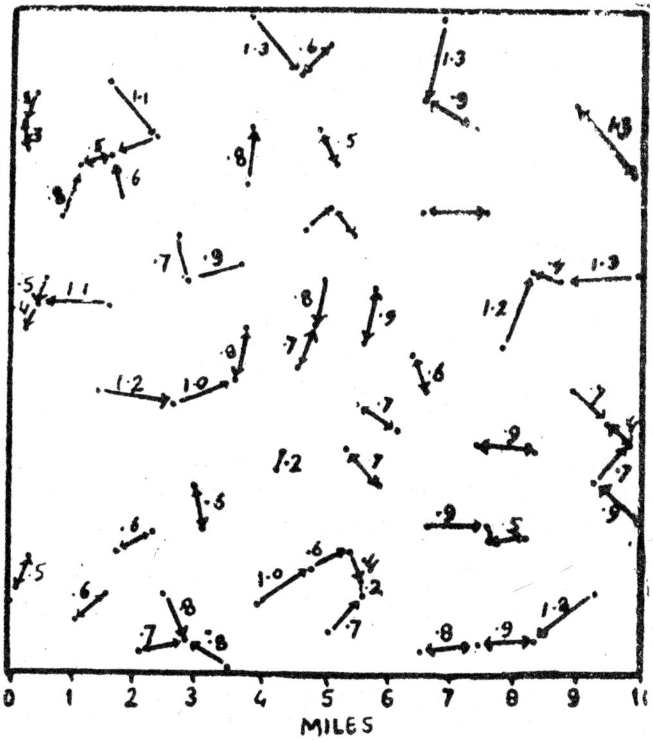

FIG. 7.4 : **Location of Rural Settlements in the Region**

$$\overline{D}\,O = \frac{44.0}{62} = 0.71$$

Since $N = 62$ and $A = 100$ Sq. miles

$$\bar{\gamma}\,E = \frac{1}{2\sqrt{\dfrac{62}{100}}} = \frac{1}{2\sqrt{.62}} = 0.635$$

and hence

$$R = \frac{\overline{D}O}{\bar{\gamma}E} = \frac{0.71}{0.635} = 1.1176$$

Since the value of R falls in between 1 to 2.15, it will show an approaching uniform distribution only when $\bar{\gamma}E$, is prominently different from $\bar{D}O$. The test of significance of $\bar{\gamma}E$ is deduced as follows:

$$S.E.\ \bar{\gamma}E\ \sigma\ \bar{\gamma}E = \frac{0.26136}{\sqrt{62\frac{(62)}{100}}} = \frac{.26136}{\sqrt{38.44}}$$

$$= \frac{.26136}{6.2} = 0.04215$$

$$\text{and } C = \frac{\bar{D}O - \bar{\gamma}E}{\sigma\ \bar{\gamma}E}$$

$$= \frac{0.71 - .635}{0.04215} = 1.77935$$

which is insignificant at 1 per cent level of significance. It can be remarked that the value of R demonstrates a random distribution.

Likewise in second grid (Fig. 7.4) the total distance between each pair of nearest point comes to 60 miles and is divided by 86 (the number of settlements in this grid) to give $\bar{D}O$ i.e.

$$\bar{D}O = \frac{60}{86} = 0.6977$$

Also

$$\bar{\gamma}E = \frac{1}{2\sqrt{\frac{86}{100}}} = \frac{1}{2\sqrt{.86}} = \frac{1}{2\times .92736}$$

$$= \frac{1}{1.8547} = 0.5392$$

$$S.E.\ \bar{\gamma}E\ \sigma\ \bar{\gamma}E = \frac{0.26136}{\sqrt{86\times .86}} = \frac{.26136}{8.6} = 0.03039$$

$$\text{hence } R = \frac{\bar{D}O}{\bar{\gamma}E} = \frac{0.6977}{0.5392} = 1.2939$$

$$\text{and } C = \frac{.6977 - .5392}{.03039} = 5.21553$$

which is significant at 1 per cent level of significance and therefore the distribution may be interpreted as approaching uniform.

REFERENCES

1. Important books that bear directly on the subject are—
Walter, Isard, *Location and Space Economy*, New York: John Wiley & Sons, 1956.
E. M. Hoover, *The Location of Economic Activity*, New York, McGraw-Hill Book Company, Inc., 1948.
A. Lösch, *Die raumliche Ordnung der Wirtschaft*, Jena: Gustav Fischer, 1934, trans, *The Economics of Location*, W. H. Woglom and W. F. Stolper, New Haven: Yale University Press, 1954.
W. Christaller, *Die Zentralen Orte in Suddeutschland*, Jena: Gustav Fischer, 1933, trans, *Central Places in Southern Germany*, C. W. Baskin, Englewood Cliffs, N. J.: Prentice-Hall, 1966.
2. L.R. Dice, "Measure of the Spacing between individuals within a population", *Contrib. Lab. Vert. Biol*, Univ. Mich. 55, 1952, pp. 1-23.
3. P. J. Clark, and F. C. Evans, "Distance of Nearest Neighbour as a Measure of Spatial Relationship in Population", *Ecology*, 35, 1954, pp. 445-453.
4. P. G. Moore, "Spacing in Plant Population", *Ecology*, 35, 1954, pp. 222-27.
5. H. R. Thompson, "Distribution of Distance to n^{th} neighbour in a Population of Randomly Distributed Individuals", *Ecology*, 37, 1956, pp. 391-94.
6. M. F. Dacey, "The Spacing of River Towns", *Annals of the Assn. of Amer. Geographers*, 50, 1960, pp. 59-61.
M. F. Dacey, "A Note on the Derivation of Nearest Neighbour Distance", *Journal of Regional Science*, 2, 1960, pp. 81-87.
M. F. Dacey, "Analysis of Central Place and Point Patterns by Nearest Neighbour Analysis", *Lund Studies in Geography*, Series 13, 24, 1962, pp. 55-75.
7. L. J. King, "A Qualitative Expression of the Pattern of Urban settlements in Selected Areas of the U.S.", *Tijdschrift Voor Economische en Sociale Geographie*, 53, 1962, pp. 1-7.
A. Getis, "Temporal Land Use Patterns Analysis with the use of Nearest Neighbour and Quadrat Methods", *Annals of the Assn. of Amer. Geographers*, 54, 1964, pp. 391-9.
B. P. Birch, "The Measurement of Dispersed Patterns of Settlement," *Tijdschrift Voor Economische en Sociale Geographie*, 58, 1967, pp. 68-75.
8. In Some literature $\bar{\gamma} E$ is also termed $\bar{\gamma}$A (i.e. actual mean distance), but since A is usually used to express area, this particular abbreviation should be avoided.
9. P. J. Clark, and F. C. Evans, op. cit., pp. 451-2.
10. L. J. King, *Statistical Analysis in Geography*, Prentice-Hall, Englewood Cliffs, N. J., 1969, p. 100.
11. D. A. Pinder, and M. E. Witherick, "The Principles, Practice and Pitfalls of Nearest Neighbour Analysis", *Geography*, 57, Part 4, 1972, pp. 277-288.
12. L. J. King, 1962, op. cit., p. 3.

13. $\bar{\gamma} E = 0.50\ (A/N)—1/2$, Haggett, *Locational Analysis in Human Geography*, Arnold, London, 1965, p. 232. In order to correct this formula and the one given by Toyne, P. and Newby, P. T., *Techniques in Human Geography*, Macmillan, London, 1971, p. 116 b., A and N should be transposed.
14. L. J. King, (1962), op. cit.
15. D. V. Lindley and J. C. P. Miller *Cambridge Elementary Statistical Tables*, Cambridge University Press, London, 1964.
16. D. A. Pinder and M. E. Withrick, 1972, op. cit., p. 280.
17. Clark and Evans, op. cit.
18. D. A. Pinder and M. E. Withrick, 1972, op. cit., p. 285.
19. For the details of the concept of confidence limits, See Gregory, S., *Statistical Methods and the Geographer*, Longmans, London, 1963, pp. 50-52.
20. L. J. King (1969), op. cit., pp. 99-100.
21. D. A. Pinder, and M. E. Withrick, "A Modification of Nearest neighbour Analysis for use in Linear Situations"; *Geography*, 60, Part 2, 1975, pp. 16-23.
22. Mahmood Aslam, *Statistical Methods in Geographical Studies*, Rajesh Publications, New Delhi, 1977, pp. 74-76.

EIGHT

O.P. Singh

QUANTITATIVE METHODS IN GEOMORPHOLOGY

GEOMORPHOLOGY, an important branch of geography deals with the evolution of land forms including the sequential developments that might have taken place in shaping them since their initial appearance above the base level of erosion. The study of landforms provides information useful to human as well as physical geographers. Now such studies are not only of academic interest but have occupied important place in applied fields, viz; civil engineering, geologic and geophysical explorations, ground water hydrology and also in regional planning and defence arrangements.

One of the recent methods of approach to geomorphological interpretation is the quantitative empirical method. This is; often, a useful preliminary method in considering problems where the number of variables and their complexity is so great that it is not possible to come to any conclusion until some of the relationships between them have been established. Some geomorphologists like Finsterwalder (1890),[1] Wentworth (1930),[2] Horton (1945),[3] Dury (1951),[4] and Strahler (1952)[5] have developed various methods of measuring the geometry of the landscape. Such works are, usually, related to slopes and drainage lines. Relationships that can be established between the variables may be referred to as the laws of morphometry (Horton, 1945, Chorley, 1957)[6]. These laws are based on measurable quantities such as stream length, stream order and slope, which can be related to one another quantitatively or dimensionlessly. This quantitative empirical approach to the problem of the comprehension of the nature of the landscape and the processes acting upon it may yield more useful results in that areas where the historical approach is of little value on account of the destruction of the evidence on which a denudation chronology must be based.

Keeping the significance of quantification to the study of landforms in view, an attempt has been made to present a comprehensive analysis of modern quantitative methods in geomorphology. Following the methods and approaches of the above modern geomorphologists, the quantitative methods of landform studies may be classified as follows :

1. Arithmetical Methods
2. Graphical Methods
3. Cartographical Methods
4. Statistical Methods

Arithmetical Methods

The arithmetical methods (Baulig, 1926)[7] deal with the mathematical analysis of topographical maps. Altimetric frequency curve, area height diagram, height-range diagram, clinographic curve, etc. are noticeable techniques in the study of landforms.

Altimetric Frequency Curve or Graph

Altimetry includes the numerical frequency of highlands including both areas or lengths and numbers (of summits, cols, shoulders, etc.). It is very useful in determining former erosion surfaces (Clarke, 1966).[8] All spot heights with or without the summit, col, shoulder, etc. points given on the map are counted with the help of grid method (Maling, 1955),[9] grouped into altitude class intervals and frequency of groups are plotted against height. Plotting of values derived from the average of highest and lowest points/contour in each grid square can also be treated in the same way. Frequency of area of summits, cols, shoulders may also be counted separately and graphed which are as follows:

(1) Hill-tops frequency curve (Hollingworth, 1938).[10]

(2) Spot-height frequency curve (Baulig, 1928).[11]

(3) Summit frequency curve (Thompson, 1936).[12]

(4) Histogram of frequency of areas of summits and benches (Thompson, 1941).[13]

(5) Shoulder, summit and col frequency histogram (Geyl, 1961).[14]

Height-Range Diagram

The diagram devised by Spark (1949)[15] is very useful for the representation of flats or levels, intersetting the fragments of former erosion surfaces or of old valley floors. Firstly, identify the flats (including also summits and cols) and tabulate them according to their location, origin, grid reference, elevation and length, and to the geological zone in which they are developed. Then, plot the height range of each flat as ordinates on a vertical seal against a horizontal seal proportional to the greatest length of the flat, thus forming a series of black rectangles.

Area-Height Diagram

It gives the visual analysis of area-height relationships. First, the area between each successive contours is noted with the help of a polar planimeter and then, the areas calculated either in terms of absolute values or in terms of percentage of total area plotted against height on suitable lines. Thus, two types of diagram may be drawn.

Hypsometric Curve : Areas enclosed by each contour are plotted against height (Hanson-Lowe, 1935)[16]. It indicates the relative significance of each altitudinal zone in a region. Major topographic breaks become obvious on this curve and repeated breaks in slope and major terraces also appear here.

Percentage Hypsometric Curve : The work of Horton (1945) of America on various relationships within a drainage basin has stimulated a study of the area-height analysis of particular drainage basins. The contribution of Strahler (1952) is most valuable in this respect. He conceived of a percentage of hypsometric method to analyse drainage basin. The values that form the basis of the graph are the area above a certain contour as a ratio of the total area of the drainage basin, as measured on the map, and the height of the particular contour as a ratio of the total height range between the lowest and highest points in the basin. The Y-axis of the graph will be the height ratio n/H, and the X-axis the area ratio a/A. The curve must always run from the top left hand corner of the graph to the bottom right hand corner. This curve particularly drawn for the drainage basins calculates a hypsometric integral which is defined as the proportion of the square graph lying below the curve. The integral indicates geomorphic development of the river.

Clinographic Curve

The clinographic curve (Hanson-Lowe, 1935) is the average longitudinal slope curve with concave-up for the land segments of the area. The total area enclosed between adjacent contours is expressed as a circle and the distance between the Y-axis of the graph and the point plotted against that contour is proportional to the radius of the circle. Successive average slopes against the height gives the clinographic curve for the area. It is highly sensitive in comparison to physographic curve so far as the interpretations of sudden changes and breaks in general relief are concerned.

Hypso-Clinographic Curve

This is the oldest method (Finsterwalder, 1890) for determination of average slope employing the hypsometric data and length of contours. The total length of contours multiplied by the contour interval and the product divided by the area, when consulted the tangent values, gives the average slope for the area. The average slope for the total area is obtained by joining the extremities of the two axes, if the numerator is plotted along the X-axis and the denominator along the Y-axis.

Slope-Height Curve

The average slope, obtained by Hanson-Lowe (1935) or Strahler's methods (1956)[17], plotted on the horizontal axis, and the contour heights on the vertical axis, gives the slope-height curve (Moseley, 1961)[18]. In this curve, uniform slopes are straight lines parallel to the vertical axis whereas a concave slope with a uniform rate of change is a straight line inclined to the left.

Terrain Representation Diagram

Another much more rapid method of illustrating a special area in relief is the terrain representation method of Robinson and Thrower (1957)[19] which is based on the inclined contour; this is the projection on a datum plane of the outline of the intersection of the ground surface with an inclined plane. The spacing of the horizontal lines used to draw the diagram controls the apparent vertical exaggeration. This spacing can be calculated by using $D = (12\,h \cot Q)/S$, where h is the contour interval, Q is the viewing angle and S the scale denominator of the map and D is the distance in inches between the parallel lines on the map. The diagram gives an oblique view of the landscape.

River Terraces Diagram

A method of representing the terrace remnants, suggested by Coleman (1954)[20], is to plot a rectangle whose height is proportional to the vertical distance between the front and back of the remnant, while rectangle is placed at the correct altitude according to the height scale on the ordinate of the graph. The abscissa is a length scale on which the value is straightened out.

Average Slope Curve

This curve can be drawn by plotting the average slope values on the horizontal and their frequencies on the vertical axes. The values of slope may be obtained either by Strahler (1953) or Wentworth methods (1930). The Strahler method is more accurate.

Graphical Methods

Graphical methods are very useful to simplify the configuration distribution of contour maps to unfold the major landform patterns of an area.

Generalised Contours

The generalised contours which touch the actual contours at the tips of spurs but are carried across valleys, help in a genetic classification of landforms, including the interpretation of cliffs and various breaks in slope. The method (Pannekoek, 1962)[21] can reveal significant stages in the development of landscape.

Profiles

Profiles of different characters are frequently used in the geomorphological interpretation of the terrain as they are useful in depicting the nature of the general reliefs and different surfaces having different slopes standing at various altitudes.

In drawing of an accurate profile is to lay a straight-edge of paper along the chosen line on the map, then mark accurately with sharp clean ticks all contour intersection, spot-heights, rivers, summits and defined points. Draw the base line of profile on a sheet of graph paper, and transfer the ticks carefully to this. Rub vertical lines at their end of the base-line, and mark of a vertical scale, which should be carefully chosen, bearing in mind the height range involved and the nature of the country. Either by following the vertical graph-lines, or by ruling perpendiculars from the ticks on the base-line, mark the position of each point according to the vertical scale, with a fine accurate cross. When all points are plotted, join them by a smooth line, not by a series of straight lines. Lastly, add significant place-names and lettered in sideways above the point on the section to which they refer (Monkhouse, 1967)[22].

Serial Profiles

A set of profiles on different frames when serially and systematically arranged to cover an area is known as serial profiles which give the visual impression of landscape like hills, spurs, knoll, valleys, etc.

Superimposed Profiles

A series of profiles when plotted on a single frame is known as superimposed Profiles (Wright, 1949) which are very much useful in determining the different erosion surfaces and in identifying plateau tops.

Projected Profiles

The total panoramic view as if seen by a plane from above in two dimensional magnitude, is well indicated by Projected Profiles (Barrell, 1920)[23] which help us in co-relating erosion level at varying altitude. In this type it must be viewed, and if, for example, this is from the west, then, the most westerly profiles in a north-south series are plotted, the next one to the east is drawn where it projects above the first and so on to the most easterly profile (King, 1966)[24].

Composite Profiles

The composite profiles carry only the uppermost line from a diagram of superimposed or projected profiles. This type of profiles, therefore, indicates only the absolute relief and the minor details illustrates the ruggedness of the skyline of an area (Shaffer, 1947)[25].

Spur Profiles

A more widely applicable method of delimiting possible erosion surfaces by drawing profiles is to construct a series of spur profiles (Brown, 1952)[26]. The line of profile must follow the highest land each spur and the resulting profiles will show clearly any spur flats suggested by the contours.

Longitudinal River Profiles

The diagram plotted height against the horizontal distance only the course of a river is known as Longitudinal River Profiles (Peel, 1941)[27] which are very useful for identification of graded and ungraded reaches of a river. The breaks of slopes in these profiles illustrate the rejuvenation heads, i.e., knick point, etc.

Compressed Profiles

A device akin to the projected profiles has been termed as a Compressed Profiles (Brown, 1952). The series of composite profiles are compressed together, and when viewed at right-angles, only those features not obscured by higher ones in the foreground are visible.

Cross Valley Profiles

These are the simple profiles constructed at right angles across the river valley. Such profiles illustrate the width of valley and the configuration of valley bottom of a river (Jones, 1924)[28].

Reconstructed Profiles

A series of such profiles are very helpful in examination of the effects of rejuvenation of a drainage system on the present form of a valley (Miller, 1941)[29].

Cartographical Method

Maps on which the surface relief is portrayed by the use of contours provide a valuable tool for geomorphologists and several cartographic analysis depends on the quality of maps and processing and plotting of data taken from contour map.

Divide Map

The map in which all water divides are shown by different categories and indicated by different weight of line is known as Divide

Map (Dury, 1952)[30] which is very useful to study the migration of water divides of a region

Cliff Map

It is useful to show the occurrences of steep cliffs and rock-faces on a map. This may be done by wedge-shaped black lines, with the thin ends pointing down the slope (Robinson, 1954)[31].

Relative Relief Map

The relative relief, obtained by computing the difference between the highest and lowest elevations within a limited area or the units grid, is one of the best methods for the measure of local roughness (Smith, 1935)[32]. Plot the relative relief values in the centre of each square or unit grid on the base map and then, draw isopleth lines. The map may be shaded into different categories if it is essential. The type of map presents 'amplitudes of available relief' of an area (Glock, 1932)[33].

For the computation of values of relative relief, Thauer (1955)[34] applied circles (with a radius of 0.5 mile) instead of grids square. This eliminates the inconsistency resulting from the use of a square and also coverage leads to the better and consistency by a systematic overlapping of circles.

In spite of the methods applied by Smith and Thauer, Hammond (1968) used actual height along each line of slopes on the map of an area. It gives a good measure of local relief.

Dissection Index Map

The picture gained from relative relief only is static, or it fails to take into account the vertical distance from the erosion base, i.e., the dynamic potential of the area studied. On the basis of this fundamental fact, Dov Nir (1957)[35]. Suggested 'Dissection Index', i.e., a ratio which always varies between zero (complete absence of dissection) and one (extreme case, vertical cliff at sea-shore) is computed and isopleths are drawn.

Stream Line Surface Map

The map of the streamline surface can be used to give an estimate of the depth of dissection of a landscape in which there is clear evidence of an erosion surface. This estimate of dissection can be obtained by first drawing in the streamlines on the map and superimposing on this the contours of the reconstructed surface, drawn on tracing paper, then the two maps can be used to plot differences of height between the two surface, this is done at the points where the two sets of contours cross. Then using the plotted points as a guide, isopleths showing line of equal depth of dissection can be drawn on the map. This method can only be used where the original surface can be reconstructed with a fairly high degree of certainty (Dury, 1951).

Slope Map

The analysis of slope and its representation on a map has been the subject of much research, particularly by American Geomorphologists. A number of methods have been developed by which slope maps may be constructed and these will be described briefly, but it should be pointed out that such maps are not of value for detailed work of slope form and analysis; for such a purpose, field surveys are essential. The slope maps, however, provide a useful visual impression and they be used merely to relate to some other geographical variables.

A method of average slope determination was used as long ago as 1890 by Finsterwalder which was as follows :

$$\text{Average Slope in Degree} = \frac{\begin{array}{c}\text{Total Length}\\\text{of Contours}\end{array} \times \text{Contour interval}}{\text{Total Area}}$$

This laborious and complicated technique was criticized by C. K. Wentworth and other geomorphologists. (Wentworth 1930) applied the following formula :

$$\begin{array}{l}\text{Slope}\\\text{(Tangent)}\end{array} = \frac{\text{Number of contour crossing per mile} \times \text{Contour interval}}{\text{3361 (constant)}}$$

The result gave the average slope in terms of the tangent of the average angle of slope which can be converted into degrees with tangent tables. Then the values plotted in the grid square and isopleths drawn.

Smith (1938)[36], on the basis of the amplitude of available relief, examined the problem in terms of Relative Relief. Still, the problem of incorporating the horizontal equivalent between the highest and lowest altitude in the squares of grid remains. It, if not taken into account, may shift the actual values of the slope upward or downward and even some scarps may be blurred up to a lesser degree of slope.

A modification in this respect was suggested by Miller (1949),[37] such as to measure the horizontal equivalent between the two stations at extreme altitudes in a square and divide the relative relief value by the measured mileage.

The approach of slope representation through relative relief was further modified by Raisz and Henry (1937)[38] by way of dividing the total area not into equal surfaces or rectangles but in unequal patches of almost similar contour spacing and then deducing the result as gradient of 0—50 feet per mile, 50—100 feet per mile and so on. This method clearly delineates the flats and scarps.

In contrast to Raisz and Henry's choropleths, Strahler (1956) devised two new types of slope map, using isopleths. Basically, a large number of slope values are plotted, derived either from using an Abney Level, or by computing them from a map coefficient scale. The tangent and sine functions are then plotted, and a series of isopleths (Isotangent and Isosine Maps respectively) are interpolated.

Still, more laborious and strenuous method was suggested by Robinson (1948)[39] who reduced the size of square to 0.01 sq. mile and then unit dots were adopted for the calculated values of squares. In theory, the dots are countable, so as to give precise information, and at the same time their density appearance gives a good visual impression. The size of the dot must be carefully chosen, in order to produce effective visual contrasts of light and shade.

Combined Map of Slope and Relative Relief

Various methods may be used in combination to illustrate some specific relationship of elements in the surface configuration. Elliot (1953)[40] developed a technique of presenting slope and relative relief on one map. He put a grid of 2½ minutes square on a map quadrangle, and determined the relative relief figure for each; he then drew isopleths to show amplitude of relative relief, following in general Smith's method (1938), but introducing some degree of subjectivity by departing from it where obvious breaks of slope occurred. He then shaded four categories in horizontal ruling. Secondly, using the same contour map base, he distinguished categories of slope, using the Raisz-Henry method (1937) of measuring contour-frequency spacings with dividers, and shaded three categories in vertical lines. When the two maps were superimposed, the combinations of the two sets of categories produced twelve composite slope and relative relief regions, obviously distinguished by degree of cross-shading; these were outlined by pecked lines.

STATISTICAL METHOD

Quantitative studies require that the form elements of a landform should be separated, quantitatively described and compared from region to region. Statistical Indices are essential because they provide the standardised measure so that the essential differences and similarities between region may be understood. Hence, statistical analysis by using numerical indices can be considered a powerful tool in investigating the fundamental problems of landform analysis.

Stream Ordering

Drainage basin is a fundamental geomorphic unit yet objectively delimited and it lends itself perhaps better than any unit area to quantitative analysis. To do the work on drainage basin, it must first be subdivided into stream segments of different order numbers based on the most widely and ordering scheme adopted by Strahler (1957)[41] from Horton (1945) in which fingertip streams are specified as order number $Nu=1$ and when two first order (i.e. of same order) tributaries join, a channel segment of order $Nu=2$ is formed and so on. This simple unambiguous Strahler system is now firmly established and this ordering system provides sequence of stream order numbers, e.g., N_1, N_2, N_3, ... N_k, which approximate an inverse geometric series with the degree of branching or stated statistically is the antilog of the regression coefficient (Chorley, 1969), i.e., $\log Nu=Q-bu$.

Roughness Index

This is defined as $R.I.=\dfrac{N\times M/4}{4}\times 10$ where N = total number of intersections of contour lines with two sets of perpendicular grids, 4 and 10 are constant values, M=distance in miles between grid lines. It expressed the total characteristics of surface configuration. The characteristics of roughness in the terrain is successfully employed by Zakrazewska (1963)[42] in delimitation of the physical boundaries.

Elevation Relief Ratio

Elevation relief ratio, defined as $\frac{E-L}{R}$ where average height, L=minimum height, and R=relief along the profile, indicates the measure of relative proportion of uplands and lowlands along a profile. The ratio value close to one indicates a profile in which most of the flat land is located on divides and valleys are narrow and the value close to zero indicates a profile of open valleys and widely spaced isolated hills.

Profile Index

This technique (Peel, 1941) determines the base level at the time of rejuvenation. The profile should not be drawn from the map but should be surveyed, or at best important points should be checked in the field. The usual formula adopted by Miller (1941) for the purpose of mathematical extension of longitudinal profiles is in the form, i.e.

$$Y=(a-k\ log)(p-X)$$

Where Y is the height of any point on the profile ; p is the total length of the stream in miles; X is the distance from the mouth of the stream; a and k are constants. The constant a is based on the height of the source of the stream, while k is connected with the radius of curvature of the concavity of the stream profile—Peel (1941) suggested 870 and 380 for the values of constant a and k respectively.

Average Height Index

It is useful to compare relief in different areas. Planimeter actual surface for the segments of the area calculated volumes above sea level by multiplying the area with mean inter-contour heights. Divide it by total area to get average height for an area (Clarke, 1967).

Drainage Frequency

Drainage frequency is defined as a number of streams per unit area of a drainage basin. It is useful for the study of texture of drainage (Horton, 1945).

Drainage Density

Drainage Density is simply a length per unit area which is expressed as

$$\text{Drainage Density} = \frac{EL}{A_b}$$

where L = total length of streams, and A_b = total area of basin (Horton, 1945). It varies according to different geological structure, precipitation, vegetation cover and rock types. Generally, low drainage density indicates old stage of the river while medium and fine density show mature and young stages respectively.

Drainage Texture

Drainage Texture is defined as the spacing of channel segments in an unit grid. It indicates the geomorphic development of the basin (Singh, 1976).[43]

Stream Gradient

Stream gradient, an important factor for studying the vertical aspect of the drainage network geometry, is obtained by dividing the stream length by the elevation difference (Horton, 1945).

Basin Slope

Simple measures of average ground slope within a basin are employed in hydrological analysis, such as (Horton, 1914).[44]

$$\text{Mean Basin Slope} = \frac{\text{Total length of contours} \times \text{Contour interval}}{\text{Basin area}}$$

Roughness Number

The number is defined as the product of relative relief and drainage frequency (Horton, 1945).

Constant of Channel Maintenance

Constant of Channel Maintenance is the area required to maintain unit length of drainage channel (Schumm, 1956).[45] The relationship between mean drainage basin areas of each order and mean channel lengths of each order of any basin is a linear function whose regression coefficient plotted (Mean Basin Areas and Mean Stream length) on logarithmic (or constant ratio) scale is equivalent to the area necessary on the average for the maintenance of unit length of drainage channel. This relationship gives an idea of the pattern of run off with the area which helps in calculating average discharge at any point in drainage basin by measuring the catchment area lying above the point.

Relief Ratio

It is the ratio of the maximum basin relief to the horizontal distance along the longest basin dimension parallel to the main drainage line (Morisawa, 1955). It is related to flood magnitude.

Elongation Ratio

The shape of any drainage basin is expressed by its 'elongation ratio; a ratio between the diameter of a circle with the same area as the basin and the basin length (Schumm, 1956).

Circularity Ratio

The ratio, a measure for the shape of a basin is expressed as

$$R_e = \frac{A_b}{A_c}$$

Where R_e=elongation ratio, A_b=area of basin, and A_c=area of the circle having the same length of perimeter of the basin (Horton, 1945).

Stream-Length Ratio

It is a ratio of average length of channels of a given order to average length of stream of next lower order (Horton, 1945).

Basin Texture Ratio

It is a ratio of maximum number of channels crossed by contour to basin parameter (Smith, 1950).[46]

Basin Area Ratio

The mean basin areas of successive streams of a given order tend to form a geometric series beginning with area of first order basins and increasing according to a constant area ratio (Horton, 1945).

Bifurcation Ratio

It is an average ratio of number of streams of a given order to number in next higher order (Horton, 1945). The value varies according to different geological structure, rainfall and geomorphic development of the basin.

Basin Slope Ratio

Average channel slope of streams of a particular order is defined as the ratio of the average vertical drop to the average horizontal distance, measured from the upper ends to the lower ends of streams of order (Horton, 1945). The ratio values differ from order to order due to variation in geological structure.

Mean Available Relief

The available relief is that part of the land that is available for erosion in any particular area. The area of the surface of each contour is measured and the volume of material between sea-level and the height concerned are calculated. The difference between these two values will give the total volume of the interfluve areas above the assumed base surface. This total volume divided by the total area will give the value of mean available relief, a measure for the comparison of degree of dissection. The volume calculations may fairly be done simply by using the formula (Dury, 1951).

$$\text{Volume} = V.I. \left(\frac{\int_0}{2} + \int_n + \int_1 + \int_2 + \ldots\ldots\ldots\ldots f_{n-1} \right)$$

where $V.I.$ = vertical contour interval; and $\int_0, \int_1, \int_2 \ldots\ldots\ldots\ldots$
= the areas enclosed by the respective contours.

Dissection Index

Dissection Index is defined as (Smet, 1954)

$$D.I. = \frac{\text{Real Area} - \text{Projected Area} \times \text{Average Slope}}{\text{Volume}}$$

It is very useful for comparison of different areas.

Basin Shape Index

Basin Shape Index (Gravelins, 1914)[47] is a ratio of average width of basin to axial length of basin where axial length is the horizontal distance along the longest basin dimension parallel to the main drainage line. Average width of basin is equal to the area of the basin divided by the axial length. Thus the index can be expressed as :

$$\text{Basin Shape Index} = \frac{\text{Area of Basin}}{(\text{Axial Length})^2}$$

Sinuosity Index

It is a ratio, derived by dividing the channel length by the valley length (Schumm, 1963)[48]. The value of 1.5 or above is as a criterion for meandering and below 1.5 as sinuous as classified by Leopold (1957).[49] A stream has a sinuosity of 1.00.

All the streams have both hydraulic and topographic sinuosity indices (*H.S.I.* and *T.S.I.*) which may be derived by the following formulas (Mueller, 1968).[50]

$$H.\ S.\ I. = \%\ \text{equivalent of}\ \frac{C\ I - V\ I}{C\ I - 1}$$

$$T.\ S.\ I. = \%\ \text{equivalent of}\ \frac{V\ I - 1}{C\ I - 1}$$

$$\text{where } C.I.\ (\text{channel Index}) = \frac{\text{Channel Length}}{\text{Air}}$$

$$\text{and } V.I.\ (\text{Valley Index}) = \frac{\text{Valley Length}}{\text{Air}}$$

(Air=the short distance between the source and mouth of the streams).

T.S.I. is outstanding during youth stage of a river when *H.S.I.* is negligible conversely, *H.S.I.* is outstanding during old age after the most of *T.S.I.* has been removed.

Symmetry Index

It is helpful to find out the nature of channel patterns, i.e. meandering, sinuous and straight. For this purpose, the calculations of symmetry of arc length, arc height and form ratio (arc length/arc height) of meanders are most essential. The following formula may be applied in this respect (Brice, 1964).[51]

$$\text{Symmetry of Arc Length} = 100 - \frac{100 \times \text{Mean Deviation of Arc Length}}{\text{Mean Arc Length}} \%$$

$$\text{Symmetry of Arc Height} = 100 - \frac{100 \times \text{Mean Deviation of Arc Height}}{\text{Mean Arc Height}} \%$$

$$\text{Symmetry of Form Ratio} = 100 - \frac{100 \times \text{Mean Deviation of Form Ratio}}{\text{Mean Form Ratio}} \%$$

More than 60 per cent symmetry shows meandering pattern of a river while sinuous and straight rivers have the values of 30—60 per cent and less than 30 per cent respectively.

Stream Meandering Index

Stream meanders have some properties like wave length or meander length $(=L)$, amplitude $(=A)$, channel width $(=W)$ and mean radius of curvature $(=rm)$. The best fitting empirical relations between size parameters for meanders are as follows (Leopold, 1964).[52]

Wave Length to Channel Width $= L = 10.9w1.01$
Amplitude to Channel Width $= A = 2.7w1.1$
Wave Length to Radius of Curvature $= L = 4.7\ rm\ 0.98$

Sampling

It is frequently necessary to collect samples for analysis. Several attempts have been carried out by geomorphologists. Strahler (1954)[53] has discussed the problems and methods of geomorphological sampling, taking an example of hill-side slope. He pointed out that although, there is a generic similarity between sets of geomorphological features, each one is an individual example and rarely are two exactly similar. He also used the measures of central tendency and variability.

Another example of the use statistics in sampling is the work of Krumbein and Slack (1956)[54] on the sampling of beach environments for the analysis of sediment characteristics. Work of this type can indicate the number of samples necessary to provide a reasonable picture of the sediment characteristics.

Significance of Results

In order to establish the validity of sampling and other techniques in geomorphology, it is often necessary to test the significance of results. Probability theory gives it at various confidence level. This type of testing is necessary to establish whether the results obtained from two different areas could be due to chance differences or whether the results indicate a true difference between the two samples from the different areas. An example of this type of analysis is discussed by Strahler (1954); the study included the measurement of stream lengths and the relationship between these and the dip of rocks in the area.

Another very useful test is the chi-square test which may be stated as

$$\chi^2 = \frac{\Sigma(O-E)^2}{E},$$

where O = observed value of the variable and E = expected value. Having applied this type of test, the observations of Andrews (1963)[55] were made by measurement of the direction and dip of the long axis of 50 or more often 100 pebbles in pits dug in the till.

Smith's work (1958)[56] on slopes of badlands illustrates how the testing method can be applied to assess whether the values of the degree of slope of two sets of the difference between the means that was applied is given by the formula

$$t = \frac{\bar{\bar{X}}_1 - \bar{\bar{X}}_2}{SP\sqrt{1/N_1 + 1/N_2}}$$

where t = student 'T', P = pooled estimate of variance, S = average of the estimated standard deviations, X = mean and N = number of degrees of freedom.

Correlation

Another field in which statistics can be of use to geomorphological analysis is in the correlation of two or more variables together and in assessing the degree to which they depend on each other. In this way, cause and effect can be studied quantitatively, the cause being the independent variable and the effect being the dependent variable. For this purpose, the co-efficient of correlation must be calculated. To measure the correlation co-efficient, the data are set out in two columns for the X and Y variables. The co-efficient can, then, be found by using the formula of Karl Pearson,

$$r = \frac{\Sigma dx\ dy - n(a_1 - X_1)(a_2 - X_2)}{n\ \sigma x\ \sigma y}$$

Where,

r = coefficient of correlation
a_1 = actual mean of X series
X_1 = assumed mean of X series
a_2 = actual mean of Y series
X_2 = assumed mean of Y series
$\Sigma dxdy$ = sum of the products of deviations from assumed averages.
n = number of pairs of items
σX = standard deviation of X series
σY = standard deviation of Y series

The value of r varies from —1 for a perfect negative correlation in which one of the factors increases as the other decreases, +1 for a perfect positive correlation, in which both factors increase together, while the value zero indicates no correlation at all between the variables. In this respect, the works of Strahler (1952, 1954) are highly appreciable. Singh (1967)[57] observed a positive correlation between two variables, i.e. slope and relative relief, and slope and absolute relief while the four landform elements, e.g. absolute relief, relative relief, slope and drainage density are negatively correlated with the settlements (Singh and Singh, 1976),[58] in Palamau Upland, a part of Peninsular India.

Conclusion

An effort has been made to present a comprehensive classification in one of the four groupings proposed in the article : arithmetical, graphical, cartographical and statistical methods. The author with their procedures and illustrations will provide the stimuli for discovery of new empirical methods bringing the relationships among the landform elements.

REFERENCES

1. S. Finsterwalder, "Uber den Mittleren Boschungswinkal and das wahre Areal, Einer Topographischen Flache", Sitz Ber. K. Beayer Akad. d. *Wissench*, Vol. 28, 1890, p. 35.
2. C. K. Wentworth, "A Simplified Method of Determining the Average Slope of Land Surface", *Ame. Jour. Science*, Vol. 20, 1930, pp. 184-94.
3. R. E. Horton, "Erosional Development of Streams and their Drainage Basins: Hydrophysical Approach to Quantitative Morphology", *Bull. Geol. Soc. Ame.*, Vol. 56, 1945, pp. 275-370.
4. G. H. Dury, "Quantitative Measurement of Available Relief", *Geol. Mag.*, Vol. 88, 1951, pp. 339-43.
5. A. N. Strahler, "Hypsometric (Area Altitude) Analysis of Erosional Topography", *Bull. Geol. Soc. America*, Vol. 63, 1952, pp. 1117-42.
6. R. J. Chorley, "Illustrating the Laws of Morphometry", *Geol. Mag.*, Vol. 94, 1957, pp. 140-49.
7. H. Baulig, "Sur une Method Altrimetrique d Analyte Morphologique Appl'quee a la Bretagne Peninsulair", *Bull. Assoc. Geog. Francais*, Vol. 10, 1926, pp. 7-9.
8. J. I. Clarke, "Morphometry from Maps" in *Essays in Geomorphology* Edited by G. H. Dury. Heinemann Education Books, London, 1966, pp. 235-65.
9. D. H. Maling, *The Geomorphology of Wear Valley*, Ph.D. Thesis, University of Durham, 1955, pp. 33-36.
10. S. E. Hollingworth, "The Recognition and Correlation of High-level Erosion Surfaces in Britain", *Quart. Joun. Geol. Soc. E.*, Vol. 94, 1938, pp. 55-85.
11. H. Baulig, "Les Hauts Niveaux d Erosion Eustatique dans b Basin de Paris", *Ann. de Geog.*, Vol. 37, 1928, pp. 289-305.
12. L. D. Thompson, "Hudson George in the Highlands", *Bull. Soc. Ame.*, Vol. 47, 1936, pp. 1831-40.
13. L.D. Thompson, Topographic Analysis of Monterey, Stauton and Harrisonbury Quadrangles", *Jour. Geol., Vol.* 29, 1941, pp. 521-49.
14. W. F. Geyl, "Morphometric Analysis and the world wide occurrence of Stepped Erosion surfaces", *Jour. Geol.*, Vol. 69, 1961, pp. 388-416.
15. B. W. Spark, "The Denudation Chronology of the Dip-slope of the South Downs" *Proceeding, Geol. Assos.*, Vol. 60, 1959, pp. 165-215.
16. L. Hanson Lowe, "The Clinographic Curve", *Geol. Mag.*, Vol. 72, 1935, pp. 180-84.
17. A. N. Strahler, "Quantitative Slope Analysis", *Bull. Geol. Soc. America*, Vol. 67, 1956, pp. 571-96.
18. F. Moseley, "Erosion Surfaces in the Forest of Bowland", Pros. Yorks. *Geol. Soc.*, Vol. 73, 1961, pp. 1025-46.
19. A. H. Robinson, and N. J. W. Thrower, "A New Method of Terrain Representation", *Geog. Review*, Vol. 47, 1957, pp. 507-20.
20. A. Coleman, "The Use of Height-Range Diagram in Morphological Analysis", *Geog. Studies*, Vol. 1, 1954, pp 19-26.

21. A. J. Pannekoek, "Generalization of coastlines and contours", *International Year Book of Cartography*, Vol. 2, 1962, pp. 55-75.
22. F. J. Monkhouse and H. R. Wilkinson, *Maps and Diagrams*, Methuen & Co., Kibdib, 1967, p. 89.
23. J. Barrell, "The Piedmont Terraces of the Northern Appalachians", *Ame. Jour. Soc.*, Vol. 49, 1920, pp. 242-58.
24. C. A. M. King, *Techniques in Geomorphology*, Edward Arnold Publisher, London, 1966, p. 237.
25. P. R. Shaffer, "Correlation of the Erosion Surfaces of the Southern Appalachians", *Journ. Geol.*, Vol. 55, 1947, pp. 343-52.
26. E. H. Brown, "Erosion Surfaces in North Cardiganshire", Trans. and papers, *Inst. British Geog.* No. 18, 1952, pp. 51-66.
27. R. F. Peel, "The North Type Valley", *Geographical Journal*, Vol. 98, 1941, pp. 5-19.
28. O. J. Jones, "The Upper Towey Drainage System", *Quart. Journ. Geol. Soc.*, Vol. 80, 1924, pp. 568-609.
29. A. A. Miller, "Attainable Standards of Accuracy in the Determination of Pre-glacial Sea Levels by Physiographic Methods", *Journal of Geomorphology*, Vol. 4, 1941, pp. 95-115.
30. G. H. Dury, "Methods of Cartographical Analysis in Geomorphology", *Indian Geog. Soc., Silver Jub. Soult.*, 1952, pp. 136-40.
31. A. H. Robinson, "Geographic Cartography", *American Geography*: Inventory and Prospect, 1954, p. 557.
32. G. H. Smith "The Relative Relief of Ohio", *Geog. Review*, Vol, 25, 1935, pp. 272-84.
33. W. S. Glock, "Available Relief as a Factor of Control in the Profile of a Landform", *Journal Geol.*, Vol. 40, 1932, p. 74.
34. A Thauer, "Relative Relief Map", *Geog. Review*, Vol. 45, 1955, pp. 53-57.
35. Dov Nir, " The Ratio of Relative Relief and Absolute Altitudes of Mt. Carmel", *Geog. Review*, Vol. 47, 1957, p. 565.
36. G. H. Smith, "The Morphometry of Landscape : Analysis of Slope, *Annals, A. A. G.*, Vol. 29, 1938, p. 34.
37. A. A. Miller, "The Dissection and Analysis of Maps", *Trans. Inst. Brit. Geog.*, 14, 1949, p. 8.
38. E. Raisz & J. Henry, "An Average Map of Southern New England", *Geog. Review.*, Vol. 27, 1937, pp. 467-72.
39. A. H. Robinson, "A Method for Producing Shaded Relief From Areal Slope Data", *Surveying and Mapping*, Vol. 8, 1948, p. 107.
40. F. E. Elliot, "A Technique of Presenting Slope and Relative Relief on one Map", *Surveying and Mapping*, Vol. 13, 1953, pp. 473-78.
41. A. N. Strahler, "Quantitative Analysis of Watershed Geomorphology", *Ame. Geophy, Union, Trans.*, Vol. 38, 1957, pp, 913-20.
42. B. Zakrazewska, "An Analysis of Landforms in a part of the Central Great Plain", *Annals, A. A. G.*, Vol. 53, 1963, pp. 536-6Q.
43. Savindra Singh, and O. P. Singh, "Morphometric controls on Rural Settlements of Palamau Upland", *National Geographer*, Vol. XI, 1976, pp. 123-31

44. R. E. Horton, "Derivation of Runoff from Rainfall Data", *Trans. A.S.C.E.*, Vol. 77, 1914, pp. 339-75.

45. S. A. Schumm, "Evolution of Drainage systems and Slopes in Badlands at Perth Amboy", *Bull. Geol. Soc. Ame.*, Vol. 67, 1956, pp. 597-646.

46. K. G. Smith, "Standard for Grading Texture of Erosional Topography',, *Ame. Journal Science*, 248, 1950, pp. 655-88.

47. F. Gravelins, *Hydrology*, Berlin and Leipzig, 1914, p. 70.

48. S. A. Schumm, "Sinuosity of Alluvial Rivers of the Great Plains", *Bull. Geol. Soc. Ame.*, Vol. 74, 1963, pp. 1089-99.

49. L. B. Leopold, and M. G. Wolman, "River Channel Patterns—Braided, Meandering and Straight", *U. S. Geol. Survey, Proff.* Paper 282, 1957, pp. 39-85.

50. J. E. Mueller, "An Introduction to the Hydraulic and Topographic Sinuosity Indices", *A.A.A.G.*, Vol. 58, 1968, pp. 371-85.

51. J. C. Brice, "Channel Patterns and Terraces of Loup Rivers in Nebraska", *U. S. Geog.* No. 18, 1952, pp. 51-66.

52. L. B. Leopold, M. G. Wolman and J.P. Miller, *Fluvial Processes in Geomorphology*, W. A. Freeman & Co., London, 1964, p. 297.

53. A. N. Strahler, "Statistical Analysis in Geomorphic Research", *Journal of Geology*, Vol. 62, 1954, pp. 1-25.

54. W. E. Krumbein and S. A. Slack, "Relative Efficiency of Beach Sampling Methods", *B.E.B. Tech. Memo.*, 90, 1956, p. 43.

55. J. T. Andrews, "The Cross Valley Morains of North-Central Baffin Island: A Quantitative Analysis", *Geol. Bull.*, Vol. 20, 1963, pp. 82-129.

56. K. G. Smith, "Evolution Processes and Landforms in Badlands National Monuments, South Dakota", Bulletin, Geographical Society of America, 69, 1958, pp. 975-1008.

57. K. N. Singh, *Southern Upland of Eastern U. P.: A Study in Landforms and Settlements*, Ph. D. Thesis, Deptt. of Geog. B.H.U., Varanasi, 1967, p. 52.

58. Savindra Singh, "On the Quantitative Parameters for the Computation of Drainage Texture, Drainage Density and Frequency: A Case Study", *National Geographer*, Vol. XI, 1976, p. 26.

NINE

Joseph Uyanga

EFFECTS OF ECONOMIC BASE DIVERSIFICATION ON POPULATION DYNAMICS

THE DEVELOPMENT and usage of terms such as regional "imbalance", "inequality", "specialization", and "lop sidedness", have been popular in the description of various unsatisfactory aspects of regional variations in development:[1] Sjaastad[2] sees the sectorial distribution of workforce—"equalising differentials", as the major aspect of regional development. In regional studies an operational need thus arises to analysis the occupational content of a region or city as a yardstick for developmental comparison. Improvement of the economic and social conditions in regions that are occupationally ill-balanced is one of the main tasks of planning.[3]

Where more than two-thirds of the regional workforce is concentrated in a single economic sector (assuming for the moment the three conventional primary, secondary and tertiary sectors), it could be argued that there is a manifest lack of balance[4]. A mining community is an example of such lop-sidedness. In an enquiry of much spatial interest, Baldwin[5] has defined this argument as one of factor proportions model in the course of explaining the failure of mining export base to adequately stimulate regional development. Besides the low possibility of generating long term linkage effects and particularly the so-called final demand linkages within a region. This argument is amplified by North in putting forth a claim that "the limited social overhead investment, lack of popular education and the general lack of urbanization in the rural areas are the direct and indirect results of the regions' absolute dependence on agricultural production".[6] It is easy to exaggerate the problem of a lop-sided economy, The kernel of the argument however, is that lopsidedness spells disaster during economic depression. Vocational opportunities are limited from the social point of view because social life and cultural activities are the contributions of individuals drawn from a single preponderant group.

The consequence of a very top-heavy labour structure for regional development is that either quantitatively (e.g. female labour) or qualitatively (e.g. lack of full, diverse or high status opportunities), the full workforce potentialities of the region will not be utilised because the economy is not sufficiently broadly-based to permit a diversified and satisfactory workforce absorption.

We assume in the present study that the existence of over one half of a region's or city's workforce in any one sector of economic activity suggests a lack of balance. We define "balance" as a situation where one element in the sectorial structure is not too predominant to cause inter-sectorial disparity. The "balance" effects on regional population changes is investigated in the first section of the paper to provide a case for the subsequent study, of the extent to which economic base diversification had affected these changes within the regional and the urban systems.

The information in Fig. 9.1 shows that the trend in regional lopsidedness in South Australia was increasing during the period 1961-71. With regards to population growth the

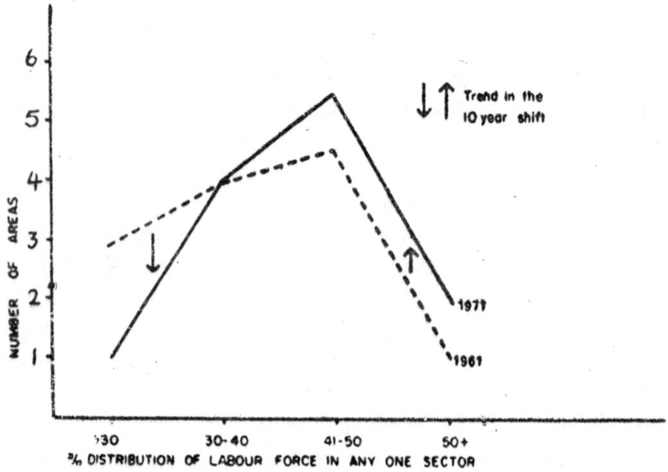

FIG. 9.1 : **Frequency of 'Balanced' and 'Unbalanced' Regions, 1961-71**

areas with less than 40 per cent of its workforce in one sector of the economy contributed 27.28 per cent if the situation in 1961 was considered or 29.38 per cent if the 1971 status was the focus. The "considerably lopsided" areas, with between 40 and 50 per cent of its workforce in any one sector suffered a loss of —5.11 per cent or —0.77, whichever year was considered, and the areas with over 50 per cent in any one sector experienced a net loss of —19.85 per cent and —18.23 per cent in 1961 and 1971. The obvious conclusion is that the more the labour force in a region was concentrated in a few sector of the economy, the more negative effects that region exerted on its population growth.

The proposition that functional diversification and differentials in South Australia regions is related to population concentrations and size was put forth by Bjorklund.[7] This has been widely accepted though no empirical test has been carried out to give credence to the operation of her concept in recent years. The following theoretical hypothesis is presented: if a significant and consistent association exists between the absolute size of population in an area and its extent of economic base diversification,[8] then, between two points in time, a change in that region's level of economic base diversification, would affect the magnitude of its population change.

Measuring Regional Diversification

There are roughly as many measures of diversification as there are students of the process. The earliest attempts by Mclaughlin (1930) was followed by Tress (1938), Florence (1943), Renwald (1949), Steingenga (1955), Rogers (1957), Devis and Hagger (1961), Conkling (1963), Parr (1965) and recently Clemente and Sturgis (1971).[9] Most of these methods—not all of which rank a town or region the same way—suffer from the shortcomings regarding the diversification at the city or regional level to be a microcosm of the nation.

Three measures are used in the present study not only to show the effect of each measure on the level of explanation of the situation in South Australia, but also to see the results of the explained, relative to each index. The first, is an expansion of Steingenga's (1965) approach.

The relative importance of each economic sector is measured on the basis of the proportion of total employment. Distribution or evenness is measured by the mean, granted that the total workforce was equally distributed among the sectors. The standard deviation from this ideal distribution indicates the disparity from evenness. The ratio of co-efficient of variation in sectorial distribution, nominated as V. The lower the V-index, the more diversified a region is. The second index, S_1, follows Gibbs and Martins (1962).[10]

The formula $1-\{\Sigma x_{ij}^2/(\Sigma x_{ij})^2\}$ measures internal specialization, where Σx_{ij} is the total employment in sector i in region j, and I is a condition of maximum diversification.

The third index, S_2, measures diversification in one region or city relative to others. That is, following Crowley (1973)[11]

$S_2 = \Sigma k_{ij}^2$ for all values of $k_{ij}>1$

$$k_{ij} = \left[\left(\frac{x_{ij}}{\sum_{i=1}^{n} x_{ij}} \right) \middle| \left(\frac{\sum_{i=1}^{m} x_{ij}}{\sum_{i=1}^{n} \sum_{j=1}^{n} x_{ij}} \right) \right]$$

where x_{ij} = employment in sector i in region j,

n = number of disaggregated sectors,

m = number of region or towns.

like S_1, the higher the S_2 index, the more relatively diversified a region was, between 1961 and 1971, in the South Australia context. It is realized that the use of the statistical subdivisions as areal units of measurement is not very appropriate. The alternative—the use of nodal or functional regions will present major difficulties and in data re-calculations. Limited justification, therefore, is found in the Australian Bureau of Statistics definitional criteria in delimiting the "statistical divisions", designed to represent closely allied and meaningful economic areas of production and community of interests"[12].

Areal Variations in Diversification

The V, S_1 and S_2 indices were calculated for two time periods in order to trace the effects of their dynamics on population dynamics, while the three measures are valuable indices of economic base diversification, several cautionary remarks concerning their use and interpretation must be made. First the value yielded is partly a function of the number of sectors categorised. Although this characteristic must be recognised, it is of little concern in the present study because here a constant number of categories are used for each region or town. Secondly, for a community to attain a maximum score of diversification, the workforce in the very minor sectors must equal that in the major ones, therefore in the polar case this is not realistic—a region with perfect or maximum diversification does not exist!.

The regional status of diversification is shown on Table 9.1, the V and S_1 indices are reasonably substitutable because the Far North which ranks highest on V also ranks highest on S_1.[13] Mallee

TABLE 9.1 : **S.A. Indices of Regional Diversification, 1961-71**

Subdivision	1961 %			1971 %			Change in Diversification		
	V	S_1	S_2	V	S_1	S_2	V	S_1	S_2
Adelaide	89	73	118	99	72	114	10	—1	4
Central Plains	92	75	54	96	73	74	—4	2	20
Yorke	116	69	47	104	75	69	12	—6	22
Kangaroo Island	103	73	99	114	72	82	—11	1	—17
North Ranges	69	78	53	81	69	66	—12	9	13
South Ranges	94	75	66	77	78	76	17	—3	10
Upper Murray	96	74	83	88	76	72	8	2	—11
Lower Murray	72	79	59	60	81	81	12	—2	22
Murray Mallee	152	58	35	145	59	54	7	1	19
Upper South East	106	72	59	102	72	69	4	0	10
Lower South East	68	80	61	70	79	74	2	—1	13
Upper Eyre	105	72	124	95	75	71	10	—1	—53
Lower Eyre	82	77	65	80	78	77	4	0	12
Spencer	73	79	92	83	77	85	—10	2	—7
Frome	88	76	75	68	89	66	20	—13	—9
Far North	47	83	72	88	76	149	41	—7	77

Note : Negative change in V means an increase in the process of diversification; positive change, of specialization.

which ranks lowest in V also ranks lowest in S_1 and even on S_2. The Lower Murray which in 1971 ranked highest on V also ranked highest on S_1. The lowest ranking for Mallee is repeated for 1971.

The least relatively diversified regions in 1961 were the Mallee and Yorke; the most relatively diversified Adelaide and Upper Eyre. By 1971 the least relatively diversified region continued to be the Mallee, followed by Frome, the most relatively diversified, Adelaide, and the Far North. The level of diversification in Adelaide tended to have declined between 1961 and 1971. Positive trends in diversification were observed in the Northern Ranges, Spencer and Kangaroo Island; negative trends in Frome, the Far North, Yorke and the Southern Ranges.

This dynamic implication that an economic space can be more or less diversified over particular points in time raises the question : If the hypothesis holds for South Australia, that changes in total population and the position of economic base diversification are somehow parallel processes, then most or all regions which had positive shifts in their diversification status should account for the largest share of the state population growth during the decade. The hypothesis is fairly supported in Table 9.2.

TABLE 9.2 : **Change in Diversification Indices and Change in Total Population, 1961—1971**

	V-index		S_1-index		S_2-index	
	Regions with Positive Shifts	Regions with Negative Shifts	Regions with Positive Shifts	Regions with Negative Shifts	Regions with Positive Shifts	Regions with Negative Shifts
Mean Population change %	16.93	—1.90	19.01	0.94	13.18	4.96

DIVERSIFICATION WITHIN THE URBAN SYSTEM

The three measures above were applied to the sectorial distribution of urban labour force. The result, (Table 9.3), shows that S_2 had a smaller range of maximum-minimum variation than the S_1 Index. In other words, South Australian urban centres are more *unlike* one another in the degree of internal diversification than they are when inter-urban relative diversification is considered[14] (Table 9.3).

High scores on inter-urban diversification went to Victor Harbour Kadina, Barmera, Strathalbyn Minlation, Gladstone, Gleve, Quorn, Naracoorte and Burra. The absence of the five largest cities (Adelaide, Whyalla, Mount Gambier, Port Pirie and Port Augusta) from this most relatively diversified towns and even (except Port Augusta) from the ten highest internally diversified towns (on V or S_1) in Table 9.3 shows that the *largest towns* in South Australia are specialised internally and relatively. The least diversified and relatively most specialised towns were Renmark, Lobethal, Leigh Creek, Woomera, Peterborough and Jamestown; the last four of the six, experienced high population losses between 1966 and 1971, just as the most specialised regions (Frome, Northern Ranges, Yorke Peninsula and the Mallee) recorded high net population decline.

The observation that a significant level of the relative index S_2 had correlation with population size (r=45), together with the small range earlier observed, suggests that to some extent economic activities in most country towns are complementary within proximate towns of approximately equal or different population size. The expected direct relationship between population size-change and economic base diversification in the urban system is supported by the positive correlation V (r= .37) and S_1 (r= .25). However the results merit further investigation due to the failure of the independent variables to explain a larger amount of the variation in each of the two dependent variables. On the basis of the generally accepted theoretical explanation, one would expect urban population size to be a stronger predictor of the workforce diversification.[15] Unlike when larger areas, the statistical subdivisions were considered, not even up to one-third of the variations in internal urban workforce diversification was explained by population size. This of course also leaves the bulk of the variation in the independent variables (population size and change), unaccounted for.

TABLE 9.3 : **S.A. Urban Centres—Population and Indices of Diversification**

(Urban Centre) Ranked from highest to lowest on V and S_1	Index of Diversification			Population Size 1971	Population Change 1966-71
	V	S_1	S_2		
(1)	(2)	(3)	(4)	(5)	(6)
Penola	66	85	40	1293	—6.51
Clare	68	84	37	2105	33.06
Port Augusta	70	83	38	12095	15.66
Ardrossan	70	83	30	714	—9.62
Crystal Brook	71	83	34	1185	—4.05
Port Elliot	71	83	44	348	—0.71
Goolwa	72	84	32	680	21.21
Murray Bridge	72	84	40	7400	14.84
Port Lincoln	73	84	39	9158	2.76
Quorn	73	83	46	1010	12.12
Mount Barker	73	84	32	2340	20.99
Kingscote	74	83	42	1007	—5.98
Kadina	74	83	51	2853	—5.59
Kingston S.E.	76	83	32	1173	10.14
Yorketown	76	83	45	704	—11,22
Burra	78	82	46	1278	—4.77
Barmera	78	82	50	1683	12.95
Border Town	78	82	40	1977	12.39
Ceduna	78	82	41	2056	41.11
Cumminua	79	82	38	736	—5.28
Balaklava	79	82	44	1111	—7.34
Gladstone	80	82	47	926	—12.06
Strathlbyn	80	82	48	1547	6.76
Moonta	81	83	42	1578	—7.29
Streaky Bay	82	82	37	913	23.88
Kapunda	82	82	39	1301	16.26
Tumby Bay	82	82	45	864	10.63
Maitland	83	82	43	1020	0.29
Jamestown	83	82	26	1339	—3.81
Angaston	83	82	31	1816	—3.76
Berri	84	81	32	2712	21.31
Mount Gambier	84	81	40	17867	3.51
Pinnaroo	84	81	42	803	—7.06
Keith	85	81	37	1211	10.39
Adelaide (Urban)	85	81	41	808466	11.23

(1)	(2)	(3)	(4)	(5)	(6)
Renmark	87	81	24	3277	7.30
Peterborough	88	81	29	3019	—3.14
Wallaroo	89	81	41	2092	—0.99
Kimba	91	80	41	704	4.14
Naracoorte	92	80	45	4399	0.43
Nuriootpa	94	80	31	2467	20.87
Minlaton	94	80	47	876	8.55
Tanunda	97	79	33	1936	—2.52
Millicent	98	79	30	5075	11.81
Victor Harbour	103	78	69	3533	12.44
Mannum	104	78	35	2036	0.10
Port Pirie	106	77	35	15506	—0.39
Cleve	107	77	46	781	15.53
Loxton	116	76	31	2658	9.83
Whyalla	149	66	33	32085	44.98
Lobethal	156	64	19	1381	25.77
Woomera	177	57	18	4069	—14.32
Leigh Creek	212	44	17		—9.24

Diversification of Economic Establishments

The number of factory and retail establishments in urban centres, in South Australia is given in the A.B.S. Divisional Statistics 1973. Two-thirds of the twenty-one towns with population above 2,000 in South Australia had the entire range of the six business categories. Food and Drink; Footwear, textile and clothing; Timber, Furniture paper; non-metallic chemicals; Basic and Fabricated metals; other manufacturing. One centre, Burra, among those with population below 2,000 had all the six categories of factories. The majority of those below 1,000 had three to four categories. In general, urban centres with population above 2000 were associated with a wider range and larger volume of activities. The association between population change and the total number of factories, was stronger the higher the urban population range. Urban centres with population between 1000 and 2000 had a stronger positive association between the number of retail establishments and population change, than the other size ranges. In all, these relations indicate roughly that as dynamic factor in urban development, in terms of the number of business establishments, population change was significant especially in the small and medium size centres.

Growth in population tended to have ensured the necessary threshold in growing country towns. Any investment decision errors probably were not as serious in their consequences as would have been in the declining towns. Probably several types of business activities were related to population size because of the "induced in industries" effects which were caused partially by an existence of local market. Some country towns could not provide some basic needs of its inhabitants. The production of goods and services which otherwise would have been imported were possible in the larger centres, thus in the larger centres there were tendencies to be increasingly self-sufficient. Therefore, the effects of income leakages were reduced. This contributed to the general observation that incomes not only increase with urban size but are more disparate and unequally distributed in the smaller centres, than the larger ones.

Despite their positive association as predicted, some of the relationships are weak enough to have resulted from chance occurrence and the r^2_s are not high. It appears again, as observed earlier that among the causes for variations in population changes industrial diversification was more important within the South Australia *regional* system than within the *urban* system. Even its low explanatory power might have been shared by other variables.

To the extent that population growth and size could have caused growth in workforce diversification and a follow-up process in industrial establishments diversification the hypothesised proposition is acceptable in the State regional structure. But owing to some regional socio-economic and political factors affecting population mobility and so, generally, population change, diversification as such cannot exclusively be said to have accounted significantly for all population growth and change in space. Given the capital intensive pattern of manufacturing and the decline of excess of supply over demand for labour which generally occurred in the State during the period under study,[16] the consistency and generality of the hypothesis lapses.

More importantly the ability of population to attract a variety of economic activities and industries and to be diversely sustained by them depends on the accelerator-multiplier effects as caused by both re-investment functions and the diversification of the existing economic activities. To a large extent the latter is normally generated by the former, yet in most South Australian country towns we have only single branch or "off-shoot" plants and factories. The profits and firm earnings are "exported" from the regions, and it is a small fraction of the proceeds from locally earned wages which are spent within the urban or regional economy. Of course, this is not powerful enough to propel and induce a highly sustained development as would occur in areas with larger population, expenditure and re-investment multiplier effects.

The Wallaroo District

This area was selected because of a lower scale of investigation. It provides an example of highly specialised region during the copper mining boom in the late 19th and early 20th centuries.

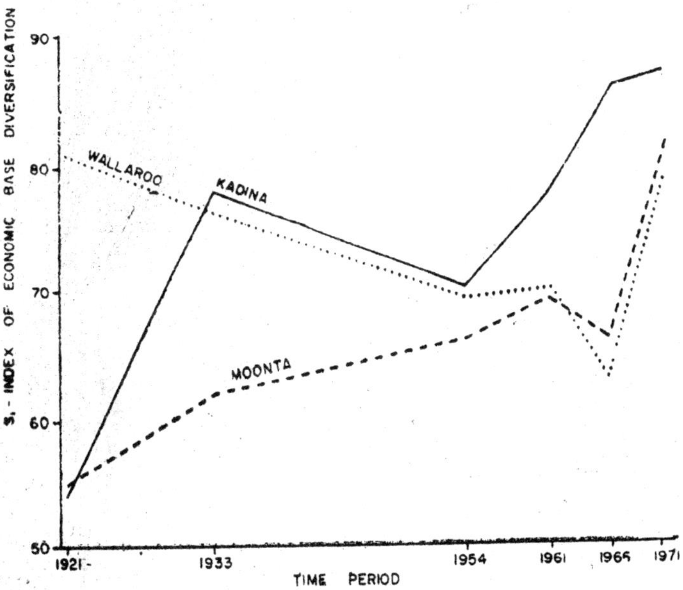

FIG. 9.2 : **The Wallaroo District—Diversification Trend, 1921-1971**

After the closure of the mines in 1923, there were various attempts at diversifying and developing new economic bases for the three towns—Moonta, Wallaroo and Kadina. The process of population dynamics can further be observed within the context of this diversification process.[17] Using the composition and sectorial distribution of the workforce at and after the closure of the mines,[18] the diversification indices S_1 and S_2 were calculated and the trend is depicted in Fig. 9.2.

This information shows that (i) in fact in their adversity the towns in the Wallaroo district had been broadening their diversity; (ii) that diversification had been progressing within the three towns at differential rates, and (iii) that Kadina is the most internally and relatively diversified of the three towns. According to our hypothesis, we should expect their population changes to reflect these differing economic base diversification process. Fig. 9.3

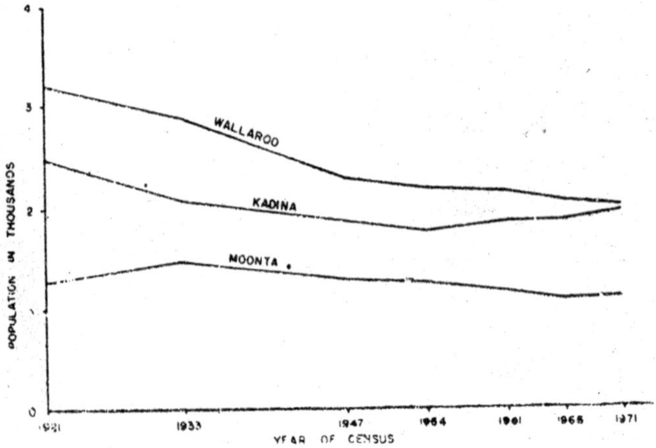

FIG. 9.3 : **Population Change in the Municipalities of the Wallaroo District, 1921-1971**

	1921 1933	1933 1947	1947 1954	1954 1961	1961 1966	1966 1971	Overall 1921—1971	Mean
Wallaroo %	—9.3	—20.7	—4.3	—6.9	—6.4	—1.00	—31.2	—0.94
Kadina %	—16.0	—9.6	—5.3	3.2	0.0	3.3	—28.2	—0.48
Moonta %	14.5	—13.3	—8.0	—5.7	—2.5	0.3	—8.0	—0.29

Note: These figures refer only to the defined municipal areas.

shows that since 1947 there had been a diminishing rate in population decline in the three towns. Kadina has grown from a position it held 36 per cent of the entire district's population in 1921 to 39 per cent in 1971 and Moonta from 18 per cent to 21 per cent, Wallaroo was gradually losing its relative grip of the total district population. Even though Kadina is not the largest town, it is the most progressive of the three, with a trade area 48 Km. radius enclosing a population of over 16,000 in the most concentrated sales region, Thompson (1955).[19] Its current rate of population growth is the highest. The writer found that almost 31 per cent of Kadina's workforce commute daily from Wallaroo (8 kilometres away) and Moonta (16 kilometres away).

Still the largest urban centre in the district, Wallaroo depends on the Port (and until 1971 the rail) facilities. The employment in its fertilizer plants fluctuates with peak despatch seasons. Wallaroo had in 1971 four times the number of factories as Kadina, and the facilities were larger.

Thus it is a good example of a town illustrating the failure of manufacturing to generate enhanced urban growth. Of the 106 persons interviewed by the writer in Wallaroo 90 (85 per cent) admitted that "many more different forms of economic activities" was the answer to their "regional problem" which 94 (89 per cent) states as "lack of local jobs" and 78 (74 per cent) as "not only the inadequacy of the employment opportunities but the *fast* decline in the existing ones." There was a general awareness of the draining effects of outmigration of young persons (especially girls) and those whose business have closed down. Apparently, "Wallaroo's future at stake"[20] aptly summarises the concern for the sapping effects of population decline, and the consequences of shopping outside the town.[21]

With regards to the whole district, the development of mining boom around copper led to an initial population and economic boom, a "burst and feast," then "famine" at the closure of the mines; subsequent population growth depended on how each town adjusted its internal economic base to support more people.

Conclusion

Within the three contexts which the question was investigated, increasing and differential economic base diversification provided the regional and urban conditions for population changes. In general the larger the region's population the higher the possibility was that employment could spread among many economic sectors; though this was not synonymous with a high level of diversification. The more diversified a region was, the more likely it was that its population change was positive, (r=0.62), and, the less the percentage workforce that was unemployed, (r=0.81).

Increased sectorial specialization had meant a decrease in the variety of economic activities and opportunities available. As the range of activities narrowed down the state urban system, the range of activities per person declined; thus a non-diversification planning policy essentially implied shifts in regional population. By specialization the less a town could provide for itself, we could assume that, the larger the monetary outflows that was.

This had implications for local multiplier effects, as a function of larger labour force participation rates, and the threshold effects of growth in local economic activities.

These conclusions, are essentially inconclusive, in that there may be significant variations in conclusions for different temporal, cultural and economic contexts, only because of definitional difficulties in the hypothesis.

REFERENCES

1. W. Baer, "Regional Inequality and Economic Growth in Brazil", *Econ. Dev. and Cultural Change*, Vol. 12, No. 3, 1969, pp. 268-285.
R.A. Easterlin, "Inter-regional Differences in Per Capita Income, Population and Total Income 1840-1900 in National Bureau of Economic Research, Conference on Research in Income and Wealth", *Trends in the American Economy in the 19th Century*, 1960, Princeton, pp. 73-140.
2. L.A. Sjaastad, "The Relationship between Migration and Income in the U.S.A.", *Papers Reg. Science Assoc.*, Vol. 6, 1960, pp. 37-64.
3. South Australia State Planning Authority, *Planning and Development Act, 1966-72 Summary*. Elsewhere such needs have been demonstrated by E. Thompson, (1966) "The Workforce Index: A Technique for Regional Planning", (*Royal*) *Australian Planning Inst. Journ.*, Vol. 4, pp. 200-204 and A.E. Smailes, (1942) "Ill-Balanced Communities: a Problem in Planning" in E.A. Gutkind, (ed.) *Creative Demobilization*, London, Vol. 2, p. 226.
4. W. Steingenga, "A Comparative Analysis and Classification of Netherland Towns", *Tijds. Voor Econ. en Soc. Geo.* June-July pp. 105, and R.J. Solomon, (1962), "Locational Emphasis of the Australian Workforce", *Econ. Geog.*, Vol. 38, 1955, p. 142.

5. R.E. Baldwin, *Economic Development and Export Growth—A Study of Rhodesia*, 1920-1960, Berkeley and Los Angels, 1966.
6. D.C. North, "Patterns of Settlements in Newly Settled Areas", *Manchester School of Economic and Social Studies*, Vol. 24, 1961, pp. 122-134.
7. E.M. Bjorklund, *Focus on Adelaide Functional Organisation of Adelaide Regions*, Dept. of Geography, University of Chicago, Research Paper No. 41, 1955.
8. See Bjorklund, op. cit., p. 40, see also R.W. Crowley, "Reflections and further Evidence on population size and Industrial Diversification", *Urban Studies*, Vol. 10, pp. 91-94 and F. Clemente, and R.B. Sturgis, (1971) 'Population Size and Industrial Diversification", *Urban Studies*, Vol. 8, No. 1, 1973, pp. 65-68.
9. Glen Mclaghlin, "Industrial Diversification in American Cities", *Quarterly J. of Economics*, Nov. 1930, pp. 131-149. Tress, R.C. (1938) "Unemployment and Diversification of Industry", *The Manchester School*, Vol. 9, No. 1, pp. 140-150. S. Florence, (1943), *Industrial Location and National Resources*, National Resources Planning Board, Washington D.C., L.T, Reinwald, "Some Aspects of Statistically Interpreting the Manufacturing Functions of U.S. Cities", Unpublished M.A. Thesis in Geography, Clark University, 1949.

   W. Steingenga, "A Comparative Analysis and Classification of Netherland Towns." *Tijds Voor Econ, en Soc. Geogr.*, June-July, 1949, pp. 105-118. A. Rogers, "Some Aspects of Industrial Diversification in the U.S.", *Econ. Geog.*, Vol. 33, No. 1, 1957, pp. 16-30. H.W. Davis, and D.F. Hagger, "Recent Industrial Changes in South Wales", *The Advancement of Science*, Vol. 18, No. 71, 1961, pp. 258-272. E.C. Conkling, "South Wales: A Case Study in Industrial Diversification", *Economic Geography*, Vol. 39, No. 3, 1963, pp. 258-272.
10. J. Gibbs, and W. Martin, "Urbanisation, Technology and the Division of Labour", International Patterns", *American Sociological Rev.*, Vol. 27. 1962, p. 669.
11. Op. cit. Ref. 9.
12. Commonwealth Bureau of Census and Statistics, *South Australian Yearbook*, 1972, p. 103.
13. This may not be so in all cases because as much as 41 per cent of the variations in V was not explained by S_1 in 1971.
14. It is noted that the drawing of urban boundaries could affect these conclusions. Renmark and Loxton for instance include a large number of farm blocks which many other towns do not.
15. Consider for example the statement, "Clearly, increased city size brings greater industrial diversification",

    W.R. Thompson, "*A Preface to Urban Economics*", Baltimore, The John Hopkins Press, (1967), p. 147.
16. Australian Dept. of Labour and National Service *Monthly Review of the Employment Situation*. See Serials from March 1969 to December 1971 for the overall situation in South Australia.

17. The History of this area can be found in Pryor, (1962) Australia's *Little Cornwall*, Riopy, Adelaide. The dynamics of the functions of the three towns is described by K.W. Thompson, "The Changes in Function of former Mining Settlements: The Wallaroo Copper Belt", *The Royal Geographical Society of Australia*, S.A. Branch, Vol. 56, 1955, pp. 47-58.
18. See J.T. Uyanga, "Population Dynamics and Regional Development: South Australia" Unpublished Ph.D. Thesis, School of Social Sciences the Flinders University of South Australia, 1974, p. 243.
19. Op. cit., p. 54 (Ref. No. 17).
20. *The Kadina Moonta and Wallaroo Times*, July 5, 1967 carried an article reporting a general meeting on that topic.
21. *Ibid*, May 16, 1968 urged the residents to "support your local business houses and tradesmen's shop locally and help employ local labour and maintain a larger population."

TEN

Ram Naresh Prasad Sharma and
Devendra Prasad Singh

STATISTICS FOR POPULATION GEOGRAPHERS

Introduction

Population geography has emerged as a distinct and well recognised branch of geography in the twentieth century. By its very definition, geography relates to the study of man and his habitat. Man, therefore, acts like a thread through almost the whole fabric of geographic thought. Trewartha[1] has lucidly stated that number, densities and qualities of the population are the point of reference from which all other elements are observed and from which they all singly and collectively derive significance and meaning.

According to Hoosen[2] geography is concerned with the problem of uneven distribution of population over the earth. It acts as a master thread, capable of weaving into a coherent pattern otherwise desperate strands of the subject and expressing its philosophical unity particularly in the context of regional geography cannot be possible.

During last few years geographers have taken interest in the study of population. Their main interest of study has been the analysis of areal variations in the distribution, composition, migration and growth of population and the way in which it relates to the nature of places. It may be postulated that the significance of population geography as a branch of geography has grown constantly with an increasing awareness of the dynamism of population[3].

Zelinsky stated that the population geography because of its being paramativable and its ability to enlighten the whole field of geographic and demographic studies should receive an autonomous status[4].

Human resources are no less important than the physical and economic resources. In fact, physical and economic resources have no meaning without human resources. Because all resources are negative unless utilised by the technological achievements of man. The appraisal of resources can only be made in the context of human wants and their capacity of utilization for which the study of population in all the depths and intensity is required. Geographers with their comprehensive knowledge of both the earth features and the inhabitants are perfectly qualified to study the people in time and space and also to portray the conclusions by suitable geographical tools. By the use of quantitative methods they can analyse and interpret the relevant statistics and draw proper conclusions in terms of the earth resources and their utilization in past, present and future. In this sphere they are better equipped than other social scientists.

The traditional subjectivity of population geography has been replaced by more precise and objective methods of measurement and analysis[5]. The collection of appropriately quantified data thus became vital and, in turn, new techniques for the accurate processing and presentation of that data have been largely adopted. Thus the quantitative revolution has been also brought to this branch of geography like other social sciences. The use of geographical techniques for the collection, processing and presentation of data became essential. These data should be scientifically collected, skilfully processed and beautifully presented, for which statistical method should be adopted. Though to many people the very mention of the word statistics is enough to cause instant depression[6]. Fortunately we have many and varied ways in which data may be statistically processed, most of them are relatively straight forward and demand only a very elementary ability. In this essay the authors intend to highlight some of the applied methods of statistics for the various aspects of population geography.

Statistics and Growth of Population

The growth of population is so rapid that it is reaching to the bursting point. Its rapidity can be easily understood by the simple fact that 3.7 human beings are born every second. This works out to 221 per minute, 318,555 per day, or more than 2.2 million per week[7]. Thus the growth of population is a very important aspect of population geography. Apart from the absolute increase or decrease per annum, one of the common measures of growth is the annual rate of increase[8]. The U.N. Demographic Yearbooks use the following formula :

$$\left({}^{t}\sqrt{\frac{P_1}{P_0}} - 1 \right) 100$$

Where P_0 is the population at the beginning of the period, P_1 is the population at the end of the period, and t is the number of years. The annual rate of increase is subject to the general qualifications for such data. It is, therefore, a useful rate and may be helpful in assessing the accuracy of vital and migration statistics.

Generally, the growth of population is the function of natural increase and migration. Birth is the natural process of population by which a new member enters the population at the age of zero.[9] Birth and death rates are the guiding factors in determining population growth. Birth rate is the ratio of the total births of a period to the total population multiplied by 1000. It is derived as under :

$$r = \frac{b}{p} \times K$$

where r is the births per 1000 persons per year, b is the total number of births during the year, p is the total population and K is the 1000. To calculate the birth rates for male and female the meaning of b is shifted to total male births, or total female births, instead of its usual meaning of total birth in the equation. The death rate is expressed in the term of equation as :

$$r = \frac{d}{p} \times K$$

where r is the death per 1000 persons per year, d is the total number of death during the period, p is the total population and K is 1000.

Natural increase is the difference between number of births and deaths and the natural increase rate is the difference between the crude birth and death rates[10]. It may be calculated as :

$$\frac{B-D}{Y}$$

where B is the total birth, D is the total death and Y is the total population for a specific year. Inspite of its popularity, this rate expresses population growth very imperfectly, because it ignores the ageing of each individual during the year and does not differentiate between the deaths of old and young persons. A population with a high proportion in the reproductive age groups might have a positive natural increase rate and yet experience low fertility and lack of replacement. Similarly, a population with declining mortality may include a preponderance of births over a long period although its fertility may be incapable to ensure replacement. Natural increase rate, therefore, should be adopted with caution.

Fertility of Population

The frequency of birth in a population is termed as fertility. It is the birth performance of a population as reflected by the number of live births. It refers to the number of children a woman has actually borne.[11] Fertility is strictly confined to females and that too during specific (15—44 or 15—49) or reproductive periods.[12] The general fertility rate can be summed up as under:[13]

$$R = \sum_{i=15}^{i=44} \left(\frac{b_i}{p_i}\right) \times K$$

where R is the total fertility rate, i is an interval of one year, b_i is the number of birth during the year, p_i is the number of women of same age and K is 1 or 1000. Thus the total fertility rate is the total number of children born to single woman or a number of women. Another method to calculate the fertility rate is as such :

$$r = \frac{b}{p} \times K$$

where r is the general fertility rate, b is the number of birth during the year, p is the population of women of age group 15—44 and K is 1000.

But both the above mentioned formulas fail to give a general picture of rate of population increase and the expectation of life at the birth is less.........[14]. A good number of children die at first year of their life. So it is better to take the survivors in the age group of zero to four. But while considering the survivors of births in the age group zero to four, the fertility is over estimated, whereas considering the actual births taken place during one year is misleading. But it is survivors who matter in the latter population and the rate of fertility can be computed as :[15]

$$R = \frac{P_{0-4}}{F_{15-49}} \times K$$

where R is the fertility rate, P_{0-4} is the number of children under four years, $F_{15=49}$ is the number of female and K is 100.

Variability Index

In his paper on population trends in India, A. Geddes has demonstrated that variability in the growth of population has a significant regional bearings. The normal type of intercensal population growth map does not reveal variability.[16] Geddes represented the variability by means of a variability index. For this purpose the actual population has been shown by plotting points to represent them on a convenient scale and joined by curves. Then the two points, first to represent the population of first year and the second to represent the population of last year, have been joined by straight line to show the population which ought to be have. The gap or deviation between the two lines has been calculated according to the scale in terms of percentage. Then he calculated the variability index as follows :

$$V = \frac{d}{i}$$

where V is the variability index, d is the deviation in percentage and i is the number of inter-censal years.

Projection of Future Population

The projection of future population depends upon the past growth of population. A number of methods so far has been employed by the demographers to estimate the future population which can be grouped as indirect and direct[17] techniques. The direct techniques are based on correlation of population numbers with social and economic indices and are more sophisticated. They are, however, difficult to apply if detailed sample survey of key areas have not been made, or if there are too many outside influences. The indirect techniques can be grouped under four heads as (a) Comparative for casting, (b) Radio and correlation method, (c) Growth composition analysis, and (d) Extrapolation,

The first method envisages the selection of models which has exhibited similar characteristics in the past. Such methods are usually suitable for developed countries. The second group of methods is an improvement upon the first, but as these methods depend upon mathematical projection for determining future ratio between the under-developed region and the method chosen, they are not superior to extrapolation method.

The third group of techniques consider natural growth, migration and territorial changes of area. The data for birth and death rates in Bihar has been found to be wholly unreliable and neither natural growth nor migration figures are available below the district level in India. Under the circumstances, the use of more sophisticated methods under this group do not seem to be justified; which leaves us with the extrapolation technique of population prejection.

However, if the present growth rate remains the same in future and no natural calamities like epidemic, drought, war etc. occur, the population growth of a region may have more or less the same ratio. Upon experiments it was found that the formula for arithmetical progression best fitted the observed growth ; which can be computed as :

$$F = P_1 + (r_i \times Y)$$

where F is the future population, P_1 is the population of first year, r_i is the rate of increase and Y is the time interval between the first and last year. We have, therefore, firstly to find out the rate of increase on the basis of following calculation :

$$r_i = \frac{P_2 - P_1}{Y}$$

where r_i is the rate of increase, P_2 is the population at last point, P_1 is the population at first point and Y is the time interval between the first and last year. Geometric progression is also one of the most popular method. It can be calculated as :

$$Y = ar^{m-1}$$

where Y is the year selected for projection, a is the first year, r is the common factor and m is the interval of years.

STATISTICS AND DENSITY OF POPULATION

The concept of population density is generally expressed in terms of areal expansion and the ratio of resources available in a region for human leaving. This ratio shows the density of population[18]. It can be represented in various ways. Generally the total population of a region is divided by the total area, which is known as general or crude density and can be calculated as such :

$$G_d = \frac{T_p}{T_a}$$

where G_d is the general density, T_p is the total population and T_a is the total area. But it is a generalised type of density and never focus the variation of population density within the unit. In the same region one area may be densely settled while others may have comparatively sparse human dwellings. Secondly, it does not reveal the clear and real pressure of population on the land because it also includes urban population which is rarely supported directly from the land. Hence it fails to highlight the internal distributional pattern and economic conditions of habitant of a region.

To avoid the demerits of inclusion of urban areas and population of crude density, the rural density can be adopted, which is calculated as such :

$$R_d = \frac{T_{rp}}{T_{ra}}$$

where R_d is the rural density, T_{rp} is the total rural population and T_{ra} is the total rural areas. Rural density of population also fails to represent the actual pressure on land. Because a large number of population get their subsistence from other than cultivation, though their percentage may be lower. But rural density includes all the rural population and even all the rural areas, whether it is under cultivation or waste or used otherwise.

A more realistic idea can be obtained by getting the agricultural density. It gives a true picture of the pressure of population on land. The agricultural density of population is the result of intrinsic interest and practical value[19]. It is more refined and presents real pressure on land because it omits non-agricultural land and people. It is based on the undermentioned formula :

$$A_d = \frac{T_{ap}}{T_{ca}}$$

where A_d is the agricultural density, T_{ap} is the total agricultural population including farm labourers and land owners and T_{ca} is the total cultivated land.

Physiological density is one of the truest indices of population concentration because it omits the non-agricultural land. It provides more refined and correct picture of population than rural density. But it fails to convey a clear idea of the actual conditions since human beings are by no means evenly distributed over the countryside[20]. It is obtained as :

$$P_d = \frac{T_p}{T_{ca}}$$

where P_d is the physiological density, T_p is the total population and T_{ca} is the total cultivated area. Physiological density is, thus, not the true representation of real human pressure on land because it includes all the population with any discrepancy of rural and urban or agricultural and non-agricultural. The pressure on land is rather exaggerated because only the cultivated land is considered for it.

In any discussion of the population problem, the food problem is of basic consideration. The population is increasing at a faster rate and the same nutritional standard is to be maintained as the one at present. The supply of food grains has to be increased at the same rate. This means that we have to find additional land for cultivation or rapidly increasing yield which is not so easy. The solution of the problem is thus more serious and difficult. The best way to observe the pressure of land is the nutritional density. The nutritional density more or less is the refined form of rural density of population which omits the non-cropped areas of the region and is often computed as :

$$N_d = \frac{T_{rp}}{T_{ca}}$$

where N_d is the nutritional density, T_{rp} is the total rural population and Γ_{ca} is the total cropped land or gross area sown.

STATISTICS AND POPULATION COMPOSITION

The composition of the population, or the population structure, refers to those aspects of population which may be measured, however inadequately[21]; e.g. age, sex, economic activities, literacy and urban and rural ratio, etc. These are also known as quantitative aspects as against qualitative aspects.

Age Structure

The age structure is an important measure of population analysis which concerns the characterisation of people according to different age groups. One cannot proceed very far in the study of population growth or migration without examination of age structure[22]. The population is often divided into three age groups i.e. (a) infants and adolescents (0—14 or 0—19), (b) young (15—59, 15—64, 19—69 or 19—64) and (c) old (over 60 or 65). Following statistics may be adopted to present a clear picture of relationships between the three age groups : (i) $\frac{c}{a}$, (ii) $\frac{c}{o}$, (iii) $\frac{c}{a+o}$, (iv) $\frac{o}{c+a}$, (v) $\frac{c+o}{a}$, (vi) $\frac{o}{a}$, (vii) $\frac{o}{c}$ where c is the children, a is the adults and o is the old. Among these the fifth (dependency ratio) and the seventh (old age index) are most common.

Sex Structure

Although numbers of the two sexes are not widely divergent, their disparity is of interest to geographers, because the two sexes play contrasting role in economy and society. The sex ratio is the proportion of men to women in the population at any given time. There are four methods to calculate sex ratios, firstly, the number of males per 1000 females; secondly, the number of females per 100 or 1000 males; thirdly males or females as the percentage to total population and lastly, the proportion of males or females as a decimal or unity.

Economic Activities

Man adopts himself according to the land and surroundings and his activities are shaped and determined as he invents processes to eke out his living[23]. His occupations are broadly divided into three groups as : (a) Primary workers get their livelihood directly from nature e.g. agriculture, forestry, mining, quarrying, etc., (b) Secondary workers process the primary products like manufacturing, construction, etc., and (c) Tertiary workers are engaged in unproductive occupations e.g. trade, commerce, transport, teaching, etc. Workers employed in these three sectors are generally represented in percentage and triangular graph is the most suitable diagram to portray them.

Non workers entirely depend on the working force for their livelihood. The heavy dependency acts as a serious drag on the production and improvement of living standards[24]. To find out the dependency ratio is a very simple statistical method for measuring the impact of age composition on the livelihood activity of the population[25]. It is assumed that the 15 to 59 years is the productive segment of the population and the infants and adolescents under 15 and persons aged 60 years or over are the dependent segment[26]. It may be said in terms of statistics as :

$$D_r = \frac{Fy_1 + Sy_1}{Fy_2 \ to \ Fn_i} \times K$$

where D_r is the dependency ratio, Fy_1 is the people under 15 years of age, Sy_1 is the people over 60 years of age, Fy_2 to Fn_i is the people between 15 and 59 years of age and K is 100 or 1000. But it is a generalised type of ratio because in underdeveloped societies children even of less than 15 years of age also work in farms, mines, industries and other enterprises. Similarly people more than 60 years of age are found engaged in various occupations. And students of higher education do not work even at the age of twentyfive or so. Hence a detailed field work is required to present an actual figure of dependency ratio.

Literacy

The acquisition as well as retention of literacy is an important catalyst for creativity in all spheres of human endeavour. The educational status of population is, therefore, one of the recognised indicators of the quality of population, while talking of literacy generally the effective literacy is concerned. It is generally computed after excluding the population under five years of age because children below it generally do not read or write. It is calculated such as :

$$E_l = \frac{T_l}{T_p - P_f} \times K$$

where E_l is the effective literacy, T_l is the total literacy, T_p is the total population, P_f is the persons below five years of age and K is 100.

Urban and Rural Ratio

Urban population greatly differs from that of rural in respect of distribution, density, mode of living, composition, economy and growth. But it is not easy to mark a distinction between rural and urban. According to sociologists urban population is town living one. They consider that their dwellings are concentrated in a recognizable street pattern. People live in some social and economic interdependence. They enjoy common administrative, cultural and social amenities. But such a distinction is vague, Presence of non-agricultural activities is the chief characteristics of towns. But it is not common in all cases in advanced countries. Because such activities frequently occur in several villages. The presence of industrial activities is another criteria but several towns have no industries and sometimes rural areas may have industrial activities though on a small scale. Monkhouse[27] considers a town as a compact settlement larger than a village with a community pursuing an urban life.

Urban-rural ratio can be represented in three ways such as :

(a) $U\text{-}R = \frac{T_{up}}{T_p} \times K$ where $U\text{-}R$ is the urban-rural ratio, T_{up} is the total urban population, T_p is the total population and K is 100 ;

(b) $U\text{-}R = \frac{T_{rp}}{T_p} \times K$ where T_{rp} is the total rural population, T_p, is the total population and K is 100 ; and

(c) $U\text{-}R = \frac{T_{up}}{T_p} \times K$ where T_{up} is the total urban population, T_p is the total population and K is 100.

According to Clarke[28] all cities have a declining population tendency towards the outer suburb. It can be stated statistically as :

$$Y = A_e^{-bx}$$

where x is the distance in kilometres from the centre of the city; Y is the density of resident population per square kilometre; b is a coefficient in measuring the rate of decline of density and A_0 is a coefficient measuring the degree of overcrowding in the centre of the city.

Statistics and Migration

Migration is a vital process which alters the size of population. Besides natural increase it is the only source of change in the population, it may be positive or negative. A migration rate is the ratio of migrants observed to the population exposed to the likelihood of migration during a specified migration interval. It may be expressed in terms of equation as :

$$m = \frac{M}{P} \times K$$

where m is the rate of migration, M is the total migration, P is the total population and K is 100.

Migration balance is to be understood by the sum of the difference between immigrants and emigrants. In terms of equation it may be presented as :

$B_m = I - E$ or $E - I$, where B_m is the balance of migration, I is the immigrants and E is the emigrants.

The significance of migrants to the total population of different regions can be brought to the notice through simple computation of size index with the help of formula[29] :

$$S.I = \frac{M_d + 100}{P_d}$$

where M_d is the migrants in the region, P_d is the population of the region.

Statistics for Population Resource Region

Many attempts have been made for such regionalisation taking basis of total weighted values for different variables or superimposing different types of regions.[30] Many sophisticated techniques, e.g. multivariate analysis, canonial ecological analysis, etc. have been used for the purpose. But they fail to give any concrete numerical representation so far as the regional context is concerned. To achieve the purpose a new method of deviation system analysis has been proposed[31] for which five variables can be taken as :

$$D_i = \frac{G_d + l_u + L + N_{aw} + D_r}{T_{nv}}$$

where D_i is the Deviation index, G_d is the general density of population, l_u is the per cent of literacy, L is the level of urbanization, N_{aw} is the non-agricultural workers, D_r is the dependency ratio and T_{nv} is the total number of variables. On the basis of results find out this way the entire region may be divided into dynamic, prospective and problem sub-regions. Such a regionalisation may be proved to be a great asset in any regional planning.

And to conclude the story of statistics for population geographers it can be said that a demographer cannot proceed any long in his objective without statistics, because this branch of geography has entered in a phase of quantitative analysis bading good bye to its entirely qualitative approach.

REFERENCES

1. G.T., Trewartha, "The Case Study of Population Geography", *A.A.A.G.*, Vol. 43, 1953, pp. 71—79.
2. D.J.M., Hoosen, "The Distribution of Population as the Essential Geographic Expression", *Canadian Geographer*, Vol. 17, 1960, pp. 10—12.
3. J.I., Clarke, *Population Geography and Developing Countries*, First edition, 1971, p. 1.
4. W. Zelinsky, *A Prologue of Population Geography*, New Jersey, 1966, p. 3
5. T. Peter, and T.N. Peter, *Techniques in Human Geography*, Macmillan, 1971, p. ix.
6. Ibid, p. 4.
7. S.F., Hartley, *Population Quantity* vs. *Quality*, Prentice Hall, New Jersey, 1972, p. 2.
8. J.I. Clarke, *Population Geography*, 1968, p. 138.
9. Ibid.
10. O. Prakash, *Population Geography of Uttar Pradesh*, Ph.D. thesis (unpublished), B.H.U. Varanasi, 1973, p. 55.
11. G.T. Trewartha, *Geography of Population World Patterns*, Wisconsin, 1968, p. 96.
12. G.W. Barclay, *Techniques of Population Analysis*, New York, (John Wiley & Sons), 1966, p. 62.
13. Ibid.
14. O. Prakash, op. cit. p. 55.

15. A. Geddes, "The Population of India, Variability of Changes as a Regional Demographic Index", *Geographical Review*, New York, Vol. xxxii, 1942, pp. 562—573.
16. F.J., Monkhouse, and H.R. Wilkinson, *Maps and Diagrams*, Methuen & Co. Ltd., London, 1971, p. 349.
17. L.N. Ram, *A Study in Regional Geography: The Subarnarekha Basin in Bihar*, Ph.D. Thesis (unpublished), Patna University, Patna, 1967, Bihar p. 244.
18. R.B., Mandal, *Nature of Population Geography, Dimensions in Geography*, Associated Book Agency, Patna, 1979, p. 10.
19. Nitya Nand; Distribution and Spatial Arrangement of Rural Population in Eastern Rajasthan, *A.A.A.G.*, Vol. 56, 1966, p. 206.
20. Ibid. pp. 206—219.
21. J.I. Clarke, op. cit., p. 62.
22. Ibid, p. 63.
23. R.P., Singh, and A. Kumar, *Monograph of Bihar*, Bharati Bhavan, Patna, 1970, p. 111.
24. Santwana Ghosh, *Population of Bihar*, Ph.D. Thesis (unpublished) B.H.U. Varanasi, 1973, p. 96.
25. Ibid.
26. Ibid.
27. F.J., Monkhouse, *A Dictionary of Geography*, Edward Arnold (Publishers), Ltd. London, 1966, p. 312.
28. C. Clarke, *Urban Population Densities*, F.R. Stat. Soc., Ser. A., 1951, p. 114.
29. A. Kumar, and N. Sharma, *Bihar's Population on Move, Dimensions in Geography*, op. cit., 1979, p. 103.
30. R.L. Singh, R.P.B. Singh & C.B. Singh, Baghelkhand Region: A Study in Population Resources Regionalization and Development Models", *N.G.J.I.* Vol. XXII, part I, 1976, pp. 7—14.
31. Ibid.

ELEVEN

J. R. Yadav

AN ANALYSIS OF EQUAL PROBABILITY (NULL-HYPOTHESIS) BY USING CHI-SQUARE (χ^2) AND t-TEST IN REGIONAL STUDIES

THE MAIN application of Chi-Square (χ^2) technique is in the testing of results from statistical significance and it also may be used to the analysis of Point Patterns. We may also adopt it in the comparison between areas of equal and unequal size also. In this method we estimate the probability that the differences we actually observe in two areas of equal size for a set of data are due to chance. If the probability[1] is very low then we may with confidence find out an explanation. However, if our estimate indicates a considerable probability that observed variations are simply on chance basis, then we seek no justification in looking for causes. Thus, chi-square test (χ^2) represents a useful method of comparing experimentally observed results with those to be expected theoretically on some formulated hypothesis. The equation for chi-square test (χ^2) is stated as follows $\chi^2 = \Sigma\left[\frac{(f_o - f_e)^2}{f_e}\right]$ in which f_o = frequency of occurrence of observed or experimentally determined facts; f_e = expected frequency of occurrence on same hypothesis. The difference between observed and expected frequencies are squared and divided by the expected number in each case, and the sum of these quotients is χ^2. The more closely the observed results approximate to the expected, the smaller the chi-square and the closer the agreement between observed data and the hypotheses being tested.[2] Contrariwise, the larger the chi-square the greater the probability of a real divergence of experimentally observed from expected results. To evaluate chi-square, we take the help of chi-square table with the computed value of chi-square and the approximate number of degree of freedom (d_f) = (r—1) (c—1) where r is the number of rows and c is the number of columns in which the data are tabulated. The chi-square has been derived from the Greek word by Professor Karl Pearson in 1900. His purpose to develop this method was to explain discrepancy between observed phenomenon and expected phenomenon.

The Null Hypothesis

It is a useful tool in testing the significance of difference in non-parametric methods between two statistics against an equal probability hypothesis, a variety of 'null hypothesis'[3]. In its simplest form this hypothesis asserts that there is no true difference between two population* means, and that the differences found between sample means is, therefore, accidental and unimportant. The difference between two statistics is tested against a 'null hypothesis'. When two or more hypotheses are formulated, it is customary to write symbolically as H_o, H_1 and H_n. It is termed "null" because it states that the two samples form part of the same population, and there is highly a probability that the observed differences are due to chance variations. The 'null hypothesis' is not confined to zero differences nor to the differences between statistics. Other form of this hypothesis asserts that the results found in an experiment do not differ significantly from results to be expected on a probability basis or stipulated in terms of some theory. A "Null hypothesis" is ordinarily more useful than other hypotheses[4] because it is exact. In chi-square test, usually three types of hypotheses are formulated for the calculation of expected frequencies. The formulated hypotheses are (*a*) Hypothesis of equal distribution, (*b*) Hypothesis of Normal distribution, and (*c*) Hypothesis of Independent distribution. χ^2 is generally appplicable in the median test. However, when N_1 and N_2 are small (e.g., less than about 10) the χ^2 test is not accurate and the exact method of computing probabilities should be used[5].

*A population in the statistical sense refers to the total number of values of the distribution with which we are concerned.

Levels of Significance and Rejection

Whether a difference is to be taken as statistically significant or not, depends upon the probability that the given difference could have arisen "by chance". Usually, a difference is marked "significant" when the gap between two sample means signifies a real difference between the parameters of the population from which our samples are derived. Before a judgement of significant or non-significant is made, some critical points must be designated along the probability scale which will serve to separate these two judgement categories. At the same time it must be emphasised that judgements concerning differences are never absolute, but on the contrary range over a scale of probability, our confidence increasing as the chance of a wrong judgement decrease.

The probability in geographical problems is generally referred to on a scale from 0 (which never occurs) to 1.0 (that certainly occurs). The researchers for their conveniences have selected several arbitrary standards—known as levels of significance—of which the .05 and .01 levels are most often used. The confidence with which a researcher rejects or retains—a null hypothesis depends the level of significance adopted The 95 per cent confidence in 'null hypothesis' is often called the .05 level of significance, meaning that the particular result has only a 5 per cent chance of occurring through random variations. Similarly, 99 per cent confidence is referred to as the .01 level of significance. This means that the result will happen by chance less than once in hundred times. Associations of numbers that are likely to occur through chance variations less than five times in a hundred are said to be statistically significant. Generally in geographical problems the .05 level of significance is regarded as sufficiently rigorous. But the actual percentage of confidence that is required for any particular problem can only be decided by researchers. This degree of confidence is known as the rejection level, i.e., the level of significance at which H_0 is decided may safely be rejected and represented symbolically by α. The .01 level of significance is more exacting than the .05 level. Fig. 11.1 shows percentage distribution of values under the normal curve. The shaded area under each tail of the curve shows 4.55 per cent of all the values two and three standard deviations from the mean. The percentage are given in more exact terms. The figure shows that 95.45 per cent of the values are within two standard deviations (plus and minus of the mean). If we select one value randomly from the distribution the probability that it would lie more than two standard deviations from the mean is 4.55 per cent. Thus, it is possible to predict the probability of occurrence of any particular value, where large samples are involved, by the use of a *Z*-score. If the value of *Z*-score is two

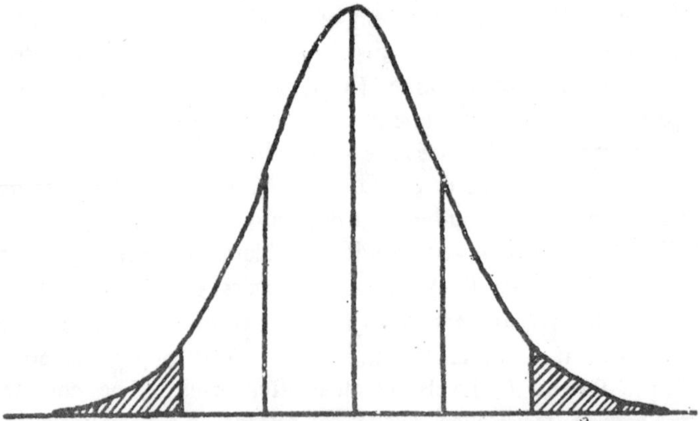

FIG. 11.1 : **Percentage Distribution of Values in Normal Value**

standard deviations or more from the mean, then the probability of occurrence is 4.55 per cent or less. Therefore, if we calculate the probability that two samples are drawn from the same population and obtain a result of 4.55 per cent, we are able to say with more than 95.45 per cent confidence that the samples are not drawn from the same population and if the probability comes to be only 0.3 per cent of a *Z*-score of three standard deviations, then we can be sure with 99.7 per cent confidence that the two samples come from the distributions so different that they must relate to different populations. In practice the decimals are usually ignored and we speak of with 95 per cent confidence, or with 99 per cent confidence.

In the case of small samples the sampling distribution will not necessarily approximate to normal and a *Z*-score cannot be used. The size of 'small' depends on the statistical test. It varies from

10 to 30 values in the sample, and is defined for each test. Over 30 values may always be regarded as a large sample.

Errors in Making Differences and Methods to Avoid Them

In testing hypotheses, usually two types of wrong inferences are drawn and they are termed as Type I error and Type II error. Type I errors are made when we reject a null hypothesis when true by marking a non-significant difference significant, although no true difference exists. Various precautions may be adopted to avoid this type of erroneous inference. If we set up a low level of significance (P greater than .05), we increase the likelihood of Type I error, whereas, if we set up a high level of significance (P less than .05), we render such erroneous inferences less likely. Type II errors are made by accepting a null hypothesis when false differences actually exists. The possibility of drawing erroneous inferences are enhanced when we set a very high level of significance. Setting a high level of significance will tend to prevent type I errors but will encourage the appearance of Type II errors. Hence it appears that a researcher must decide which kind of wrong inference he would rather avoid, as apparently he can prevent one type of error only at the risk of making the other more likely. Perhaps the wisest step to avoid errors in drawing inferences is—when significance is doubtful or uncertain, it demands more evidences. This gives the data a chance to refute (or fail to refute) the null hypothesis. Additional data, repetition of the experiment and better controls will often make possible a correct judgement.

Limitations of the Chi-Square (χ^2) Test

Though, Chi-Square test is very useful in geographical problems but it has certain limitations and following precautions should be taken while using this method. Firstly, the number of observed frequencies (f_o) and expected frequencies (f_e) in one cell should not be less than 5, otherwise values of chi-square test are not useful. Hence, the numbers should be increased or "Yates correction for continuity method"[6] should be used. Secondly, the number of observed frequencies and expected frequencies should be large or minimum, it should be 30 and by doing so, the probability increases from the normally distributed values of f_o and f_e. Thirdly, in chi-square test only absolute numbers should be used and use of percentage should be avoided. If use of percentage cannot be avoided, then chi-square values should be changed into absolute numbers. Fourthly, on the basis of chi-square test, we find out only "association" in variables, but it is difficult to find causation from such association. Fifthly, usually 'value of co-efficient of contingency' derived from chi-square test, is limited. For example, the value of 2 × 2 in contingency table is not more than .707. In the same way this value for 3 categories is .816 and for 4, 5, 6, 7 and 10 categories, it is not more than .866, .894, .913, .926 and .949 respectively. For these limitations, the comparative studies become difficult.

The t-Test as a Tool in Testing the Validity of Null Hypothesis

The validity and reliability of chi-square technique may be tested by using *t*-test. In *t*-test we study significance of difference between two means in normal distributions. Actually analysis of variance is also an extended form of *t*-test, where we study significance of difference among three or more means groups. The *t*-test may be computed as $t = \frac{M_1 - M_2}{\sigma d}$, where M_1 and M_2 = Means ($\bar{x}$) sample of number 1 and 2 and σd is the standard error of difference (s.e.d.). Haggett adopted the formula in the spacing of towns in two regions of Eastern Texas and Iowa-Missouri[7] as follows $t = (\bar{x}_1 - \bar{x}_2)/(\text{S.D.}\sqrt{1/n_1 + 1/n_2}$, $df = n_1 + n_2 - 2$. where ($\bar{x}_1$ and $\bar{x}_2$ = means sample of number 1 and 2, S.D = Standard deviation, n = number, and df = degree of freedom.

Testing of Hypothesis : An Example of Azamgarh Area

The Chi-Square test (χ^2) has been tested in the "Synchronization of Rural Period Markets in Azamgarh area" (Fig. 11.2) which is a part of middle Ganga valley with 158 Periodic market centres in an area of 5,744 Km^2 of six sub-divisions—providing socio-economic services to 28,57,484 inhabitants. The synchronization[8] of rural period markets in the study area has inverse relationship of spacing within spatio-temporal frame and are organised to achieve an 'optimal' sequence to ensure that market centres located close together in time, are much widely spaced to attract 'threshold population'. While the markets which meet on the same day or adjacent days have competed and clashed with the same people at same time and in the same area. It has been noticed in the study area that such markets meeting on the same day or adjacent days are widely spaced in temporal frame. In this paper three formulated hypotheses, using the chi-square test (χ^2) are as (a) the first hypothesis is that the number market meetings in Azamgarh area depend upon wide extension of the area, (b) the second hypothesis is that the number of market meetings depend on seven days of a week, and (c) the third hypothesis is that the number of market meetings depend on wide area and week days both.

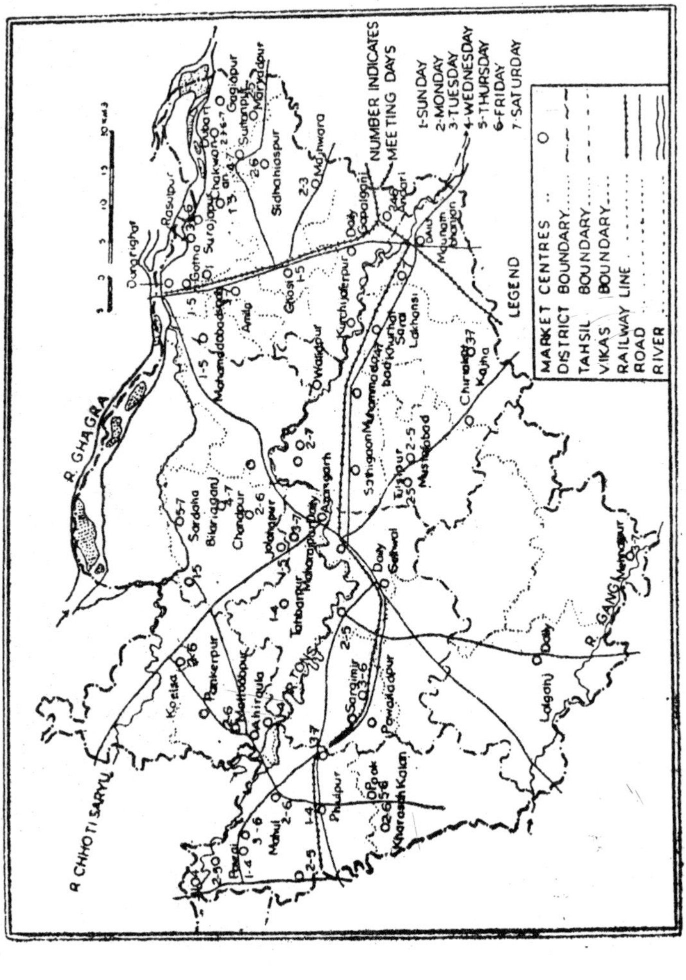

FIG. 11.2 : District Azamgarh—Periodic Market Centres

In the present analysis each day (i.e. if a market meets on Monday and Friday) both days have been scored. Thus, these are frequencies of market meetings, not market centres.

TABLE 11.1 : **Frequency of Market Meetings on Days of the Week in Azamgarh Area**

Days of a week	Sub-Divisions 1 Sagari	2 Ghosi	3 Azam-garh	4 Moham-dabad	5 Phul-pur	6 Lal Ganj	Total	Mean No. of market days
1. Sunday	7	13	5	4	13	2	44	7.3
2. Monday	7	18	9	10	13	1	58	9.7
3. Tuesday	4	23	9	6	8	2	52	8.67
4. Wednesday	4	8	5	6	13	2	38	6.33
5. Thursday	7	16	8	7	8	1	47	7.83
6. Friday	7	15	9	6	14	1	52	8.67
7. Saturday	6	18	3	9	6	2	44	7.33
	Σ 42	111	48	48	75	11	335	
Mean No. of market days	6.0	15.86	6.86	6.86	10.71	1.57	47.86	$\chi^2=57.14$ $df=(7-1)\ (6-1)=30$ $P < .01$ Significant
Chi-Square*	5.12	8.63	8.96	4.55	7.02	22.86	57.14	
No. of market centres	22	56	18	20	39	3	158	

*Chi-Square related to third hypothesis.

TABLE 11.2 : **Areawise Market Meetings in Azamgarh Area**

	Sub-Divisions						Remarks
	1 Sagari	2 Ghosi	3 Azam-garh	4 Moham-dabad	5 Phul-pur	6 Lal Ganj	
f_o	42	111	48	48	75	11	$\bar{x} = \frac{335}{6} = 55.83$
f_e	55.83	55.83	55.83	55.83	55.83	55.83	χ^2* = 102.72
$(f_o - f_e)$	13.83	55.17	7.83	7.83	19.17	44.83	df = (6—1) = 5
$(f_o - f_e)^2$	191.27	3043.73	61.31	61.31	367.49	2009.73	$P < .01$
$\frac{(f_o - f_e)^2}{f_e}$	3.43	54.52	1.10	1.10	6.58	35.99	Significant at .01 Level

*Chi-Square computed for first hypothesis.

TABLE 11.3 : Daywise Number of Market Meetings in Azamgarh Area

	Days of a Week							Remarks
	1 Sunday	2 Monday	3 Tuesday	4 Wednesday	5 Thursday	6 Friday	7 Saturday	
f_o	44	58	52	38	47	52	44	$\bar{x} = \frac{335}{7} = 47.86$
f_e	47.86	47.86	47.86	47.86	47.86	47.86	47.86	χ^2* = 5.535
$(f_o - f_e)$	3.86	10.14	4.14	9.86	0.86	4.14	3.86	df = (7—1) = 6
$(f_o - f_e)^2$	14.90	102.82	17.14	97.22	0.74	17.14	14.90	$P > .05$
$\frac{(f_o - f_e)^2}{f_e}$	.31	2.15	.36	2.03	.015	.36	.31	Insignificant at .05 Level.

*Chi-Square related to second hypothesis.

TABLE 11.4 : **Mean Differences between Areas Based on *t*-Test**

	Sagari	Ghosi	Azamgarh	Mohamdabad	Phulpur	Lal Ganj	Remarks
Sagari	×	9.86*		2.00	4.71	4.43	df = (N—1) + (N—1) = (7—1) + (7—1) = 12
Ghosi	9.86*	×	9.00*	9.00*	5.15	14.29*	S.D.** = 4.813 $\sigma d = S.D. \sqrt{1/n_1 + 1/n_2}$
Azamgarh		9.00*	×		3.85	5.29	$\sigma d = 4.813 \sqrt{1/7 + 1/7}$ $\sigma d = 4.813 \sqrt{.14 + .14}$
Mohamdabad		9.00*		×	3.85		$\sigma d = 4.813 \times .53$ $\sigma d = 2.55$
Phulpur	4.71	5.15	3.85	3.85	×	9.14*	(a) Significant Difference level on .01 level at 12 df = 3.054×2.55 = 7.787
Lal Ganj	4.43	14.29*	5.29	5.29	9.14*	×	(b) Significant Difference level on .05 level at 12df = 2.179×2.55 = 5.56

*Significant at .01 level.

**Standard Deviation (S.D.) $= \sqrt{\Sigma \frac{fx2}{N}} = \sqrt{\frac{1204.75}{52}} = 4.813$

TABLE 11.5 : Mean Differences between Days of a Week based on *t*-Test

	Sun.	Mon.	Tues.	Wed.	Thurs.	Fri.	Satur.	Remarks
Sunday	×	2.34	1.34	—1.0	0.50	1.34		df = (N—1) + (N—1) = (6—1) + (6—1) = 10
Monday	2.34	×	1.00	3.34	1.84	1.0	2.32	S.D.* = 4.813 $\sigma d = S.D. \sqrt{1/n_1 + 1/n_2}$
Tuesday	1.34	1.0	×	2.34	0.84		1.34	$\sigma d = 4.813 \sqrt{1/6 + 1/6}$ $= 4.813 \sqrt{.16 + .16}$
Wednesday	1.0	3.34	2.34	×	1.50	2.34	1.0	$= 4.813 \times .56$ $= 2.70$
Thursday	0.50	1.84	0.84	1.54	×	0.84	0.50	(a) Significant difference on .01 level at 10df = 3.169×2.70
Friday	1.34	1.0		2.34	0.84	×	1.34	= 8.6 (b) Significant difference on .05
Saturday	×	2.34	1.34	1.0	0.50	1.34	×	level at 10df = 2.228×2.70 = 6.2 Result = Insignificant

*Standard Deviation (S.D.) = $\sqrt{\Sigma \frac{fx^2}{N}} = \sqrt{\frac{1204.75}{52}} = 4.813$

To test the above three hypotheses Chi-Square test[9] has been computed against 'equal probability hypothesis—a variety of null hypothesis' which explains that frequencies of markets in different six areas and on different 7 days of a week are expected equal number of times on chance basis. The first null hypothesis has been rejected at .01 level as $P < .01$ is significant on $\chi^2 = 102.72$ at $5df$ in the case of area having six sub-divisions (see Table 11.2). Similarly, the second null hypothesis—a variety of equal probability has been retained or accepted at .05 level as $P > .05$ is insignificant on $\chi^2 = 5.536$ at $6df$ in the case of seven days of a week. In the case of third hypothesis an "independent hypothesis—a variety of null hypothesis" has been rejected at .01 level as $P < .01$ is significant on $\chi^2 = 107.72$ at $30df$ in case of area and week days both. The concentration of market meetings on any particular day is by chance and it depends upon wide areal extension.

Further detailed analysis has been done in order to find out 'significant differences' in market meetings between areas with the help of t-test. In the case of first hypothesis there are 15 paired comparisons among 6 sub-division areas.

The table number 11.4 represents greater column means over row means. Thus, the significant differences are marked between 10 paired sub-division areas at .01 level. The fact remains that Ghosi has significantly greater number of market meetings than Sagari, Azamgarh, Mohamdabad and Lal Ganj sub-divisions, Azamgarh has significantly greater number of market meetings than Ghosi, Mohamdabad than Ghosi, and Phulpur than Lal Ganj sub-division (see Table 11.4).

In case of second hypothesis, an analysis of insignificant result has also been done to know if there is any single pair of days which may be said to differ significantly.

There are 21 paired* comparisons between 7 days of a week. The largest mean difference is observable between Monday and Wednesday only—being 3.34 greater in favour of Monday. However, this difference is insignificant. Consequently it can be asserted that different number of market meetings on different days in a

* The number of paired comparisons are obtained as $\frac{N(N-1)}{2}$. The Number of days in a week are seven, so, the number of days comparisons are 21. Similarly, 15 number of paired comparisons are in areas as number of sub-divisions in the area are 6.

week is only chance and days have no relevance in this context. Thus, it may be concluded that area is significant at .01 level by using t-test in the testing of first hypothesis and days are insignificant, hence second hypothesis has been accepted and retained in the analysis of t-test.

REFERENCES

1. Patrick McCullagh, "Science in Geography 4, Data Use and Interpretation", Oxford University Press, London, 1974, p. 6.
2. Leonard Zobler, "The Distinction between Relative and Abstract Frequencies in using Chi-Square for Regional Analysis", *Annals Ass. Ame. Geogr.*, Vol. 48 (1958), pp. 456-57.
3. Davis Peter, "Science in Geography 3, Data Description and Presentation", Oxford University Press, London, 1974, pp. 29—31.
4. R.E. Quandt, "Statistical Discrimination Among Alternative Hypotheses and Some Economic Regularities", *Journal of Regional Science*, Vol. 5, 1964, pp. 1—23.
5. H.M. Walker, and Lev. J., *Statistical Inference*, Henry Holt, New York, 1953, p. 103.
6. L.J. King, *Statistical Analysis in Geography*, Prentice Hall, Inc. Englewood Cliffs, N.J., New Delhi, 1969, pp. 69 and 77.
7. P. Haggett, *Locational Analysis in Human Geography*, St. Martin's Press Inc., New York, 1965 also, A.N. Strahler "Statistical Analysis in Geomorphic Research" *Journal of Geology*, Vol. 62, 1954; pp. 1—25.
8. Hill Polly and R.H.T. Smith, "The Spatial and Temporal Synchronization of Periodic Markets: Evidence From Four Emirates in Northern Nigeria", *Economic Geography*, Vol. 48, 1972, pp. 345—355.
9. J.R. Mackay, "Chi-Square As a Tool for Regional Studies", *Annals Ass. Ame. Geogr.* Vol. 48, 1958 ; p. 164 ; also J.R. Mackay and B.J.L. Berry "Comments on use of Chi-Square", Annals. *Ass. Ame. Geogr.*, Vol. 49, 1959, p. 89.

SECTION TWO

APPLIED STATISTICS

TWELVE

R. B. Mandal

PHYSICAL GEOGRAPHY

IN THE SPHERE of physical geography statistical techniques have been applied on advanced level especially in slope analysis and measuring the intensity of streams. Further this study highlights the problems of concentration of sediments, hydrological character and land utilization, chemical conversion of inorganic and organic matter, computation of contour entropy, susceptibility and potential of glacier, precipitation probabilities as indices of climatic variation and several others which are directly or indirectly associated with the rainfall statistic and hypso-clinographic curve of various erosional surfaces.

Bivariate Analysis

Geographers tools in deriving the laws that relate the change of one quantity with respect to change in another related quantity is known as *bivariate analysis.*[1] The term bivariate means simply "two-variable", that is two measurable physical properties engaged in simultaneous variation. In weather science, atmospheric pressure decreases with altitude. Pressure in millibars is scaled on one axis of the graph; altitude in Kms on the other. Here one can say that pressure depends on altitude. Pressure is one variable quantity, altitude is the other. Their simultaneous variation is systematic and is represented by a smooth curve. A rather simple mathematical equation describes the simultaneous variation of pressure and altitude; it conforms to a well known law of physics.

Independent and Dependent Variables

In bivariate analysis it is essential to assign the effect of one variable to the other. In the example cited above altitude is considered the "cause" and pressure the "effect".

In statistics the cause is referred to as the independent variable while the effect constitutes the dependent variable (it depends upon the other variable). In mathematics the independent variable is designated by the symbol X while dependent variable by Y. The mathematician states "Y" is some function of "X". The equation or analysis shows that the relationship between X and Y is referred to as regression analysis and equation as model.

The role of bivariate analysis is played by time which is mostly scaled on the horizontal axis of the graph. In a hydrograph runoff, the dependent variable changes with time. Obviously, it would be absurd to say that time depends upon runoff.[2]

Hydraulic and Topographic Sinuosity Index[3]

$$HSI = \frac{C_1 - V_1}{C_1 - 1}$$

$$\text{and } TSI = \frac{V_1 - 1}{C_1 - 1}$$

where, HSI is hydraulic sinuosity index per cent.

TSI is topographic sinuosity index per cent.

C_1 is channel index.

V_1 is valley index.

Measuring Drainage Density with a Dot Planimeter

Drainage intensity (1) can be conceived in terms of a numerical count ratio—an area ratio, or a refinement of the latter. Thus :[4]

$$1 = n/N = a/A = wL/A \qquad \ldots(1)$$

Where N=total number of planimeter dots in study area, n=total number of dots-intersecting drainage lines, A=total study area in square inches, a=total area in square inches lying within one dot radii of stream lines, w=width in fraction of an inch of two dot radius and included drainage line, and L is the total length of drainage lines in inches within the study area. The equality $n/N = Wh/A$ is of particular interest as it permits the conversion of drainage intensity ratios to drainage density values. The transformation is :

$$L = \frac{nA}{Nw} \qquad \ldots(2)$$

Concentration Time of Sediment

It is a measure of the time for storm water to concentrate within a basin, and was developed some thirty years ago by C.E. Ramser (1927). It has been calculated here from the expression[5]

$$T_c = 0.0078 \sqrt{\frac{L3}{H}0.770}$$

where T_c is the concentration time (minutes), L is the maximum length of travel (feet), and H is the difference in elevation between the most remote point and basin outlet (feet). Length of mean watershed area, gross watershed area, is an important variable restricting pairs of watersheds to areas less than 100 acres. The sample was made amenable to a series of empirical formulae developed for small watersheds but as yet untested for larger size units.

Index of Vegetation Resistance

The expression used was :

$$V_i = \frac{w}{h}\left(\frac{c+g}{2}\right)$$

where V_i is an index of vegetation resistance, w is an estimate of vegetation weight in tons/acre, h is mean canopy height in feet, c is percentage crown cover and g is percentage ground cover. Vegetation was either weighed in the field (Brown, 1954) or estimated from trunk diameter (timler cover).[6] The system of combining these values into a single index was suggested by the findings of Kramer and Weaver and Osborn (1954).

Elevation Relief Ratio

The ratio is derived from the following equation :[7]

$$ER = \frac{E-L}{R}$$

where ER is the elevation relief ratio, E is the mean elevation of randomly chosen points within the sample, R is the relief of the area, because the average elevation minus the lowest elevation can approach the relief in value but never reach it, the elevation relief ratio must always be greater than 0 and less than 1. A value near zero indicate terrain flat with some isolated peaks, while a value near 1 indicate a similar area but with some deep canyons instead of isolated peaks.

Specific Relief Factor

The measure of the expected height of the tallest terrain features present and is useful primarily for the comparison of relief extremes between areas. In general terms it can be reduced to the following equation.[8]

$$\frac{\text{An Map } X}{\text{An Map } Y} = \frac{\text{Maximum Relief } X}{\text{Maximum Relief } Y}$$

where *An* is the specific relief factor.

Model for Maximum Probable Precipitation

The Hershfield's Statistical model for obtaining maximum probable precipitation (PMP) of a point station is based upon the formula.[9]

$$X_{pmp} = X + \text{K}.\sigma$$

where K and σ are the mean and standard deviation of series of extreme annual rainfall values of a station for one day duration and K is the frequency factor. Hershfield obtained the envelope value of 15 for K on the basis of world envelope rainfall data. The *PMP* values given in the generalized chart can be used for the design of such hydraulic structures where no risks can be undertaken due to failure of these structures.

Hydrological Character and Land Use

An elaborate discussion on correlation of different hydrological character and land use items of the sub-divisions have been made after a study of the precipitation, drainage, floods and droughts, river sub-basins and ground-water properties and extraction in the region. Each sub-division has a pattern difference or a difference in intensity of monopattern form which has been determined by a deviation from a standard set for the region. The standard is obtained by applying the formula.[10]

$$S = \frac{(a_1 + a_2 + a_3 + \ldots\ldots\ldots) \times 100}{t}$$

where S represents the standard, a_1, a_2, a_3, ... areas under different items at 't' the total area of the region.

Index of Aridity

Various formulae have been devised to define the boundary of an arid climate, mostly empirical, involving relationships between temperature, precipitation or evaporation, values are plotted and isopleths are drawn. Formulae include.[11]

$$1 = \frac{P}{T + 10} \qquad \ldots\ldots 1$$

where, 1 is the index of aridity, T is the mean annual temperature in °C and P is the mean annual rainfall in mm. Rain factor

$$\text{Moisture} = \frac{P}{T} \qquad \text{(R. Lang)} \ldots\ldots 2$$

Rain mainly in winter $R = T$ 3
Rain evenly distributed throughout the year $R = 1+7$.
Rain mainly in summer $R = t + 14$.
Where t is the mean annual temperature in °C, if the annual rainfall (R) in cm is less than $2t$ and greater than t, the climate is regarded as arid (W. Koppen and R. Geiger).

Index of Temperature Efficiency

An empirical index devised by C.W. Thornthwaite, in conjunction with a precipitation efficiency 1, as the basis for this first scheme of climatic classification. T.E. is defined for each month as[12]

$$\frac{T-32,}{4}$$

where T mean monthly temperature in °F.

Rain Factor

An index devised to express a relationship between precipitation and temperature, so as to give some indication of climatic aridity which might be used in the limitation of climatic regions devised by (R. Lang) :

Annual Precipitation in mm
r.f. = Mean Annual Temperature in °C.

Precipitation Efficiency Index

An index devised by C.W. Thornthwaite for the basis of his first climatic classification system from the formula.[13]

$$i = 11.5 \left(\frac{P}{T-10}\right) \frac{10}{9}$$

where P is the monthly mean precipitation in inches and T is the monthly mean temperature.

The Rate of Chemical Conversion of Inorganic and Organic Matter in Landscapes

A significant part of the solid and liquid products of technogenesis reach the surface of the soil and into the soil itself. They thus became part of the cycle of soil processes and their rate of chemical conversion will depend on the energetics of soil formation.

According to U.R. Volobuyev (2) the energy balance of soil formation can be presented as follows.[14]

$$Q = w_1\ tw_2 + b_1 + b_2 + e_1 + e_2 + g + v_1$$

where Q is the amount of energy that enters into soil formation and is spent as follows :

- w_1 for physical weathering.
- w_2 for chemical decomposition of minerals.
- b_1 for biochemical reactions of transformation of organic and inorganic compounds.
- b_2 energy accumulated in living matter and in the soil.
- e_1 energy lost in evaporation from the surface of the soil and plants.
- e_2 energy lost in transpiration.
- g energy lost in process of mechanical migration through the soil.
- v energy lost in the process of heat exchange between soil and atmosphere.

Compilation of Contour-Entropy Maps

This technique helps to establish the correspondence of phenomena for any map contour by means of entropy, the main function in information theory. Previous writings on the application of information theory to geography and cartographic investigations have shown that entropy is a useful measure of the homogeneity of phenomena.[15]

$$H(A) = \sum_{i=1}^{n} w_i \log_2 w_i,$$

where w_i is the shape of i phenomena in system A on the share of the area occupied by i-contour of the map.

The entropy function possesses are important properties that can be used for estimating the degree of contour correspondence on a map. Let a contour (say a vegetation contour) b_i entirely within contour b (a soil contour). We can then say that a correspondence entirely to b, and the entropy of contour a in that case would be zero. Now let us assume that contour a lies partly in contour b_1 and partly in contour b_2. In that case, the two phenomenon are not in total agreement and entropy increases. As more and more contours fall within the area a, agreement between the phenomena will decline and entropy will increase further. Even when the number of b_i contours remains the same, entropy may change according to the area occupied by each.

1. (Top row)—entropy of contour a increases with number of contours b_i within its limit.

2. (Bottom row)—entropy of contour increases with change in properties between areas occupied by contour b_i. (The number of b_i contours remaining constant, $u=3$). When contour a is almost entirely within contour b_i and only to a small extent in contour b_2 and b_3 entropy will be small, but when the share of the contours is equal ($wb_1=wb_2=wb_3=1/3$), then the entropy function will reach a maximum for the given number of contours.

Entropy used as an indicator of contour correspondence is thus a function both of the number of contours b_i and of the area within each contour.

Coefficient of Variability

The mean value of rainfall at a station is the mean of actual annual totals, which are widely dispersed on either side of the mean. The used statistical formulae are :[16]

Method I

$$C \text{ of } V\,(\%) = \frac{\sigma}{M} \times 100$$

where, σ=standard deviation and M is the mean value.

Method II

$$C \text{ of } V\ (\%) = \frac{\text{Interquartile Range}}{\text{Median}} \times 100$$

Drainage Density

$D = \frac{\Sigma L}{A}$, where ΣL is the total length of streams in a drainage basin of 'A' areal unit.[17]

Stream Frequency

$E = \frac{N}{A}$, where N is the total number of streams in a drainage basin of A areal unit.

Index of Susceptibility and Potential of the Glacier

The energy or potential of the glacier is indicated by the area enclosed by the plotted oscillation line and the abscissa : the susceptibility (E) by the ratio of the difference between the length (D) of the line joining the plotted points of the tongue position and the length (d) of the base line to the number of years considered, conveniently reduced to 10-year intervals.

$$E = \frac{10(D - d)}{\textit{No. of years}}$$

An index of susceptibility is derived from the geometry of the glacier and is the ratio of the product of the slope over the whole glacier (A) and its surface length ($c+m$) (as an indicator of gravity and distance) to the slope of the tongue, times its length (as a measure of resistance to movement).[18]

$$\text{Index } E = \frac{\tan A \times (c + m)}{\cot \beta \times C}$$

After ranking the data for different mountain groups and climatic conditions, Paschinger finds a correlation of index E with the number of peaks in the movement graph.

Precipitation Probabilities as Indices of Climatic Variation

This paper uses precipitation probability values derived from estimators of the gamma probability function as developed by

Thom (1) where the frequency distribution for the random variable X is given by :

$$f(X) = \frac{e^{-x/\beta} X^{m-1}}{\beta my\ (M)}$$

where X = Precipitation amount $\beta = \bar{X}/m$
m = gamma parameter as found from the equation.
$12\ [\ln \bar{X} - (1/N)\ \Sigma \ln X]\ m^2 - 6m - 1 = 0$
$Y(m)$ = gamma function.

$$\int_0 X_e - x\ X^{m-1}\ dxi,\ 0 \leqslant x \leqslant \infty,\quad m > 0$$

We do not present the details of the development of this method. The reader is referred to References 19, 20, 21 which have shown that the gamma probability function gives a good fit for precipitation in a data series.

Two parameters of the gamma probability function must be estimated from the above equations. The shape parameter gamma is inversely related to skewness in that small gammas indicate that a few large values cause positive skewness, while a large gamma indicate that the probability function approaches normality. The scale parameter data indicates the range or dispersion of the curve".[22]

Principal Component Analysis

(*CPA*) Principal Component Analysis was used to simplify the data and ordination plot of profiles on the component planes were compared with the surveyors original classification. The object is to resolve several more or less correlated variates into a set of new uncorrelated variates or components of the type[23] :

$$Z_r = \sum_{r=1}^{P} W_{ir}\ X_i\ (i,\ r = 1, 2, \ldots\ldots, \mathrm{P})$$

Where X_i is an original variate, Z_r stands for the *rth* component and w_{ir} is the weight of the *ith* variate on the *rth* component. The exponents are extracted such that the first component accounts for the maximum variance that can be expressed in one dimension, the second for the maximum of the remainder and so on.

Slope Morphology

The area analysed by Wentworth's method properly represents

represents the slope morphology[24]. The technique is very useful in making quantitative generalization for the study area.

With the originai formula of Finsterwalder's $\left(\frac{I\ (L)}{A} \text{ in ft.}\right)$.

Wentworth substituted the number of lines per linear mile at right angles to their direction for the total length of the contour lines and thus made a formula of S_m IN/5280, where S_m is mean slope, N is number of contour lines per mile, and I is contour interval. He further corrected this formula for contour crossings at oblique angles and determined that the average space between contours along any random line is greater than the average space normal to the slope in the ratio 1 : 0.6366. Thus he gave the final form to this formula :

$$S_m \quad \frac{I\ (N)}{5280\ (.6366)} = \frac{I\ (N)}{3361}$$

Dynamics of Slope Forming Processes

The slope map, consisting of lines of equal, sine of slope angle, tells the distribution of the part of gravitational acceleration, '*g*' contributing to the down slope force at any given point on the surface. The critical value of '*f*' in Bengham's formula varies on the nature and mechanism of the rockmass of the region. The Bengham's Law is as follows: (Strahler, 1952a, p. 926).

$$d_u/d_y = \frac{F_s - f}{n}$$

where du/dy is the rate of Shear (rate of change of velocity u, with respect to depth y), F_s is shearing force, f is Bengham's yield stress and n is consistency of the "viscosity in a fluid" solid.[25]

Stream Orders

A map bearing a network of stream channel, can be sub-divided into a channel segments according to orders of magnitude assigning a sequence of numbers, to the orders. Each finger tip channel is designated as a segment of the first order. At the junction of any two first-order segments, a channel of the second order is produced and extended down to the point where it joins another second order channel, where upon a segment of third order results and so forth. However should a segment of the first order join a second or third order segment, no increase in order occurs at that point of junction. The trunk stream of any watershed bears the highest order number of the entire system. Channels of the first and second order usually carry flowing streams only in wet weather.[26]

The number of segments of each order were then counted to yield the figure (Table 12.1). The order of a stream segment is designated by the symbol, *u*, the number of segments of a given order by the symbol number. Consider the ratio between the number of stream segments of any given order to the number of segments of the next higher order, a proportion designated the bifurcation ratio (symbol R_b).

TABLE 12.1 : **Order of a Stream Segment**

Stream Order, u	No. of stream segments, N_u	Bifurcation Ratio, R_b
1	139	
2	46	3.02
3	11	4.18
4	3	3.66
5	1	3.00

(Data from K.G. Smith, 1958).

The bifurcation ratio between successive orders is then defined as

$$R_b = \frac{N_u}{N_u+1}$$

In the example from big Badlands there are just over three times as many first-order segments as second order; over four times as many second order segments as third order; three and two-third times as many third-order as fourth order; and three times as many fourth-order as fifth-order. The differences in these bifurcation ratios can be attributed to chance variations in the shape of any stream network. An average of the four bifurcation ratios is close to 3.5, which is a good representative value for the series.

Studies of many stream networks conform the principle that in a region of uniform climate, rock type, and stage of development, the bifurcation ratio sends to remain constant from one order to the next values of bifurcation ratio between 3 and 5 are characteristic of natural stream systems.

Law of Stream Numbers

The number of stream segments of successively lower order in a given basin send to form a geometric series, beginning with a single segment of the highest order and increasing according to a constant bifurcation ratio. For Example, if the bifurcation ratio is 3, and the trunk segment is of the sixth order, the number of segments will be 1, 3, 9, 27, 81, and 243.

The relationship between orders and numbers that follow a geometric progression conforms to a mathematical model known as negative exponential function. The formalized statement of Horton's law of stream number is then as follows:—

$$N_u = R_b\ (K - u) \text{ and } R_b = \frac{N_o}{N_o + 1}$$

where N_o is the number of streams of Order o in a basin. Symbols N_u, u, and R_b are as defined in an earlier paragraph. Symbol K is defined as the order of the main trunk stream which designates the segment of the highest order. There is of course, only one trunk segment. The value of K for the Allegheny River basin is 7. It should be kept in mind that the above equation has meaning only for integral values of u (for example, 1, 2, 3, 4 etc.) because there are no intermediate values of order under the definition that has been adopted for order of a stream segment.

TABLE 12.2 : **Allegheny River Drainage Basin Characteristics**

Stream Order	Number of Segments	Bifurcation Ratio	Mean length of Segments Miles L_u	Cumulative mean length Miles L_u	Length Ratio R_l	Average Watershed Area, Sq. miles A_u
1	5966	3.9	0.09	0.09		0.05
2	1529	4.0	0.3	0.4	3.3	0.15
3	378	5.7	0.8	1.2	2.7	0.86
4	68	5.3	2.5	3.9	3.1	6.1
5	13	4.3	7	11	2.8	34
6	3	3.0	20	31	2.9	242
7	1		8+(not complete)			550 (not complete)

Data by Marie E. Morisawa, 1959.

A simple test of the exponential equation can be made as follows. Assume an ideal stream net with a bifurcation ratio of exactly 3. Let the highest order, K, be 5.

Order, u	Number of Segments, N_u	Bifurcation Ratio, R_b
1	81	
2	27	3.0
3	9	3.0
4	3	3.0
K=5	1	3.0
	N_u=121	

Suppose that we wish to determine the number of segments of the second order (N_2), knowing only that the bifurcation ratio is 3 and that $K = 5$. Substituting in the equation:—

$$N_2 = 3^{(5-2)}$$

$$N_2 = 3^{(3)}$$

$$N_2 = 27.$$

Horton further observed that the total number of stream segments of the entire drainage basin can be expressed as follows:

$$\Sigma N_u = \frac{R_b^{K-1}}{R_b - 1}$$

The symbol ΣN_u means "the sum of segments within each order" (the Greek letter sigma has been adopted to instruct the reader to sum all terms). Testing this equation against the ideal set of numbers in which R_b equals 3, we obtain

$$\Sigma N_u = \frac{3^5 - 1}{3 - 1}$$

$$\Sigma N_u = \frac{243 - 1}{2} = \frac{242}{2}$$

$$\Sigma N_u = 121$$

Stream Lengths

Referring to the drainage network map, (Fig. 12.1) the mean length of stream segments, in miles, increase by a ratio of roughly three times with each increase in stream order. This proportion of length increase is known as the length ratio (Symbol, R_l), and tends to be approximately constant for a given drainage system. Chance variations to be expected in the configuration of any drainage system which will produce inequalities of observed length ratio from one order to the next.

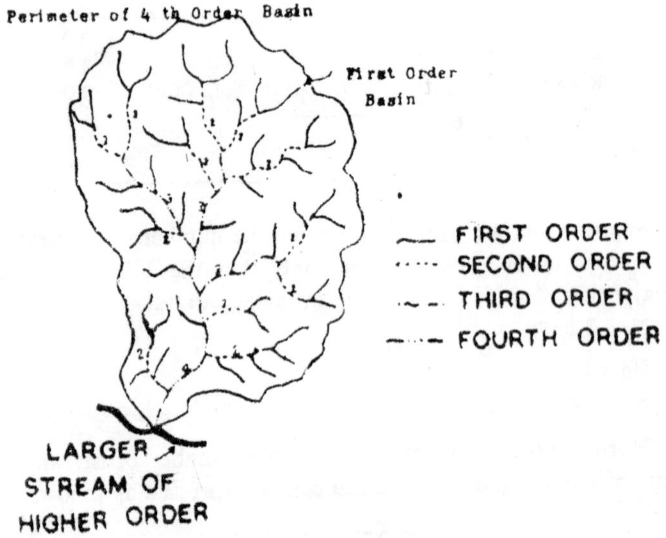

FIG. 12.1 : **Orders of Magnitude Assigned to the Segments of a Branching Stream System (after Strahler)**

Definition of length ratio resembles that of bifurcation ratio and is as follows:

$$R = \frac{\bar{L}u}{\bar{L}u-1}$$

The symbol $\bar{L}_u$ represents the mean length of all stream segments of order u. In the practice of morphometry, a distance measuring instrument (not measure) is run over all segments of a given order on the map and their total distance read off. This total length is then divided by the number of segments of that order, yielding the mean length. Stated in rigorous fashion:

$$L_u = \frac{\Sigma L_u}{N_u}$$

where ΣL_u means "the sum of lengths of all stream segments of order u".

Study of many drainage systems led Horton to formulate a law of stream lengths, which with necessary modifications may be stated as follows:—

> "The cumulative mean length of stream segments of successive orders tend to form a geometric series beginning with the mean length of the first-order segments and increasing according to a constant length ratio. The word "cumulative" in this law indicates that the mean lengths are progressively added (cumulated) starting with the second order. For order 2, mean lengths of first and second orders are summed, for order 3, mean lengths of orders 1, 2, and 3 are summed, and so forth."

As with the law of stream orders, the law of stream lengths can be given mathematical expression by an exponential regression equation. Figure 12.2 is a graph in which cumulative mean stream lengths are plotted on a constant-ratio (logarithmic) scale on the vertical axis, stream order is plotted on a arithmetic scale on the horizontal axis. If the plotted points fall on a nearly straight line, the validity of Horton's law of stream lengths can be regarded as strongly supported.

Horton's law of stream length is stated mathematically by the following equation:

$$\bar{L}_u = \bar{L}_1 \, R_l^{\,u-1)}$$

where $\bar{L}_1$ is the mean length of the first-order segments, the other symbols having been previously defined.

Basin Areas

Turning next to the areas of drainage basins, we can study the relationship between mean area of basin of a given order (Symbol, A_u) and the order itself. Horton's law of stream length has been paraphrased into a law of basin areas, stated as follows: the mean

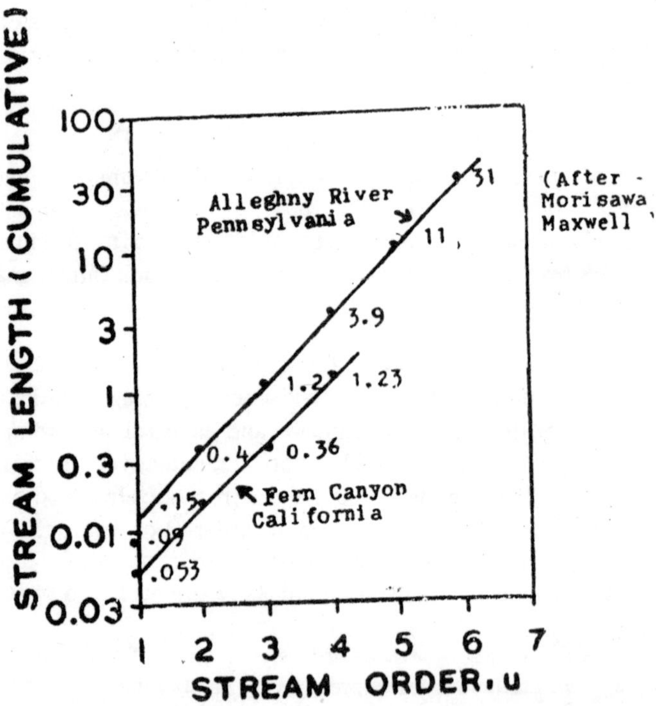

FIG. 12.2 : **Mean Length of Streams of each Order (cumulative) Plotted against Order Products a Straight Line Regression of Positive Exponential Form**

basin areas of successive stream orders tend to form a geometric series beginning with mean areas of the first-order basins and increasing according to a constant area ratio. The definition of area ratio, R_a, is

$$R_a = \frac{\bar{A}_u}{\bar{A}_{u-1}}$$

where $\bar{A}_u$ is mean area of basin of order u. By anology with the law of stream length the law of basin area is as follows:—

$$\bar{A}_u = \bar{A}_1 R^{(u-1)}$$

The symbol $\bar{A}_1$ denotes mean area of the first-order basins.

Law of Allometric Growth

Any measurable property of a fluvial system, such as stream

length or basin area is comparable to an organ of a growing animal and increases according to the allometric growth law. We have seen that in Horton laws relating mean stream length to order, a constant ratio of increase in length holds for increase in order. In this case, successive orders signify increasing time, as inferred from observations of growth of a youthful stream network. Similarly, increase in basin area with order can be thought of as a constant ratio increase with time.

Stated mathematically, the relationship between mean basin area and mean stream length (cumulative) is

$$\bar{A}_u = \overline{aL_u}^b$$

where a is a numerical constant and b is an exponent. This regression equation represents a power function.

Stream Flow and Basin Area

If stream systems were filled with gauges at lower end of each channel segment of each order, the investigation could proceed according to basin areas by order. In practice, gauges are situated at various points on streams. Therefore, we can only relate stream discharge to the total contributing area of watershed above the gauge.

Figure 12.3 shows the observed relationship of average discharge, $\bar{Q}$, to drainage area, A, for the Potomac River basin. Each point represents one gauge. Obviously, gauges in the headwater regions show as points at the lower left, those for downstream lie at the upper right, while the individual points show marked departures from the filled straight line. The trend is obvious. The straight line is described by the power equation.

$$\bar{Q} = aA^b$$

where a is a numerical constant and b is an exponent. Because the fitted straight line runs at 45° across the graph, it can be said that the value of b is exactly 1. The meaning of the exponent of 1 is that the discharge increases in direct proportion to the area. Actually, values of b differing somewhat from 1 are observed in other drainage basins.

One practical use of the mathematical equation relating stream discharge to basin area is that it enables the hydrologist to estimate mean discharge at any point in the system by measuring the watershed area lying above that point. Such knowledge would be essential in designing hydraulic structures, such as dams, bridges and irrigation diversions.

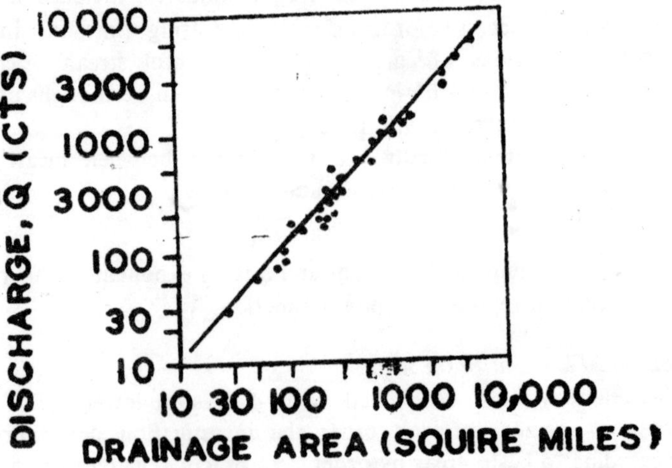

FIG 12.3 : **Relationship of Mean Stream Discharge to Drainage Area for all Gauging Stations in the Potomac River Basin (after J.T. Hack)**

Stream Slopes

On the basis of data, such as shown in Figure 12.4 Horton formulated a law of stream slopes, stated as follows : the mean slope of stream segments of successively higher orders in a given basin tend to form an inverse geometric series decreasing according to a constant slope ratio. Stated as an equation, the law of stream slopes is as follows :

$$\bar{S}_u = \bar{S}_1 \; R_s^{(u-1)}$$

The symbol R_s represents the slope ratio, and is defined as $R_s = [\bar{S}_u \neq (\bar{S}_u - 1)]$. Slope ratios must be less than 1. Values from 0.3 to 0.6 are typical. Individual slope ratios differ from order to order due to variations in the resistance of the rock beneath the stream channel. In Horton's law of stream slopes a single constant value of slope ratio is assumed.

The Law of Stream Slopes

After discussing the law of stream numbers and stream lengths, the law of stream slopes is the third law of Horton which concerns average channel slope which states that "there is a definite relationship between slope of the streams and stream order, which can be expressed by an inverse of geometric series law". The formula is,[27]

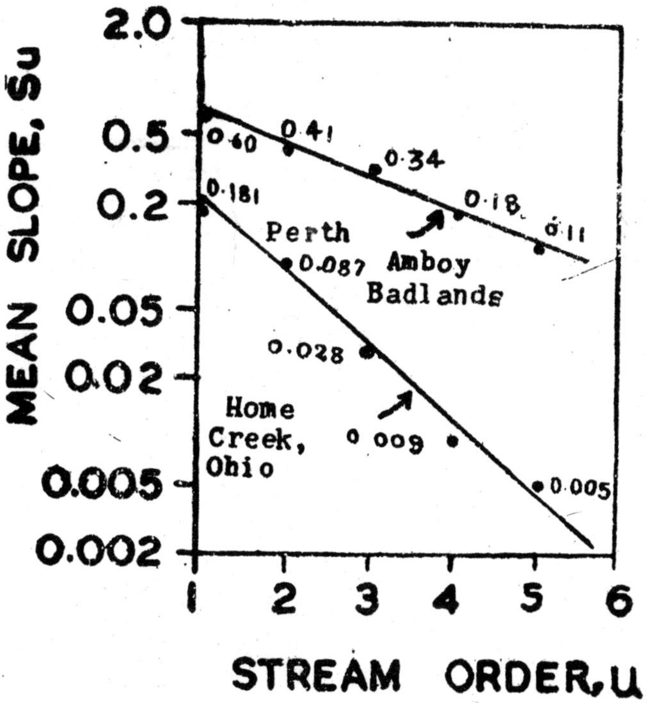

FIG. 12.4 : **Mean Slope of Each Order—Straight Line Regression of Negative Exponential form (after Schumm and Morisawa)**

$$S_o = \frac{S_1}{r_s^{\,0-1}}$$

where S_o is the average slope of stream order O; S_1 is the average slope of the first-order streams; r_s is the slope ratio or the ratio of the average slope of streams of given order of the average slope of streams of the next higher order S_0+1, $rs = \frac{S_o}{S_0+1}$

The consistency of this law is again verified by the Perth Amboy stream slopes in New Jersey.

Horton's three laws may be summed up in a single equation which he called the formula of drainage composition

$$D_d = \frac{l_1 \, r_b^{S-1}}{A_s} \times \frac{P_s - 1}{P - 1}$$

where D_d is the drainage density, defined as the ratio of all streams in a drainage basin to the drainage area, measured in per square mile, l_1 is the average length of streams of first order, r_b is the bifurcation ratio, *i.e.*, the ratio between the number of streams of one order to the number of streams of the next higher order, A is the total drainage basin area, having a main stream order of S, and P is the ratio $\frac{r_l}{r_b}$, or ratio of length ratio to bifurcation ratio.

Schumm[28] suggested a fourth law in the manner of Horton's equation of drainage composition. He called it the law of stream areas, which states that "the mean drainage basin areas of streams of each order tend closely to approximate a direct geometric series in which the first term is the mean area of the first order basins".

Symbolically this is expressed as follows :

$$\bar{A}_o = \bar{A}_1 \, R_a^{\,0-1}$$

where $\bar{A}_o$ is the mean drainage basin streams of order O_1, $\bar{A}_1$ is the mean drainage basin of first-order streams, R_a is the area ratio, *i.e.*, the ratio of the average basin area of the stream of given order A_o to the average area of basins of the next lower order $A_0 - 1$,

$$R_a = \frac{A_o}{A_0 - 1}$$

Hack[29] tested Horton's law of drainage composition for basin in Maryland and Pennsylvania, and verified it by deriving equations for length and drainage area of the principal stream of any order S. Thus given :

$$P = \frac{r_1}{r_b} \text{ and } D_d = \frac{l_1}{a_1}$$

$$A_s = \frac{l_1 r_b^{s-1}}{\frac{l_1}{a_1}} \times \frac{P^{s-1}}{P-1}$$

$$\text{Since } A_s = a_1 r_b^{s-1} \times \frac{P^{s-1}}{P-1}$$

$$\text{and } L_s = l_1 r_1^{s-1}$$

$$\text{Therefore } A_s = a_1 r_b^{s-1} \times \frac{P^{s-1}}{P-1}$$

where L_s and A_s are length and drainage area of the principal stream of any order S.

Morisawa[30] in her empirical analysis of the geometry of 15 watersheds in the Appalachian Plateau found that the geomorphic characteristics conform to Horton's equation of drainage composition, but only in horizontal properties, and not in vertical properties, owing to variations in geologic and lithologic structure.

An interesting property, related to drainage networks is that of texture. Smith in his comparative study of erosional topographies for 54 basins, found that "texture ratio" could be expressed:

$$T = \frac{N}{P}$$

where T is the texture ratio, N is the number of crenulations on a selected contour, and P is the length of the basin perimeter, and it is a logarithmic function of the ratio of the sum of all stream length to the drainage area

Relations between River Discharge and Basin Characteristics

The inter-relationship between stream discharge and the geometric properties of watersheds can be expressed by simple power or exponential function and it can be summarized by a series of equations :

$$W = aQ^b \quad \text{...(1)}$$
$$D = cQ^f \quad \text{...(2)}$$
$$V = kQ^m \quad \text{...(3)}$$
$$L_d = pQ^j \quad \text{...(4)}$$
$$S = tQ^z \quad \text{...(5)}$$
$$N = rQ_y \quad \text{...(6)}$$

Where W is the width of the channel in feet, D is the depth in feet, V is the velocity in feet per second, L_d is the suspended sediment load in tons per day, S is the stream slope in feet per foot, N is the roughness or friction factor defined as ratio of grain size to the depth of flow, and Q is the discharge in cubic feet per second, a, c, k, p, t, and r are numerical constants, and b, f, m, j, z and y are numerical exponents.

Leopold and Miller[31] found that stream order O was related to the number of streams (N), stream (L), stream slope (S), drainage area (A_d) and discharge (Q) all of which seem to be consistent with Horton's scheme of classification. These relations may be expressed in the following forms :

$$O \propto \log N \qquad \ldots(7)$$

$$O \propto \log L \qquad \ldots(8)$$

$$O \propto \log S \qquad \ldots(9)$$

$$O \propto \log A_d \qquad \ldots(10)$$

$$O \propto \log Q \qquad \ldots(11)$$

Several inter-relations among the above equations may be noted. One relation is that between stream length and drainage area:

$$\text{Since } L \propto Q^j$$

$$\text{and } Q^j \propto A_d^k$$

$$\text{therefore } L \propto A_d^{jk} \qquad \ldots(12)$$

Another example of a similar inter-relation is that between stream slope and discharge:

$$\text{Since } S \propto A_d^k$$

$$\text{and } A_d^k \propto Q^j$$

$$\text{Therefore } S \propto Q^{jk} \qquad \ldots(13)$$

It is evident that stream length, drainage area, discharge, and stream slope are all inter-related and can be expressed in exponential forms. When plotted on log-log paper these relationships of drainage network factors form straight lines, indicate that there is not only inter-relations but collinearity.

Meanders Runoff at the Time of Deglaciation

Dury's theory[32] of underfit streams argues that modern underfit streams can only be accounted for its channel forming discharges were once much greater, and induces regional changes of precipitation, precipitation intensity and temperature to supply the greater discharges. He asserts that glaciation, lithology and other factors are, in general, not adequate as bases for explanation of the phenomenon. Kennedy however, in a criticism of the 'bank full' concept, finds Dury's detailed reconstruction of climate unconvincing. Is there, perhaps, a less elaborate explanation of these supposed high discharges.

Changes of Regime due to Change of Temperature

Streams with snowmelt regimes in which most of the annual discharge runs off in an intense but brief spring flood, carry higher discharges more often than they would if their regimes were more regular. The one year flood in such a stream would be much smaller if all the precipitation falling on its catchment were redistributed promptly by evaporation, groundwater recharge or runoff, instead of most of it being stored as snow and released over a short period during the spring.

The mid latitudes were colder at the end of the Plestocene than they are today, and the proportion of total precipitation falling as snow was greater than now. 'Mid latitude' refers here to regions which experienced glacial or periglacial conditions or which were characterized by winter snowfall during the late plestocene. The extent to which the 'snowmelt effect' altered the magnitude-frequency distribution of discharges probably varied widely but calculations presented here show that the effect might have been sufficient to fill now-underfit channels in many mid-latitude regions.

A Model Ice Cap

There is, however, another possible source of water needed to feed these channels. In comparison with its life span the dissipation of a warm ice cap was a short affair, of North American deglaciation, for example implies that at times the Laurentide ice front must have retreated at rates well in excess of $200 my^{-1}$, with rates exceeding $1000 my^{-1}$ for the same major salients.

Consider a simple model of decaying ice cap which rests on a flat floor and is circular with perfect radial drainage, there is always net ablation everywhere on its surface and its profile is given by

$$Z = h\,[1-(r/R)^2], \qquad \ldots(1)$$

where h = ice thickness at centre of ice cap (m),
M =annual volume of melted water resulting from ice cap retreat (m^3),
S_v = annual water equivalent snowfall (m^3), derived from
S = annual water equivalent snowfall (m),
L = fraction of $(M+S_v)$ lost by evaporation or infiltration.

Q will be reduced to $\bar{Q}_{r1}$ mean total discharge in m^{3s-1} per km of ice front assuming T_q = 100 days, where T_q is duration of the runoff season, M and S_v, and their corresponding mean total discharges $\bar{Q}M$ and $\bar{Q}S$ will also be given per km of ice front.

M is derived as follows. The total ice cap volume

$$V = \int_0^n \int_0^{2\pi} r^2 \, dQ \, dZ, \qquad \text{...(2)}$$

Where Q is the polar co-ordinate in the horizontal plane. Substituting for r, and solving (3)

$$V = \pi \frac{R^2 h}{2} \qquad \text{...(3)}$$

and

$$\Delta V = V_1 - V_2 = \frac{\pi}{2} [R_1^2 h_1 - R_2^2 h_2] \qquad \text{...(4)}$$

Subscripts 1, 2 denoting situations before and after any one year's melt. Assuming ablation in the vertical (Z) direction to be everywhere constant and equal to $(h_1 - h_2)$, vertical ablation at the centre of the ice cap, then

$$h_2 = \frac{R_2^2 \, h_1}{R_1^2} \qquad \text{...(5)}$$

Substituting for h_2 in (5)

$$\Delta V = \frac{\pi}{2} h_1 \left[R_1^2 - \frac{R_2^4}{R_1^2} \right] \qquad \text{...(6)}$$

and for a unit length of ice front

$$M = \frac{1}{2\pi R} [0{\cdot}917 \, \Delta V] \qquad \text{...(7)}$$

Where $\bar{R} = (R_1 + R_2)/2$ (Km), and 0·917 in the density of ice.

TABLE 12.3 : **Annual Melt from a Decaying Ice Cap**

h_1 (m)	w (m)	R_1 (km)	A_1 (km^2)	M (m^3 km of $front^{-1}$)	$\bar{Q}M$ (m^3S^{-1} km^{-1})
		50	25	$11{\cdot}445\times10^6$	1·3247
	50	400	200	$11{\cdot}460\times10^6$	1·3264
		1000	500	$11{\cdot}461\times10^6$	1·3266
		50	25	$45{\cdot}576\times10^6$	5·2750
250	200	400	200	$45{\cdot}815\times10^6$	5·3027
		1000	500	$45{\cdot}836\times10^6$	5·3051
		50	25	$11{\cdot}292\times10^7$	13·069
	500	400	200	$11{\cdot}441\times10^7$	13·242
		1000	500	$11{\cdot}454\times10^7$	13·257

$W = R_1 - R_2$ = annual amount of retreat.
A_1 = area drained by 1 km of ice front, assuming perfect radial drainage.

Discussion

The model is not to be constructed as a serious attempt to replicate conditions in any actual instance of deglaciation, particularly since it includes no consideration of the physical properties of the ice cap. At present it gives no more than a crude approximation to the volume of runoff during deglaciation. The calculations, however, are not demanding, and Q^r can be computed easily for different values of variables. Examples in tables were chosen merely to cover a wide range of situations which might have prevailed at the end of Plistocene, although some, perhaps, are extremes.

Calculation of Limestone Removed by Solution

Corbel (1959) has devised a formula to calculate the number of millimetres of limestone per 1,000 years removed by unaided solution only in a limestone region. The formula is as follows.[38]

$$\text{Erosion } (X) \text{ in } m^3/\text{year/km}^2 = \frac{4\ ET}{100}$$

Where $E =$ height of water in decimetres
$T =$ amount of $CaCO_3$ in mg per litre.

Corbel's formula gives the following result for northwest Yorkshire, at the present rate of solution of limestones and with the present rainfall : solution of the rock is equivalent to a lowering of the surface by:

0·04 mm per year by solution at the surface,
0·043 mm pér year by solution underground.
= approximate 0.083 mm per year total lowering.

The rate of lowering of surface of the limestone deduced in this way is approximately 41 mm in 1,000 years.

Micro-Erosion Process and Slope Development

Various research workers have attempted to relate rates of erosion to factors causing erosion. W.D. Ellison[34] has shown experimentally the enormous Kinetic energy exerted by rainfall defined by the following equation :

$$E = KV^{4 \cdot 33} \, d^{1 \cdot 07} \, I^{0 \cdot 65}$$

where $E =$ soil intercepted in splash samplers during @ 30 minute period in grams.
$V =$ velocity of drops in feet per second.
$d =$ diameter of drops in millimetres.
$I =$ intensity of rainfall in inches per hour
and $K =$ constant.

The detaching force for a given rainfall is the product of Kinetic energy and duration of rainfall. W.H. Wischmeier and D.D. Smith (1958) have reported a rainfall erosion factor (R) in which :

$$R = \frac{\Sigma EI}{100}$$

where $E =$ storm energy in foot-tons per acre-inch.
$I =$ maximum 30-minute intensity in inches per hour.

The factor R is used in U.S. Weather Bureau's so-called 'universal equation' (U.S. Dept. of Agriculture, 1961):

$$A = R.K.LS.C.P.$$

where $A =$ Average annual soil loss in tons per acre,
$R =$ Rainfall erosion factor,

LS = Slope Length and Steepness factor,
C = Cropping and Management factor, and
P = The supporting conservation practice.

In this study, volumetric investigations of soil erosion were not possible because of the problems encountered with the sediment trap.

Continuity Equation and Transport Laws

Equations governing the transport of debris down to a slope or river profile are as follows :

(A) The Continuity Equation[35] :

$$M+D = \frac{\delta y}{\delta t} \qquad ...(1)$$

where M is the rate of mechanical lowering,
D is the rate of chemical (dissolved) lowering,
y is the elevation,
and t is time elapsed.

(B) The relationship between mechanical lowering and mechanical transport.

$$M = \Delta . S \qquad ...(2)$$

where Δ indicates the vector divergence, and S is the vector sediment transport. This relationship can be expressed in simpler form if horizontal distance X is measured along the line of greatest slope, in which case :

$$M = \frac{\delta S}{\delta X} - \frac{S}{P} \qquad ...(3)$$

Where P is the radius of curvature of the contours measured as positive where contours are concave outwards, as in hollows.

In considering the continuity equation as two-dimensional, it is assumed that the lines of greatest slope do not migrate laterally through time. This is a reasonable assumption for spur tops and valley axes, and where contour lines are straight.[36]

Emberger's Formula of Bioclimatic Types

Considering the meagre meteorological data that was available during his African (Emberger, 1955) and Mediterranean (1956) bioclimatic studies Emberger devised this simple formula with wide applications and reliable results. After giving a careful thought to the absence of detailed meteorological data he put precipitation and temperature together characterizing a bioclimatic region. The following three basic factors enter into the composition of this formula :

1. Rainfall is expressed as the mean annual total precipitation in mm.
2. Temperature is expressed as $(M+m)$ where M is the mean of the maxima of the hottest month and m is the mean of the minima of the coldest month in °C.
3. Evaporation.

In absence of direct measurement of evaporation the factor is used as $(M—m)$ and it is found that the higher the evaporation more the value. It has been concluded that the higher the value of mean extreme temperature $\frac{(M+m)}{2}$ and of the extreme amplitude $(M—m)$, the drier is a country. In humid regions the ratio between quotient and precipitation and the extreme temperature and extreme amplitude is smaller. The ratio termed as Pluviothermic Quotient is as follows.[37]

$$Q = \frac{P \times 100}{\frac{(M+m)}{2} \quad (M-m) \times 2}$$

To avoid decimals as far as possible, the formula is multiplied by 100 and is divided by 2.

Two concepts which are of fundamental importance in this method are as follows :

1. The aridity-humidity scale is indicated by the Pluviothermic Quotient. Driest stations have the lowest I value of quotient: (Meher Homji, 1968).[37]
2. The temperature scale is indicated by the mean of the minima of coldest month in 'Bio-climatic types' of Kumaon Himalayas according to the Emberger's formula. The results of the Emberger's formula of Q has been given below on the basis of 15 years data (1957-72) regarding temperature and rainfall.[38]

TABLE 12·4 : **Pluviothermic Quotient Values**

Station	Quotient	Mean Minima	Mean Maxima	P mm	$\frac{M+m}{2}$	$M-m$	Effective Temp. Index
1. Bazpur	87	7·5	40·3	1373	23·9	32·8	153·4
2. Chakrata	332	1·7	23·9	1855	12·8	22·2	103·3
3. Dehra Dun	182	6·1	36·2	2324	21·15	30·1	144·5
4. Haldwani	132	6·5	39·25	1978	22·9	32·7	169·8
5. Joshimath	133	2·8	27·9	1031	15·1	25·6	112·0

Source : Indian Meteorological Reports and Field data, 1974.

Summer Monsoon Rainfall and Rank Statistic Test

The Mann-Kendall Rank statistic test recommended by (World Meteorological Organization) climatology commission working group on climatic fluctuations (1966) has been applied for testing randomness against trend. In this test, each term of rainfall series is replaced by its rank. The first term of this series is compared with all the subsequent terms and n_1, the number of terms with rank higher than the first term is noted. Starting with 2nd, 3rd,......(N—1) the term, the process is repeated and $n_2, n_3 \ldots {}^nN-1$ are obtained, N being the total number of terms in rainfall series. The test statistic r is given by: [39]

$$T = \frac{4 \sum_{i=1}^{N-1} n_i}{N(N-1)} - 1$$

Distribution of r is Gaussian, provided $N > 10$, and the expected value and variance of T are given respectively by

$$E(T) = 0$$

$$\text{Var}(T) = \frac{4N+10}{9N(N-1)}$$

T has been calculated for each station in the basic network and its significance has been tested. August rainfall series at three stations, September rainfall series at two stations and seasonal rainfall series at three stations show significant increasing tendency. However, these stations are randomly distributed over the area and moreover, the percentages of significant cases are not different from those that may be found by chance. The hypothesis of randomness of the monthly and seasonal rainfall series over South-East Asia, is therefore, not contradicted.

Intensity of Slope[40]

$$\text{Intensity of Slope} = \frac{\text{Max. Altitude} + \text{Minimum Altitude}}{2}$$

Grids of one inch have been prepared for extracting data from one inch map. Maximum and minimum heights of each grid have been recorded. They have been added and then divided by two. Each calculated figure represents the average slope per square mile. All these figures have been plotted within the centre of each grid in half inch map to prepare choropleth.

The Percentage Hypsometric Curve

The percentage hypsometric curve involves two ratios of relative height $\left(\frac{h}{H}\right)$ and relative area $\left(\frac{a}{A}\right)$. On the ordinate is plotted the ratio of height (h) of any contour above base to the total height (H), while on the abscissa is plotted the ratio of area (a) between any given contour and lowest contour to the total area (A).

Projected Area and Real Area

The relationship of real area to projected area helps the study of dissection. It is the basis of the dissection-index proposed by Slaucitajs (1936) and termed relief entwicklung:

$$\frac{\text{Real area} - \text{Projected area}}{\text{Projected area}} \times 100$$

De Smet (1951) finds this index incomplete without reference to volume: in the classical scheme of the erosion of an uplifted peneplain. Volume should diminish constantly, whereas real area and average slope should pass by a maximum when the lost remnants of the old peneplain disappear. Accordingly, De Smet proposes this more complex index of dissection :

$$\frac{\text{Real area} - \text{Projected area} \times \text{Average slope}}{\text{Volume}}$$

This is the most satisfactory of several indices of dissection.

Altitudinal Distribution

Altitudinal distribution can be expressed by a number of common indices, using hypsometric data. It is, for instance a simple matter to determine average height.

$$\sum \frac{\text{Percentage inter contour areas} \times \text{Mean inter contour height}}{100}$$

As the area contained within the hypsometric curve and its axes is proportional to the volume of relief, average height may be obtained by dividing this area by the maximum abscissa.

Clinographic Curves

Curves of the clinographic type have been applied with varying success to demonstrate average slope between contours. They include :

(a) The clinographic curve of Finsterwalder;
(b) The hypso-clinographic curve of Finsterwalder;
(c) The hypsographoid of Uhorczak/Romer;
(d) The clinographic curve of Hanson-Lowe,
(e) The mean slope curve of Strahler and the average profile of De Smet;
(f) The curve of average inter-contour widths of De Smet;
(g) The slope-height curve of Moseley.

All the relevant methods are depend.nt upon contour lengths or inter-contour areas, rather than on slope facts delimited by field analysis.

Clinographic Curve of Finsterwalder (1890)

Average slope of the total area, however, may be obtained by dividing the area contained within the curve by the area of the region, and by consulting a table of tangents. Finsterwalder and Peucker both expressed the angle of average slope as:

$$\frac{\text{Counter interval} \times \text{Total length of Contours}}{\text{Total area}}$$

This method of slope analysis is more complex than Wentworth (1930) method. Wentworth method requires the calculation along the lines of grid squares of the number of contour crossings per mile. A second count, along a grid oblique to the first, provides additional data. The following formula is then applied to the mean of the two sets of readings :

$$\frac{\text{Average number of contour crossings per mile} \times \text{Contour interval}}{\text{3361 (a constant)}}$$

In Wentworth method, one of rapid sampling, provides a considerable measure of accuracy.

Peguy (1942, 1947) used contour lengths and average slope in his indices *d* aeration:

$$l = \frac{D}{d \,.\, \tan \alpha} \qquad \text{where } D = \text{length of longest contour}$$

$$d = \frac{\text{Sum of contour lengths}}{\text{number of contours}}$$

tan α= tangent of average slope. But without inclusion of area and volume, this is not a really satisfactory index of dissection.

The Hypsoclinographic Curve of Finsterwalder

De Smet (1954) shows that the hypsoclinographic curve of Finsterwalder does not show precisely the average slope of each inter-contour area. Inter-contour angles of slope are not obtainable by Finsterwalder formula:

$$\frac{\text{height} \times \text{length of lower contour}}{\text{inter-contour area}} \quad \left[\frac{h.L_1}{A_1}\right]$$

but by the following formula:

$$\frac{\text{height} \times \text{mean length of two contours}}{\text{inter-contour area}} \quad \left[\frac{h.L_1 + L_2}{A_1}\right]$$

De Smet suggests that the vertical axis should be divided as follows :

$$h\left[\frac{L_1 + L_2}{2}\right], \; h\left[\frac{L_1 + L_2}{2} + \frac{L_2 + L_3}{2}\right],$$

$$h\left[\frac{L_1 + L_2}{2} + \frac{L_2 + L_3}{2} + \frac{L_3 + L_4}{2}\right]\ldots\ldots\ldots$$

He also suggests that, as the slope of the area above the highest contour cannot be measured by Finsterwalder's formula, an approximation could be found by dividing the length of the highest contour by two. But in no circumstances can the hypsoclinographic curve be regarded as an average profile, as the axes represent areas not lengths.

The Hypsographoid of Uhorczak/Romer (1931)

Uhorczak, a pupil of Romer, wished to produce a graph which gave a more exact picture of average profile than the hypsographic curve. He envisages representation of relief by a series of concentric circles proportional in size to the areas above contours. He constructed hypsographoids with altitude on the central vertical axes and with radii of circles arranged proportionally as horizontal lengths, the ends of these being joined. Uhorczak produces hypsographoids for continents, oceans and major drainage-basins, comparing them with hypsographic curves. By calculating the volume of a hypsographoid as πr^2 (area of base) multiplied by average height, and by representing it as a cone, Uhorczak then establishes average slope. The hypsographoid, however, neither facilitates visual impression of area-height relationship, nor demonstrate a correct average profile, except for conical reliefs. The reason for these disadvantages will be considered in the discussion of Hanson-Lowe's clinographic curve (1935), which is merely one-half of the hypsographoid.

The Clinographic Curve of Hanson-Lowe (1935)

Hanson-Lowe devised his clinographic curve to demonstrate the average angle of slope of inter-contour areas in the form of an average profile. He regards contour as a series of concentric circles, finding the radius of the area above a given contour by[41]

$$\sqrt{\frac{\text{area above contour}}{\pi}}$$

The average distance between two contours is, therefore, the radius of the lower contour minus the radius of the higher contour, and the slope of each inter-contour area becomes

$$\tan \alpha = \frac{\text{contour interval (in feet)}}{\text{inter-contour distance (in feet)}}$$

The inter-contour angle of slope are then drawn on a graph, beginning at the highest point and working down to the base. In most cases, all angles must be multiplied by a constant to enable the curve to be easily drawn, although rescaling of vertical data can be effected merely by graphical means. The length of the horizontal axis is equal to the radius of the whole area. Unfortunately, by ignoring the length of the contour, the clinographic curve of Hanson-Lowe maximizes inter-contour width and thus minimizes inter-contour angle of slope.

The Mean Slope Curve of Strahler (1952) and Average Profile of De Smet (1954)

Strahler finds the mean width of each inter contour area by dividing this area by the mean length of the two adjacent contours. The tangent of the angle of slope is the contour interval divided by the mean inter-contour width. The mean slope of the highest inter-contour area cannot be accurately obtained because there may be several distinct summits, but an approximation is possible when half the length of the highest contour is taken as denominator. For studies of small drainage basins Strahler suggests identical scales on the horizontal and vertical axes. He expresses the relationship of hypsometric curve to the mean slope in the following equation:[42]

$$\frac{l}{L} \tan Q = K \tan \alpha$$

where Q = Slope of hypsometric curve,
α = mean ground slope,
l = contour length at given relative height,
L = length of longest contour in basin,
K = constant.

Comparison of the Hanson-Lowe and Strahler methods for the islands of Gozo and Gueransey reveals marked differences in results. Most emphatically, Strahler method gives by far the more accurate representation. Table 12.5 shows that with Hanson-Lowe method every inter-contour width becomes longer and therefore every angle lower than with the Strahler method. Differences are not constant. Slopes obtained by Strahler method are from $1\frac{1}{4}$ to more than 5 times as steep as those obtained by the Hanson-Lowe method. Although more laborious than the latter, Strahler method must be preferred.

De Smet proposes a curve, similar to the mean slope curve, which he terms the average profile. He used inter-contour widths cumulatively on the horizontal axis, but suggests no means of calculating inter-contour angles of slope. Such calculations are however unnecessary for the construction of an average profile.

The average slope for a region or drainage basin can be easily calculated from mean slope data[43] :

$$\frac{(\alpha_1 \times A_1) + (\alpha_2 \times A_2) + (\alpha_3 \times A_3) + \dots\dots\dots + (\alpha_n \times A_n)}{A_1 + A_2 + A_3 \dots\dots\dots + A_n}$$

Where α = slope, and A = area of inter-contour areas. This formula is more precise than the method advocated by Finster-walder and Peucker.

TABLE 12.5 : **Hanson-Lowe Method for Inter-Contour**

	Gozo				Guernsey			
	Inter-contour distance (ft)		Angle of slope		Inter-contour distance (ft)		Angle of slope	
Method	Hanson-Lowe	Strahler	Hanson-Lowe	Strahler	Hanson-Lowe	Strahler	Hanson-Lowe	Strahler
0-50	285	169	1°58′	16°26′	2598	1042	1°6′	2°45′
50-100	380	191	7°30′	14°40′	739	299	3°52′	9°30′
100-150	486	200	5°53′	14°4′	334	362	3°26′	7°52′
150-200	760	258	3°46′	10°58′	1051	362	2°43′	7°52′
200-250	1156	313	2°29′	9°5′	1911	565	1°30′	5°3′
250-300	1859	397	1°32′	7°11′	2476	693	1°9′	4°7′
300-350	2402	445	1°12′	6°25′	5122	1411	0°30′	2°2′
350-400	2323	409	1°14′	6°52′				
400-450	2529	486	1°8′	5°53′				
450-500	1251	326	2°17′	8°43′				
500-550	1109	401	2°35′	7°7′				
550-600	195	155	14°20′	17°51′				
600	407	508	4°38′	5°37′				

Source : After J.I. Clarke, Essays in Geomorphology, p. 255.

REFERENCES

1. A.N. Strahler, *Physical Geography*, Wiley Eastern University Publication, Bombay, 1971, p. 665.
2. Op. cit.
3. M. Taher, "Fluvial Process and Geomorphology of the Brahmaputra Plain", *Geographical Review of India*, Vol. 36, No. 1, 1974, p. 42.
4. J.J. Donahue, "Measuring Drainage Density with a Dot Planimeter", *The Professional Geographer*, Vol. XXVI, No. 3, 1974, pp. 317-318.
5. C.E. Ramser, "Runoff From Small Agricultural Areas", *Journal of Agricultural Research*, 14, 1927, pp. 797-823.
6. P. Haggett, "Land Use and Sediment Yield in an Old Plantation Tract of the Serra Do Mor, Brazil", *The Geographical Journal*, Vol. CXXVII, Part I, 1961, p. 57.
7. R.R. Drumond, & H.W. Dennis, "Qualifying Relief Terms", *The Professional Geographer*, Vol XX, No. 5, 1968, p. 329.
8. Op. cit., p. 330.
9. S.C. Ambastha, "*Hydrology and its Impact on the Land Utilization of Magadh, Bihar*", First Indian Geographical Congress, New Delhi, 1972, Section Climatology.
10. S.C. Ambastha, Op. cit.
11. F.J. Monkhouse, "*A Dictionary of Geography*", Edward Arnold, Bristol, 1970, p. 21.
12. Op. cit.
13. Op. cit.
14. M.A. Glazovskaya, "Technobiogeoms : The Initial Physical", *Geography*, Vol. XIV, No. 4, 1973, p. 216.
15. A.M. Borlyant, "Maps of Interrelations and Their Application in Geographic Research", *Soviet Geography*, Vol. XIII, No. 7, 1972, pp. 472-482, (479).
16. F.J., Monkhouse, *A Dictionary of Geography*, Edward Arnold, Bristol, 1970, p. 75.
17. N.K.P. Sinha, *Geomorphic Evolution of Northern Rupununi British Guiana*, Unpublished Ph.D. Thesis, McGill University, 1966, p. 95.
18. V. Paschinger, (Abstract Review), *The Geographical Journal*, Vol. 131, Part 3, 1965, pp. 391-392.
19. H.C.S. Thom, *A Note on the Gamma Distribution*, Statistic Laboratory Iowa State College, 1947, (Manuscript).
20. D.G. Friedman, and B.E. James, "Estimation of Rainfall Probabilities", University of Connecticut, *Agricultural Experiment Station Bulletin*-332, 1952, p. 24.
21. G.L. Barger, R.H. Shaw & R.F. Dale, "Gamma Distribution Parameters, Agriculture and Home Economics Experiment Stations", Iowa State University, 1959, p. 183.
22. V.L. Eichenlanle, N.D. Stommen, & D.G. Dickason, "Precipitation Probabilities as Indices of Climatic Variation Over the Eastern United States, "*The Professional Geographer*, Vol. XXIII, No. 4, 1971, p. 301.

23. F.M. Courtney, & R. Welester, "A Taxonometric Study of the Sherborn Soil Mapping Unit", *Transactions of the Institute of British Geographers*, No. 58, 1973, p. 117.
24. S.C. Kharkwal, "Slope Studies on A Himalayan Tarrain", *The National Geographical Journal of India*, Vol. XVII, Part 1, 1971, p. 2.
25. S.C. Kharkwal, Ibid, p. 14.
26. A.N. Strahler, *Physical Geography*, Wiley Eastern University Publication, Bombay, 1971, pp. 482-497.
27. S.T. Wong, "A Multivariate Statistical Model for Predicting Mean Annual Flood in New England", *Annals of the Association of American Geographers*, Vol. 53, No. 3. 1963, pp. 298-311.
28. S.A. Schumunm, & J.C. Maxwell, "Quantitative Geomorphology of the San Dimas Experimental Forest California", Columbia University, Department of Geology, Technical Report No. 19, Office of Naval Research, Project NR 389-042, 1955, p. 14.
29. J.T. Hack, "Studies of Longitudinal Stream Profiles in Virginia and Maryland", (Washington D.C., U.S. Geological Survey Professional Paper, 294-B, 1954, p. 65.
30. Morisawa, and Lucien M. Brush, Jr., "Drainage Basins, Channels and Flow Characteristics of Selected Streams in Central Pennsylvania", Washington D.C., U.S. Geological Survey Professional Paper, 282-F, 1961.
31. L.B. Leopold, & M.J. Thomas, "The Hydraulic Geometry of Stream Channels and Some Physiographic Implications, Washington D.C., U.S. Geological Survey Professional Paper, 252, 1953, pp. 16-21.
32. J.G. Cogley, "Meanders on Runoff at the Time of Deglaciation", *Area*, 1973; Vol. 5, No. 1, pp. 33-37.
33. G.H. Dury, *Essays on Geomorphology*, Heinemann, London, 1966, p. 205.
34. W.D. Ellison, "Some Effects of Raindrop and Surface Flow and Soil Erosion and Infiltration", *Trans. An. Geophys. Un.*, 26, pp. 415-29. (2) W.H. Wischmeir, and D.D. Smith, "Rainfall Energy and Its Relationsl to Soil Loss", *Trans. Am. Geophys. Un.*, 39, pp. 285-291.
35. D. Brunsden, Slopes Form and Process, *Institute of British Geographers*, No. 3, January, 1971.
36. M.J. Kirkley, "Hillslope Process-Response Models Based on the Continuity Equation". *Slope Form and Process* (Edt. & Brunsden), No. 3, 1971, (Of. Institute of British Geographers), pp. 16-17.
37. Homji, V.M. Meher, "Application of Pluviothermic Quotient Xerothermic-Hygrotherm.c Indices to the Indian Subcontinent", *The Indian Forester*, Vol, 94, No. 5, p, 353.
38. J.S. Chauhan, "Application of Pluviothermic Quotient of Emlerger in the Central Himalayas"; *Geographical Review of India*, Vol. 36, No. 4, 1974, pp. 316-317.
39. D.A. Mooley, "Climatology of the Asian Summer Monsoon Rainfall—Controls and Concentration", *Geographical Review of India*, Vol. 37, No. 1, 1975, p. 9.

40. Anil Kumar, *Geographology of Simdega and Its adjoining Area, Bihar*, Vol. 1, Unpublished Ph.D. Thesis, Bhagalpur University, 1968, p. 81.

41. J. Hanson-Lowe, "The Clinographic Curve", *Geological Magazine*, 72, 1935, pp. 180-184.

42. A.N. Strahler, "Hypsometric (area altitude) *Analysis of Erosional Topography, Bull. Geol. Soc. Arer.*, 63, 1952, pp. 1117-42.

43. C.K. Wentworth, "A Simplified Method for Determining the Average Slope of a Land Surface", *American Journal of Science*, 20, 1930, pp. 184-94.

THIRTEEN

R. B. Mandal

REGIONAL GEOGRAPHY

THE FIELD of regional geography includes several aspects related to spatial distribution of any item on the earth. The wider use of statistics in regional analysis is necessary for scientific development of the subject and to create research frontiers to its manifold objects for the regional scientists.

Measurement of Angular Momentum

"It is common knowledge that the greater the distance from the sun, the slower the planet's movement along its orbit. But the decrease in velocity is in the ratio of $1\sqrt{R}$, i.e., slower than the increase in distance R. As a result the angular momentum per unit of mass (specific angular momentum) increases in ratio to the square root of the orbital radius ($R : \sqrt{R} = \sqrt{R}$). For movement along an elliptic orbit[1] with major semi-axis equal to a and eccentricity equal to e, the specific angular momentum is in direct ratio to the value $\sqrt{a\,(1-e^2)}$."

Computation of Deflated Income per Head

The deflated income per head for each country (YA) is computed by the following formula[2]

$$'YA = \frac{1}{3} \sum_{i=1}^{3} \frac{Y_1}{N_1\, I_1} \left[\frac{\sum\limits_{j=1}^{12} b_j}{\sum\limits_{j=1}^{12} b_j \left(\dfrac{\sum\limits_{i=1}^{n} P_{ij}\, A}{\sum\limits_{i=1}^{3} P_{ij1}} \right)} \right]$$

Where Y_1 = national income in each country's currency.

N_1 = Population.
I_1 = Index of general prices (1973-74=100)
b_j = Weight of the jth commodity and j represents 1, ...(12)
P_{ija} = Price in local currency of jth commodity for country A.
P_{ij_1} = Same as P_{ija} except that it represents the Indian price.

i represents 1 = 1971, 2 = 1972, 3 = 1973.

Computation of Exchange Ratio's among National Currencies

If we convert all national incomes into Indian rupees and multiply by the ratio of Indian price level to the price level of each country in turn, the exchange rate "cancels out". For example, if the national income of India = Y_1 in rupees and that of any other country A = YA in A's currency while the exchange rate between rupees and A's currency = r, we could convert YA to rupees by multiplying r. If, we multiply YA by the ratio of the Indian price level to that of A in rupees, we obtain :

$$r\,YA \left[\frac{\sum_{i=1}^{n} b_i\, P_{i1}}{\sum_{i=1}^{n} b_i\, P_i\, Ar} \right]$$

Where b_i = Appropriate weight for the ith commodity, and P_{i_1} and P_iA = the price of commodity i for India and A respectively. The r's cancel out and we are left simply with :

$$YA = \left[\frac{\sum_{i=1}^{n} b_i\, P_{i1}}{\sum_{i=1}^{n} b_i\, P_iA} \right]$$

Which is independent of the exchange ratio.[3]

Conversion of Fahrenheit Degree Temperature into Centigrade Degree Temperature

A graduated scale of temperature on which 0° represents the melting point of ice, 100° the boiling point of water at sea level. Officially centigrade thermometer is known as Celsius since 1948.

Method I : For $86°F$ The centigrade value is :

$$°F \text{ to } °C = (F°—32°) \times \frac{5}{9}$$

$$°C = 86° - 32° \times \frac{5}{9}$$

$$°C = \frac{54° \times 5}{9} = 30°C$$

Method II– Formula is :

$$°F \text{ to } °C = \frac{C°}{100°} = \frac{F—32°}{180°}$$

$$\frac{°C}{100°} = \frac{86° - 32°}{180°}$$

$$\frac{°C}{100°} = \frac{54°}{180°}$$

$$°C\ 180° = 100° \times 54°$$

$$°C = \frac{100° \times 54°}{180°} = 30°C$$

Method III

$$°C \text{ to } °F = \frac{9 \times °C}{5} + 32°$$

The Adjusted R

The concept of "Adjusted R" test is to consider the magnitude of coefficients of correlation when adjusted for changes in degrees of freedom. When a variable is added into the analysis, it may be expected to raise the coefficient of correlation. However, it also decreases the degree of freedom. The "adjusted R" answers the question, is an increase in the coefficient significant in the light of the loss of degree of freedom ? The resultant 'adjusted R is compared with the preceding "adjusted R". If it is larger, the variable retained; if it is smaller, the variable does not add significantly to the explanation and is debted. (Table 13.1).

The formula for its calculation is as follows :

$$\text{The adjusted } R = 1 - \frac{1—R^2 (N-m)}{N-1}$$

Where N = number of cases (Census tracts).

m = number of partial regression coefficients plus the constant a.

TABLE 13.1 : **Addition of Variables and Test with Adjusted R**

No. of variables	Variable Added	r^2	r	Adjusted R
1	9	0.6952	0.8338	0.8338
2	21	0.8599	0.9270	0.9277
3	20	0.8771	0.9365	0.9371
4	19	0.8936	0.9453	0.9712
5	6	0.9031	0.9503	0.9740
6	17	0.9056	0.9516	0·9749
7	18	0·9114	0.9547	0.9767
8	15	0.9165	0.9573	0.9782

Source : After Coulson, M.R.C.

The Quadrat Method

In the quadrat method, the plane is divided into a number of equalized cells, and the number of points occurring in each cell or in randomly selected cells is noted. Tests are made about the hypothesis of random like nearest-neighbour analysis. If the observed frequency of points occurring in cells does not approximate the expected distribution which is derived from the observed density function, then the distribution can be thought of as either grouped or dispersed depending on the direction in which the observed value differ from the expected values.

Test of significance can be made using the chi-square static because the distribution of chi-square approximates the ratio of the sum of squares to the variance.

Therefore, the equation :

$$X^2 = \frac{\Sigma (X-\bar{X})^2}{S^2} \quad ...(1)$$

but,

$$X^2 = \frac{\Sigma (X-\bar{X})^2}{\bar{X}} \quad ...(2)$$

The ratio of the mean to the variance would equal 1 in random distribution, and values higher than 1 would indicate a tendency toward evenness, and values lower than 1 would indicate a tendency toward groupedness.[4]

$$R = \frac{\bar{X}}{\Sigma \frac{(X-\bar{X})^2}{N}} \quad \ldots(3)$$

The values for R and chi-square were obtained and are summarized in Table 13.2.

Empirical Aspects of Regional Growth

An alternative dependent variable, growth in State final demand 1958-63 can be derived, using estimates prepared by Polenske et al. for the Harvard Multi-regional Input-output study. These estimates are built up from the following equation:[5]

$$SFD = C + I + V + X^F + S + G$$

where,

SFD = State final demand,
C = Personal consumption expenditure
I = Gross private capital formation.
V = Net change in inventories.
X^F = Gross foreign exports.
S = State local government expenditure.
G = Federal Government Expenditure.

Both consumption expenditures and private capital formation refer to total State purchasing regardless of place of production or residence (location) of purchasers.

A Model of the Use of Territory

With the qualitative content of the relationship of elements in the use of territory, we will now attempt to build a model of the phenomenon under study.

The linkage between the need for a certain product and the area used for that purpose may be expressed by several equations:

$$P_i = a_i S_i \quad \ldots\ldots(1)$$

Where P_i is the needed amount of ith product; a_i is the unit area productivity of territory for ith product; S_i is the area used for the production of the given product.

Since in actual practice, a particular product is produced by the use of a number of segments (types of territory) that vary in productivity, the expression CW should be written more properly as :

TABLE 13.2 : **Summary of Quadrat Method Data and Measures**

Year	Frequency of Points of per Cell						Total Points	Total cells	Mean point per cell	*R*	Probability of greater χ^2 than that observed
	0	1—	2—	3—	4—	5—					
1900	89	13	3				19	105	0.181	0.882	0.16
1910	112	20	4	3			37	139	0.266	0.696	0.0005
1920	261	43	12	2	3	1	110	322	0.342	0.762	0.0005
1930	315	66	18	5	3		129	407	0.317	0.679	0.0005
1940	389	82	15	5	2	2	145	495	0.293	0.640	0.0005
1950	433	76	15	5	1		125	530	0.236	0.746	0.0005
1960	586	64	4				72	657	0.110	0.999	0.50

Source : Getis, A., 1964, p. 397.

$$P_i = a'_e S'_e + a''_i S''_i + \dots\dots\dots + a'''_i S'''_i = a'_i S'_i \quad \dots(2)$$

Where a'_i, a''_i ...... a'''_i are the unit area productivities of different types of territory used for the production of ith product and S'_i, S''_i, ... S'''_i are the areas of those types of territory.

The overall used for territory required to perform all kinds of (m) functions would be as follows, on the basis of (1) and (2):

$$\sum^{m} P_i = \sum^{m} a_i S_i = \sum^{m} \sum^{n} a'_i S'_i \quad \dots(3)$$

The multiplicity of uses to which various types of territory might be put by different engineering system can be represented in the form of an m by n matrix, in which m columns would represent the set of types or segments of territory. The specific content of the matrix would depend on the purpose of investigation and the availability of data.[6]

Extension of One-Dimensional Mathematical Treatment to Two-Dimensional Statistics

Extension of the usual one-dimensional mathematical treatment to two-dimensional geographical statistics has led to some interesting and cartographically convenient concepts. If the coordinates of n basis statistical points (e.g. settlements) are denoted by (X_i, Y_i) and any quantitative property (number of inhabitants, etc.) by P_i, we can use the weighted means:

$$\bar{X} = \frac{\sum_{i=1}^{n} X_i P_i}{\sum_{i=1}^{n} P_i} \qquad \text{and } \bar{Y} = \frac{\sum_{i=1}^{n} Y_i P_i}{\sum_{i=1}^{n} P_i}$$

as the coordinates of the centre of the distribution (the centre of gravity). Similarly, the standard distance,

$$d = \frac{\sum_{i=1}^{n} P_i [\,(X_i - \bar{X})^2 + (Y_i - \bar{Y})^2]}{\sum_{i=1}^{n} P_i}$$

$$= \sqrt{\delta X^2 + \delta Y^2}$$

corresponds to standard deviation and is easily calculated with the aid of the weighted standard deviations of the X and Y coordinates S_x and S_y.

Calculating :

$$T_g\ 2\alpha = \frac{C_{ov}\,(X, Y)}{\delta x^2 - \delta y^2}$$

We obtain the direction X of major elongation of the distribution, with maximum deviation along it. δ maximum and minimum at right angles to it. Moreover,

$$\delta\, x^2 + \delta\, y^2 = \delta^2 \max + \delta^2 \min = \alpha^2$$

The main statistical characteristics of a geographical distribution can now be represented graphically by a cross centered at $(\bar{X}, \bar{Y})$, with arms of length 2δ max and 2δ min at an angle of α to the major axes, and with a diagonal

$$\alpha = \sqrt{\delta^2 \max + \delta^2 \min}$$

standrad distance (i.e. dispersion). This distribution cross can be printed on the maps, and will show at a glance quantitative difference in synchronous distributions or change in time-sequence series, etc.[7]

Scale of Aerial Photographs

Like maps, the aerial photographs do not have a uniform scale. In fact the scale on a photograph varies from point to point depending on the variation of relief and tilt of the camera axis (Fig. 13.1).

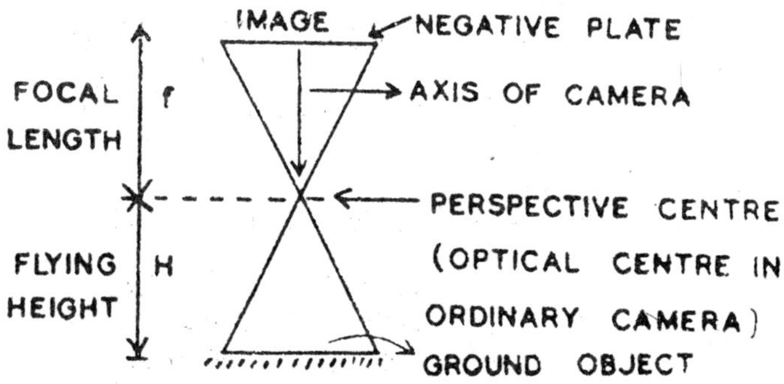

FIG. 13.1 : **Relief and Tilt of the Camera Axis**

$$\frac{\text{Image Distance}}{\text{Ground Distance}} = \frac{\text{Focal Length}}{\text{Flying Height}} = \frac{f}{H}$$

The average scale of an aerial photograph is obtained from the relationship—Scale of Photograph : focal length of the camera used/ flying height of the aircraft above the mean terrain. In fact, the scale of the photograph in the ratio between the image distance and the ground distance. To take an example, the average scale of the photograph taken at 5,000 ft. above the mean terrain with a 6-inch focal length lens is

$$\frac{0.5}{5,000} = 1 : 10,000.$$

In relation to the corresponding map, the scale of a photograph may be determined from the relation:

$$\text{Scale of Photograph} = \text{Scale of Map} \times \frac{\text{Distance on Photo}}{\text{Distance on map}}$$

In aerial photograph, the variation of scale is caused by two factors i.e. effect of relief and tilt displacement. In the central projection of an aerial photograph the images of ground positions are shifted or displaced because of terrain relief. In a truly vertical photograph, the displacement of images is in a direction radial from the centre of the photograph. This radial displacement of images due to relief is responsible for scale difference within a photograph and this fact is to be considered at the time of interpretation and mapping. The magnitude of radial displacement of images due to relief is given by the equation

$$m = \frac{rh}{H}$$

where m = relief displacement of top of an object with respect to its base

r = radial distance from centre of the photograph to the base of the image.

h = height of the object.

H = flying height of the aircraft above mean terrain level.

A Method for Determining Elevation Changes on Aerial Photographs

Once the scale of photographs (S) and focal length of Camera (f) have been determined, the flight altitude (H) of the aircraft may be calculated according to the following equation:

$$H = \frac{f}{S} \qquad ...(1)$$

This displacement value of the water-tower (d) or the relative change of the two-point image on two successive photographs, must now be determined. Accurate measurement of the water tower extremities (top and bottom photo identities) is obtained as a distance measured along the base lines of both photographs. The values may then be substituted in the height formula below:

$$h = \frac{\text{H}}{b + d} \cdot d \qquad ...(2)$$

Where h is the height of the water tower, H is the height of altitude, d_i is the value $(Xb_1 + Xb_2)$ subtracted from $(Xt_1 + Xt_2)$ and b is the average length of the photo air bases

$$\left(\frac{b_1 + b_2}{2}\right)$$

All measurement should be converted to decineal portions of a feet to facilitate ease of computation. A simple check of the accuracy of the height measurement (h) may be obtained by solving for H in the derivative formula :

$$H = \frac{h\,(b—d)}{d} \qquad ...(3)$$

If the flight altitude (H), as determined here, results in a value within 0.10% of the original flight altitude derived in equation (1), may be assumed that the vertical measurement is also accurate to within 10% of its true value. This degree of accuracy represents the maximum error allowable with photography of a scale of 1 : 20,000. Due care in pin-picking points and conducting measurement will reduce the margin error.[8]

Friction of Distance

The unconstrained gravity model usually relates the volume of interaction between two places to three independent variables as follows:

$I_{ij} = f\,(O_i \,.\, D_i \,.\, d_{ij})$

Where I_{ij} = the volume of interaction from i to j.

O_i = an index of the total volume exported by place.

$i_j \,.\, D_j$ = an index of the total volume imported by place.

J_i and d_{ij} = the distance from i to j.

Three-Dimensional Vector Technique for Analysing Till Fabrics

The output from the computer program gives a resultant vector or mean orientation, a value for dip which together with the preferred orientation, provides the point of balance, vector strength R and the value for R-N where N is the number of observations in the sample. The R value can be tested against the null hypothesis for randomness by using a formula proposed by Watson (1956) which is successfully approximated by the equation:

$$R95 = \ldots\ldots \sqrt{\frac{NX^2_{95}}{3}} \qquad \ldots\ldots(1)$$

Where X^2 95 is the value for chi-square with three degree of freedom and N is the number of observations in the sample. For N equal to 50 the value for R should be above 11.39 at the 95 per cent level. Steimmets (1962) discussed fully the use and implication of these results. It should be pointed out, however, that R does not necessarily measures dispersion around the mean and thus imposes same disadvantages upon analysis.

In calculating the 95 per cent field of confidence surrounding the mean orientation and centre of balance, the second program uses the output of the first. Further, it calculates the statistic referred as to the estimate of precision parameter. The equation used for this is :

$$K = \frac{(N-1)}{(N-R)} \qquad \ldots\ldots (2)$$

This statistic K, measures homogeneity of variance and can be tested by the maximum F ratio where the degree of freedom are $2(N-1)$ and a ll samples have an equal number of observations.

The next step tests whether or not the mean directions of populations are identical. The statistic with degrees of equal to

$$\frac{(P-1)}{(P-1)} \frac{(q-1)}{(N'-q)} \text{ is}$$

$$F = \frac{(N'-q)}{(q-1)} \cdot \frac{\Sigma R_i - R}{N'-R_i} \qquad \ldots\ldots(3)$$

Where N is the total number of observations, q is the number of samples. ΣR_i is the sum of the R values for each sample. R the combined value for N observations, and P number of dimensions.[9]

Diffusion of Fowl-Pest Disease

Two new parameters need to be added to the nature of the surface being affected by the epidemic. Distance from the original source of the infection (D_1), and distance from the sea in all directions (D_2) are probably related to the diffusion of the virus (S) in the following way where K_1, K_2 and K_3 are constants to be defined:[10]

$$S = K_1 \left(D_1^{K^2} / D_2^{K^3} \right)$$

Generalization in Isopleth Mapping

An isopleth line on a map is determined by interpolation and extrapolation of values derived for different control points located at different parts of the map. Naturally the more numerous and closely spaced control points are more accurate in an isopleth line. As the alignment of a section of an isopleth line is fixed with reference to the value of control points, the strength (S) of the isopleth at any position varies inversely as the mean distance (d) and mean departure (V) of control values from the isopleth value at that position. Let (V) be the value of an isopleth line drawn by interpolation of control values V_1, V_2, V_3, ... V_n. Their respective distances from a point on the line being d_1, d_2, d_3, ... d_n. Then the mean distance of the control point is given by ;

$$d = (d_1+d_2+d_3 \ldots\ldots\ldots\ldots + d_n)/n = \sum \frac{d}{n}.$$

and the mean departure of value is given by:

$V = (A_n \; \mho \; V)$

Where n denotes the number of control points.

The strength (S) of an isopleth at any of its point, therefore, varies inversely as dV. Conversely, the degree of relative generalization (G) which is reciprocal of strength, that is $\frac{1}{S}$, varies directly as the mean distance and the mean departure of the control values lying in the immediate vicinity of the said point on the isopleth. The relation may be expressed as :

$$G = \frac{1}{S} = KdV \qquad \ldots\ldots (1)$$

Where K is a constant. In the above equation G may range from zero upwards depending on the value and distance of control points. Since the distance of an isopleth line from its control point can never be uniform. It is apparent that wide local variation or generalization is a necessary property of an isopleth line which is one of its serious drawbacks.

With some modification of the equation (1) it is possible to determine in general terms the degree of generalization of entire isopleth line, instead of its individual points. If n is the total number of central points used to fix the alignment of an isopleth line of length l, the degree of generalization involved is given by :[11]

$$G = \frac{Kdvl}{n} \qquad \text{... (2)}$$

Where d is the mean distance of the control points from the isopleth line and v is the mean departure of control values from the isopleth value.

The nature of the values of the control points is also an important factor in the process of cartographic generalization. The values of some control points on a map may be more reliable than those of some others. In those cases where the values of a control point is derived by taking the average of a large number of observations (m) the equation may be modified in the form

$$S = \frac{Kdvl}{nm} \qquad \text{... (3)}$$

When the value of control points is not an observed value, in such cases it established that the generalization at each control point is directly proportional to the area of the corresponding spatial unit based on which the control value has been derived. The relation here may be expressed in the form

$$G = Ka \qquad \text{... (4)}$$

Where a is the area of spatial unit, and K is a suitable constant.

Generalization in Choropleth Mapping

In case of uniformly spread all over the area the degree of regional generalisation in choropleth mapping can be expressed in the form;

$$G = \frac{Ka}{E} \qquad \text{... (5)}$$

Where K is a constant.

The evenness of distribution E is a complex function of spacing (S) of a distribution with reference to total area, size (r) of a distribution in relation to total distribution in relation to the centre of gravity of the geographical area. The function may be expressed in the form :

$$E = f(S_{rc})$$

The exact nature of these functions and their parameters are yet to be worked out. According to J.K. Wright, the coefficient of absolute areal evenness of any distribution is given by

$$E = \frac{1-M}{Q}$$

Where M shows the amount of hypothetical displacement to achieve perfect areal evenness and Q is the amount of displacement that would obtain with medium unevenness of distribution.

Generalization in Mapping Absolute Distributions

The probability of quantitative generalisation in a proportional symbol varies inversely as its size. That is, larger the size, less is the possibility of error of approximation. The empirical relation may be expressed in the form :

$$G = \frac{K}{S} \quad \ldots (6)$$

Where S is the size of the symbol and K is a suitable constant. In case of circular or spherical symbols the equation may be expressed in the form:

$$G = \frac{K}{r^2} \text{ or } \frac{K}{r^3}$$

where r is the radius of circle or sphere as the case may be. Similarly, for line symbols indicating quantities by their width or length the amount of generalization is obtained by :

$$G = \frac{K}{l}$$

where l indicates linear dimensions of width or length of the symbol depicted on the map.

If v is the quantitative value of a point symbol on the map, the probability of maximum error or the amount of maximum possible local generalization is given by

$$G = \frac{\pm V}{2} \quad \dots (7)$$

If the total number of operations involving use of such a symbol is n, total generalization for the entire mapped area will be equal to the algebraic sum of all errors. The relation may be expressed in the form:

$$G = \frac{\pm V}{2} \text{ up to the } n^{\text{th}} \text{ term,}$$

The value of G varying between 0 and $\pm\, nv_2$ for the mapped area as a whole.

Locational Generalization

Geographical location of point symbols on a thematic map is also subject to generalization. The maximum possible geographical displacement of a point symbol from its true position is equivalent to half of the length (l) of the point symbol. The empirical relation for medium possible genęralization may be expressed in the form:

$$G = \frac{Kl}{2} \quad \dots (8)$$

Where K is a constant. In case of dots or circular symbols of radius (r) the equation is:

$$G = Kr_4$$

Similarly in case of a line symbol the amount of maximum locational generalisation will vary as half the width (w) of the line or

$$G = \frac{Kw}{2} \quad \dots (9)$$

Therefore, probability of error in plotting point-symbols and line symbol can be reduced by using relatively smaller sized point-symbols and finely drawn line symbols.

Generalization in Thematic Classification

If the size of the class or the class interval of the thematic scale in relation to the total range which is the determining factor here. The empirical relation between the degree of generalization and the class interval (i) may be expressed in terms of function of 'i' in the form.[12]

$$G = Kf(i) \qquad ...(10)$$

Therefore, larger the value of interval, the greater is the degree of generalization involved. The absolute amount of the generalization can be worked out in terms of the actual value of the observed data and the interval of the class in which the observed data fall. The amount of generalization of an observed value (V) varies between $V \pm i_2$ and $V \pm i$.

Characteristics of Thematic Scales on Maps

In the conventional system of classification the Pth class limit (l_p) of the Thematic scale on a map is given by the general expression;

$$l_p = V \pm \left(P - \frac{(q-1)n}{m} - 1 \right) \sum_{i=1}^{m} i_q$$

Where V indicates initial class boundary and m represents the broad groups or parts into which the Thematic scale (comprising n classes) is divided, n being divisible by m. The class size or class interval in each broad group is indicated by i_q, where by definition $i_q = 0$

When $p >$ or $< \frac{(n+m)}{m}$ as the

case may be and where

$q = 1, 2, 3, \ldots\ldots\ldots\ldots m$, when $\frac{m + qn - 1}{m} \leqslant P \leqslant \frac{qn}{m}$.

It is possible to set a scale in such a fashion as to put the class intervals gradually increasing or decreasing in perfect progression. In this system of sliding scale the class limits are spaced out in the following manner.

$V (V \pm i_1), (V \pm i_1 \pm i_2), (V \pm i_1 + i_2 + i_3), \ldots\ldots$, and

$(V \pm n \sum_{i=1}^{n} in)$,

Where V indicates the initial class boundary. The value of the Pth class limit here will be given by[13]

$$l_p = V \pm \sum_{i=1}^{P} i(j-1).$$

Where $i_o = o$

It is also possible to divide the Thematic scale in such a manner as to obtain a predetermined frequency (f) of the thematic distribution in each class.

The Pth class limit is obtained from expression

$$l_p = (P - l) f.$$

The greatest advantage of this system of classification of the thematic scale is that the degree of generalization in each class is more or less uniform throughout the map.

Concept of Continentality

This concept has been derived from a hypothetical situation of a plane surfaced land mass of regular shape where latitude has no relevance and where the geometrical centre of the land mass would have the highest continentality, on the other a vision of continentality as the degree to which any part of that area exhibits a climate typical of being inland and a way from the sea's influence.

Most of the attempts at a continentality index employ climatic criteria but lay different emphasis on the significance levels of certain parameters, especially temperature range. In addition, the indices of Zenker (1888), Hann (1903), p. 215, Gorc Zynski (1920), Johansson (1931), Conrad (1946), and even Ivanov (1959) attempted to detect the theoretical core of continentality using one inductive approach considering first the raw data and then compensating for external factors, finally arriving at an accurate index with no criticism. The decrease of thermal amplitude from high to low latitudes due to the inclination of the earth's axis was the first to be compensated for with almost all indices taking note of it, followed by certain remodifications to overcome the arithmetic difficulties that arise in low and high latitudes.[14]

TABLE 13.3 : **List of Zenker Type Indices of Continentality**

Originator	*Index*
Zenker	$K = \frac{100\,A}{\phi}$
Zenker[15]	$K = \frac{6A}{\phi} = 20$
Gorczynski[16]	$K = \frac{1.7\,A}{\sin\phi} - 20.4$
Johansson[17]	$K = \frac{1.6\,A}{\sin\phi} - 14$
Conrad[18]	$K = \frac{1.7\,A}{\sin(\phi+1^{\circ})} - 14$
Ivanov[19]	$K = \frac{100\,(A+AD+0.25D)}{0.36\,\phi+14}$

Where A = Annual temperature range (°C),
K = index of continentality.
ϕ = latitude.
AD = daily temperature range.
D = Saturation deficit

Method of Regional Differentiation

The method of work on regional differentiation according to employment travels and manufactured goods shopping travels was adopted depending on the usability or rather on the availability of original sources of information. The amount of information concerning industrial goods shopping travels is substantially smaller. The only more or less comprehensive material is represented by the results of a spatial research by the State Institute for Regional planning. On the total some of the retail trade turnover in 1967 of stores selling industrial goods, conducted in 220 centres and in all the districts and regions of the CSR. By processing this information we have been only able to calculate the likely numbers of inhabitants of the areas gravitating to the respective centres[20] :

$$\frac{M_i}{\frac{V_o + V_k}{2}} = Pi, \qquad \ldots(1)$$

Where M_i = the retail trade turnover of centre "i" in 1967.

V_o = the average per capita spending on industrial goods in 1967 by inhabitants of the district where the centre is located.

V_k = the average per capita spending on industrial goods in 1967 by population of the region where the centre is situated.

P_i = the number of inhabitants of centre "i".

Thus, the number of "fully served inhabitants" was calculated for the surrounding area of each centre. In view of the existence of additional industrial goods store outside the "selected centres" a certain part of the population staged out of the specified centres and their gravitating areas. The selection of centres was based on the following principle : the minimum number of inhabitants of an employment or a service gravitating area of the centre should be, 5000.

A General Measure of Diversity

The Factors of Diversity

Diversity can be expressed as a function of two variables (1) No. of categories or classes (N), and (2) the evenness of distribution among these categories expressed as per cent (E).[21]

$$D = F(N, E)$$

Where D is Diversity. First of all in many situations the determination of N (the number of categories) is not as concise or easily accomplished as by use of prescribed census classification. Really, in some problem it is not possible to determine N. Secondly, the first few categories embrace, in the usual case, the region in which the actual curve departs most from the diagonal and, therefore, the addition of the tail end of the Lorenz curve actually decreases the sensitivity of the area ratio. The last advantage lies simply in the saving of time.

A General Index of Diversity

Starting with equation (1)

$D = F(N, E)$: It is proposed that N should be introduced linearly, so that with E held constant a ten category distribution would be considered twice as diverse as one of five categories. And E will be expressed as

$$E = \frac{A_i}{A_a} \quad ...(2)$$

Where A_i is the area under the first-three points of the ideal or diagonal Lorenz curve and A_a is the area under the first-three points of the actual curve (Fig. 13.2).

Equation (1) can then be written

$$D = N = \frac{A_i}{A_a} \quad ...(3)$$

In order to evaluate the ratio A_i/A_a considered the rank percentages as P_1, P_2, P_3. These are the increments in ordinate values of the actual Lorenz curve at segment junctions. It can easily be shown that :

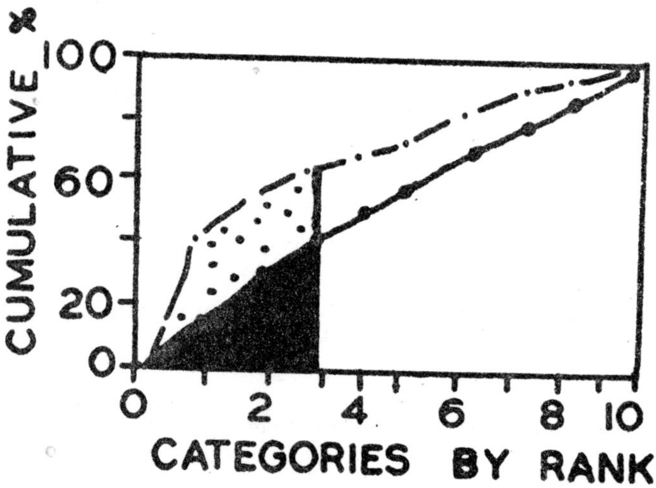

FIG. 13.2 : **Realistic Lorenz curve and its ideal counterpart**

$$A_a = \frac{5P_1+3P_2+P_3}{2}$$

While the area under the ideal curve is

$$A_i = \left(\frac{1}{2}\right) \left(\frac{100}{N} . 3\right) . 3$$

The second term being the ordinate value of the triangle, the final factor 3 being the abscissa value and the ½ factor for triangularity.

Inserting these expressions for A_i and A_a in equation (2) we have :

$$E = \frac{A_i}{A_a} = \frac{900/N}{5P_1+3P_2+P_3} \qquad ...(4)$$

Inserting this in equation (3) and arbitrarily introducing a factor of 10 to produce a convenient scale of magnitudes, we have :

$$D = (N)\frac{900 \times 10}{N(5P_1+3P_2+P_3)}$$

or

$$D = \frac{9 \times 10^3}{5P_1+3P_2+P_3}$$

Attributes of the Index

It is a simple algebraic expression through which diversity is obviously known. The idea of the above three percentage distributions (be it in number of employs, value added, or whatever, we are not concerned here) are all that is necessary. A most interesting feature of the final form is the absence of the N factor. The problem of its determination has been eliminated.

A second interesting attribute of this index is that perfect diversity for N category is $10N$, so that for six categories perfect diversity is 60, for 10 it is 100 and so forth.

Stereoscopic Profiling of Land-Units from Aerial Photographs

(a) *Planimetric Scale C* i.e. representative fraction in the horizontal Plane)[22].

$$SP = \frac{d}{D};$$

Where d=measured distance in approximately the same horizontal plane between two plotted points of the stereo-model. The points were on or close to the profile. D = The map or ground distance between the same two points and expressed in the same units as d.

(b) *Vertical Scale*

$$Sa = SP\left(Ck - \frac{1}{K} \quad \frac{N_1 . n_2}{e} + \frac{1}{K} \cdot \frac{b}{e}\right); \text{ where}$$

$$K = \frac{\text{Principal distance of instrument}}{\text{Principal distance of camera}}$$

e = Interpupillary distance (eye Base)

n_1, n_2 = Distance between principal points in the stereo-model.

b = Machine Base.

The Construction of Point Scales

The points as distributed on the scale must follow a certain pattern if the interval is not conform, the points should increase or decline in terms of a particular mathematical function. A point scale that does not follow such a function and contains variable intervals is not a logical scale, and its use in statistics may lead to serious error (say, in calculating the number of events under each point).

The technique of constructing a scale with increasing intervals is relatively simple. Let us designate the desired number of points as n and the length of a segment on the scale from 0 to i (where $i = 1, 2, \ldots\ldots\ldots n$) as L_i. The expression L_i is related to n by same kind of function, usually an exponential. In that case $L_i = n^x$.

The problem is to calculate the power X. If we convert to logarithms, we get

$$X = \frac{\log L_i}{\log n}$$

For example, where $i = n = 4$ and $L4 = 80$,

$$X = \frac{\log 80}{\log 4} = 3.16.$$

The exponent usually turns out to be a fraction. We then calculate the intervals in terms of distance from the initial point on the scale :

$$L_3 = 3^3 . 16 = 32.4$$

$$L_2 = 2^3 . 16 = 8.9$$

$$L_1 = 1^3 . 16 = 1.0$$

The points are then plotted on scale *V*. That was the method used in the construction of the Beaufort scale. It should be noted that the intervals $L_i - L_i - 1$ will depend on the selected dimensionality. When the magnitude *L* is taken *i*, *n* another dimensionality, the intervals too will be different.

The quantitative equivalent of points on the scale should be expressed with sufficient accuracy to ensure that a particular magnitude will be allotted to the çored point on the scale, especially near the boundary between two points. In this case, the numerical intervals would be as follows :

Point 1 = 0.0-0.9
Point 2 = 1.0-8.9
Point 3 = 9.0-32.4
Point 4 = 32.5-80.0

Or, if the numerical equivalents are rounded to the nearest unit

Point 1	0-1
Point 2	2-8
Point 3	9-32
Point 4	33-80

A scale with decreasing intervals is constructed in the same manner, except that the segments are plotted from the end. For the construction of a scale with a bunching of intervals in the centre. We need to calculate the intervals for each half of the scale by taking *n* equal to 0.5 plus the number of classes (since one half of the zero interval must be taken into account), as example cited here $n = 5.5$ and $L_n = 40$. If the difference between points becomes too abrupt, their number can be increased (compare *V* and *E*). A scale can also be constructed by the use of another function, say, sum of the arithmetic progression :

$$L_n = \frac{n}{2}[2a + r(n-1)]$$

Or the same of the geometric progression

$$L_n = \frac{a_1(1 - q_n)}{1 - q}$$

In which two out of three variables are given: in the first case, the magnitude of the first interval a_1, the difference of the progression r as well as the number of points n; in the second case, the denominator q is given instead of the difference.

Correspondence between Point Scale and Classified Phenomena

If we want the point values to reflect the quality of the study phenomenon, we must first be clear about the nature of the phenomenon before applying the appropriate arithmetical operations.

The problem come down to constructing the mathematical model of a particular system, i.e. to applying the conceptual formula.[23]

$$Y = f(K_1\ AK_2\ BK_3\ C_1, \ldots\ldots\ldots K_n - 1\ KK_nL)$$

Where Y is the overall suitability of the selected conditions for the purpose of which the point scale is being constructed; A, B, $C_1 \ldots\ldots\ldots\ldots\ldots K$, L are the factors on which the attainment of the purpose depends; K_1, K_2, etc. are coefficients, n is the number of factors being considered.

Geography of Consumption

Since the income for the Bureau of labour statistics total sample group for each city vary, some method for normalising expenditures to a standard identical income was necessary. The income of the sample group for each city were averaged and each city's sample population income was compared to this average. The difference from the average income was computed and the normalised expenditures for each category were then calculated on the basis of the following formula ;[24]

$$RE = K \cdot X, \text{a}/1{,}000$$

Where RE = revised expenditure for a given item.

K = a constant which reflects the varying propensity to consume goods and services as a result of additional small increments to ones income.

X = expenditure for a given group of goods and services.

a = a city's income differences from that of the sample group average (\$ 6,200).

Allocated differences are indeed pronounced.

Magnitude of Changing Patterns in the Mineral Economy

The magnitude formula may be expressed as[25]

$$M = \frac{E_n + V_s}{2}$$

Where M is the magnitude of mining E_n is the total number of employees as a percentage of the U.S. total, V_s is the value of shipments as a percentage of the U.S. total.

A Cartographic Process of Radar Photogrammetry

One approach, reflecting transposition of standard parameters is the measurement of distance between two identifiable points on a radarscope photograph and the corresponding points on a verified source such as map, aerial photograph, or radarscope photographic scale can be found through the following formula:[26]

$$S_r = \frac{(D_s)\ (S_s)}{D_r}$$

D_s = Distance of Source
S_r = Scale of radarscope photograph
S_s = Scale of source
D_r = Distance on radarscope photograph.

To achieve higher accuracy, the process can be repeated for several points on each radarscope photograph and the total figures combined to reach an average figure.

Mathematical Method of Regionalization

Knowing their extreme values U_n *Max* and U_n *Min* on the earth, it is possible to express them in percentage using the parameter

$$t_n = 100\ \frac{U_n - U_n\ min}{U_n\ max - U_n\ min}$$

to guarantee full comparability. The natural boundaries of simple regions on maps showing the distribution of the values of measurable features with the so-called megi-gradients they always run between two isolines the minimum distance between them being ΔS, and their value is:

$$S_m = 8848\left(\frac{\Delta t_n}{\Delta S}\right) = \frac{884800}{U_n\ Max - U_n\ Min}\left(\frac{\Delta U_n}{\Delta S}\right)$$

On the maps showing the non-continuous distribution of features the values of the parameters refer to strictly limited areas. The values of these boundaries are determined by difference between parameters of the neighbouring areas:

$$\Delta t_n = (t_{n_1} - t_{n2}) = \frac{100}{U_n\, Max - U_n\, Min}\ (\Delta U_n)$$

Therefore, system of natural boundaries may be delimited according to the measurable values on any map.[27]

Method of Proportionate Distribution

The assumption here is that each urban centre exerts some influence as a service or market centre on the nearby countryside or upon another urban centre within its orbit. The degree or extent of influence is greater near the urban centre than farther from it progressively decreasing from the centre outwards being minimal at the periphery of its area of influence.

The area of influence is assumed to be circular for the sake of convenience in mappings and proportional circles are drawn for all the towns is the area under study, with radius:

$$R = \sqrt{\frac{T \times A}{TTU}}$$

Where A = Total area of the state,
U = Total Urban Population in the state,
T = Population of each town.

The larger the circle the greater the degree or extent of influence.[28]

This method insures a more precise delimitation of the areas of town concentration and scatter, and the areas inadequately served by towns.

Composite Index of Urbanity

It can be expressed by the formula :

$$I = \frac{dA}{dP+dA} \cdot IP + \frac{dP}{dP+dA}\ IA$$

Where dA and dP are measures of deviations of IA and IP from unity i.e. the national value:

Where IP = Index of urbanity sector (urban to total population) dimension.

IA=Index of urbanity, Spatial (urban population to total area) dimension.

I = Composite index of urbanity.

Coefficient of Evenness for Aerial Distribution

Wright (1937) had indicated the possibility of the use of the concept of '*Lorenz Curve*' to measure the degree of evenness of the distribution of certain population on earth's surface. He rightly pointed out that the ratio of population to area or to any other locational measure, for example, the density of population is a simple measure of the degree of evenness or of association of the spatial distribution of population on the basis of the concept of 'Lorenz Curve' he formulated the "coefficient of evenness for aerial distribution", which may be expressed as :

$$R = \frac{2\sum_{j=1}^{n}\left(A_j \sum_{i=1}^{j} P_i\right) - \sum_{j=1}^{n} A_j P_j}{\left(\sum_{j=1}^{n} P_j\right)\left(\sum_{j=1}^{n} A_j\right)}$$

Where P_i and A_i are the population and the area of the ith subdivision of a region (of which the degree of evenness of the distribution of population R, is to be determined), and the suffix "i" in the region is so chosen that all $\frac{(P_i)}{A_i}$ are monotonic increasing, n being the total number of such values in the region. When A_i's are equal we get the corresponding value denoted by T, which on simplification becomes :

$$T = \frac{2\sum_{j=1}^{n} (n + 1 - j)\ P_j - P}{nP}$$

Where the total population $P = \sum_{j=1}^{n} P_j$

and P_j's are monotonic increasing. Two spatially separated regions may be delimited by the same range of density values (by some population range when subdivisions are equal in area, but may have different values of R or T,

($O < R, T \leqslant 1$), since R or T takes into account the order of $\frac{(P_i)}{A_i}$ is or P_i's and the number of subdivisions, n, in a region.[29]

The Functions of Place

The concept "function of a place"[30] might be interpreted as a place, i.e. a particular segment of geographical space, performs (or may perform) a particular function in the life (or in the activities) of society and thus satisfies (or may satisfy) a particular need of society or of man.

In a static situation (when $P_1, P_2 \ldots\ldots P_i =$ constant, where $P_1, P_2, \ldots\ldots P_i$ are social needs), as both needs and technology develop slowly, the social function of a place depends on its capacity to satisfy certain social needs. Since the totality of properties of the particular place depends on its position in multi-dimensional space, the social function of ith place (ϕi) may be viewed as a methematical function of its position [$x_i\,(n)$] in n dimensional space.

$$\phi\, i = f\,[X_i\,(1)),\ X_i\,(2) \ldots\ldots\ldots,\ X_i\,(n)] \qquad \ldots(1)$$

Strictly speaking, the parameters of multidimensional space should also include time as well as the function performed by the particular place since these are significant characteristics of the place. However, for the purpose of our analysis, both function and time factor are treated separately and are, therefore, excluded from the totality of properties of the place.

The mathematical expression contains yet another significant simplification. It assumes that the social needs ($P_1, P_2, \ldots\ldots\ldots P_i$) corresponds to a certain extent to the particular properties ($a_1, a_2 \ldots\ldots\ldots aK$) of each place. Actually, in view of the potential multitude of needs that may be satisfied concurrently or alternatively by particular combinations of properties of each given place, the extent to which these aspects correspond is one of the unknowns rather than a given parameter.

The expression reflects a static situation while the tasks of prediction, planning, designing require a dynamic approach. In this approach the asynchronous character of changes in each of the properties of place as well as various types of needs and their demand upon these properties determine the variability of the functions of place through time,

Coefficient of Spatial Association (CS)

Let us assume two sets of co-ordinates of a series of points in two dimensional space :

$A(X, y/X_1, y_1, X_2y_2, \ldots\ldots\ldots X_n, Y_n)$

$B(X, y/X_1, y_1, X_2, y_2 \ldots\ldots\ldots X_m, y_m)$

Then a measure of spatial association is given by the ratio of the average distance to nearest neighbour within the sets and between the sets.

Let dA_i be the distance of X_iA to its nearest neighbour within A and dB_i likewise for X_iB. Then the average within set nearest neighbour distance for all elements of A and B is given by :

$$\left(\sum_{i=1}^{n} dA_i + \sum_{i=1}^{m} dB_i\right)\Big/(n+m) = a \qquad \ldots(1)$$

Note that A and B do not have to contain the same number of elements. Let the nearest element within B for X_iA be located at distance dAB_i and dBA_i likewise for X_iB. The average nearest neighbour distance to a number of another set for all elements of A and B is given by :

$$\left(\sum_{i=1}^{n} dAB_i + \sum_{i=1}^{m} dBA_i\right)\Big/(n+m) = b \qquad \ldots(2)$$

Expressing (1) as a ratio of (2) gives a measure of spatial association called (S) such that :

$$S = \left(\sum_{i=1}^{n} dB_i \Big/ \left(\sum_{i=1}^{n} dAB_i + \sum_{i=1}^{m} dBA_i\right)\right) = a/b \qquad \ldots(3)$$

S has a range $0 \leqslant CS < \infty$, as $CS \rightarrow 0$,

The two sets of point patterns become increasingly disassociated until at $CS=0$. As $CS \rightarrow \infty$ the data sets are increasingly associated.

A number of such transformations exist e.g.

$$f(X) = 1-2\ e^{-0,\ 6931X} \qquad \ldots(4)$$

$$f(X)-(X-1)\ 1\ (X+1) = (a-b)/(a+b) \qquad \ldots(5)$$

Where $X = a/b$

Here transformation (5) is used, which gives a range $1 < f(X) < +1$. So far it has been implicitly assumed that association between sets *A* and *B* is mutually of the same order. As has already been suggested, this is not necessarily the case if we define a coefficient of partial association (*CS* such that :

$$CSa = (\Sigma dA_i - \Sigma dAB_i)/(\Sigma dA_i + \Sigma dAB_i) \qquad ...(6)$$

$$CSb = (\Sigma dB_i - \Sigma dBA_i)/(\Sigma dB_i + \Sigma dBA_i) \qquad ...(7)$$

dAB_i and $dBA_i \neq 0$

Where *CSa* represents the asociation of set *A* and *B* and *CSb* likewise the relationship of *B* to *A*, then rarely does *CSa*=*CSb*. Neither is *S* equal to the average of these two functions except in special cases e.g; $CS = +1, -1$), This observation does not invalidate *CS* as a general measure of spatial association between two sets of point patterns. *CS* always lies between *CSa* and *CSb*. *CS* has the following properties and characteristics:

(1) *CS* is scale free.

(2) The number of elements in sets *A* and *B* may be different.

(3) No arbitrary real framework is imposed which might give different value according to the size of the nest or its disposition.

(4) No surface have to be calculated.

(5) Both overall and partial association values of the type described previously can be calculated.

(6) There is a boundary problem in that the nearest-neighbour to an element both within and between sets, may lie outside the region or area being examined. This problem is similar for most measures.

(7) A conceptual problem may arise in considering each object as a point in space, but the significance of this observation varies according to the size of the area being considered. Thus a shop in the context of a *CBD* may be satisfactorily regarded as occupying a point in space.

(8) *CS* has problems in coping with highly skewed sets of data, but again this falling is characteristics of most measures of association.

(9) No test of significance of *C* value is available,

TABLE 13.4 : **Analysis of CS Values**

Crude *CS* value	Transformed *CS* Value	Degree of Association or Disassociation
$<\frac{1}{3}$	<-0.5	High Disassociation
$\frac{1}{3}$—$\frac{2}{3}$	-0.2 - $+0.2$	Moderate Disassociation
$\frac{2}{3}$—1.5	-0.2 - $+0.2$	Intermediate Relationship
1.5—3	0.2-0.5	Moderate Association
>-3	>0.5	High Association.

A number of transformations can be made to the inputs of *S* and to the method of calculation. First distance could be either route distance or airline distances, and may be transformed by some distance decay function e.g. $f(d_i) = d_i^2$. Second each X_iAB can also be weighted, but manipulation is more difficult than for distance transformation. High order nearest neighbour distances may be used as a basis for calculating *CS* if it is thought that they are more accurate by portraying the latent characteristics of population. Thus the values of *CS* is a pure non-expressing a spatial relationship between two areally distributed sets of points.[31]

Map Lettering from Lantern Slides

One should know the size of the room and the size of the screen. Let us assume that the rear seats are 60 ft. from the screen, which is 10 square feet, and we made the map one feet wide or high. The question is : how far has the map to be held from our eyes to cover the screen.[32]

$$X : W = D : W \text{ and } X = \frac{WD}{W}$$

$$\text{In our case } X = \frac{1 \times 60}{10} = 6 \text{ ft}$$

Where X is the distance from map to eye;

W is the width of map,

D is the distance from eye to screen.

In placing the map in front of the man in the near seat, at $60 : 10 = 6$ ft distance, the map would just cover the width of the screen. Thus our smallest lettering should be visible from 6 feet.

Vertical Exaggeration of the Map

It is a human nature to be more sensitive than to its horizontal measure. We usually overestimate the steepness of slopes, and artists draw mountains steeper than they really are.

To bring out the ups and downs to correspond letter with our mental concept, we exaggerate the vertical component. The degree of exaggeration depends on the scale, on the nature of the country, and on the purpose of the profile. In a hilly country a rough rule is :

$$VE = 2\sqrt{\text{Miles per inch}}$$

Thus we exaggerate on a map of the scale:

1 mile 1 inch	2 times.
4 ,, ,, ,,	4 times.
16 ,, ,, ,,	8 times.
64 ,, ,, ,,	16 times.

In mountainous regions we exaggerate less, in plain regions considerably more.

Slope Measurement on Land

Slope can be expressed in several ways. From a contour map the easiest way is to read the steepness in feet per mile. For instance, on a 1 mile/inch map with 20 feet contour interval a 200 feet 1 mile slope will have the contours 0.1 inch apart. Engineers like to express slope in percentage. Thus, with a 200 feet/mile slope

$$\frac{X}{100} = \frac{200}{5280} = 3.78 \text{ per cent}$$

For computations, the simple fraction called the gradient is used :

$$\frac{\text{Vertical distance}}{\text{Horizontal distance}}$$

$$\text{In this case } \frac{200}{5280} = \frac{1}{26.4} \text{ or } 0.0378.$$

The same as one-hundredth of the percentage. The surveyor measures the slope in degrees. The gradient of 1° is 1/57.3. A 100 ft/mile slope is about 1° 05′. This figure can be used as long as the angle is not steep. Thus a 200 mile/inch slope is 2°1′ but at a slope of over 20°, the gradient has to be calculated by the function $\tan \phi = \frac{h}{D}$, where $\tan \phi$ is the angle of slope, h is the height, and D is the horizontal distance.

Horizon Measurement

From Mount Washington, N.H. 1 can you see the light of the new television tower as great Blue Hills, near Boston. This and similar questions are often asked to a cartographer, and he should be prepared to give the answers. First, let us calculate the radius of the visible horizon from a given height. A trigonometric equation gives the answer :[33]

$$r = \sqrt{[(R+h)+R][(R+h)-R]}$$

$$r = \sqrt{(2R+h) \times h}$$

As h is relatively small, approximately

$$r = \sqrt{2Rh}$$

Where r = Radius of the visible horizon from a given height.
R = Radius of the earth.
h = height of the hill from the horizon.

When expressed in miles.

$$r = \sqrt{\frac{7916 \times \text{height in feet}}{5280}}$$

$$= \sqrt{\frac{3}{2} \text{height in feet}}$$

as 7916/5280 is almost exactly 3/2.

Our equation can be improved if we correct for refraction. Refraction is caused by the decreasing density of air with altitude which enables us to see beyond the geometrical horizon. Its value varies with the weather, averaging 1.065 r. Thus corrected for refraction, when expressed in miles

$$r = 1.065 \sqrt{\frac{3}{2} \text{height in feet.}}$$

The Economic Potential Model

The fundamental concept underlying the potential model is that the degree of economic interaction between regions is a function of the level of economic activity in each area and the distance separating each area from the others. The relative attractiveness of each region is governed by its capacity to act as a market supplier of inputs for economic activities and in turn the region of greatest attractiveness will be that where distance costs are little to all possible markets and source of supply. In this way, the model derives a spatial patterns of the relative potential for economic growth of each region.

The model as applied in the New Zealand situation was based on that developed by Clark, Wilson, and Bradley (1969, 197-212) and is expressed by the following formula :[34]

$$P_i = \frac{I_i}{M_i} + \sum_{j=1}^{n} \frac{I_j}{M_j + T_{ij} + F}$$

Where P_i = Economic Potential of region i; $\frac{I_i}{M_i}$ =Influence of local market and transport costs in region i 1 to m=19 regions of New Zealand $m+1$ to $n=5$ major New Zealand markets (U.K., U.S.A., European Economic Community, Japan, Australia)

I_j=Regional income of region j

M_j=Minimum cost of transport within region j

T_{ij}=Transport cost of moving goods between regions i and j,

F=Tariff factor taking into account international trade between New Zealand regions and overseas countries.

Potential of Place

Potential of place[35]$=\Sigma P \cdot d$.

Where P=Population at another point.

d=distance of that point from A.

A variation on this method has been developed by Warntz and Neft and involves the calculation of a dynamical radius which is given by the formula:

$$r = \Sigma \frac{P \cdot d^2}{P}$$

Where P=population at another place.

d=distance of that place from point being measured.

P=total population of the study area.

Using the same formula it can be seen that

$$r = \sqrt{\frac{(400 \times 5^2) + (50 \times 10^2) + (10 \times 20^2) + (40 + 4^2)}{700}}$$

$$= 5.3 \text{ Kms.}$$

Thus within a radius of 5.3 Kms. of place A live 91 per cent $\left(\frac{640}{700} \times 100\right)$ of the total population of the region under study.

Calculation of Population Potential

Settlement	Population	Distance (A)d	P.d
A	200	—	200
B	400	5	2000
C	50	10	500
D	10	20	200
E	40	4	160

Change in Livestock Units and the Acreage of Improved Pastures

The variables used in this analysis are change in number of livestock units and change in the acreage of improved pastures over the period 1950—65.[36]

$$X = I.P / L.\ U_o$$

$$Y = LU_1 / L.\ U_o$$

Where=$I.P$ is the increase in the area of improved pastures 1950—65.

$L.\ U_o$ is the number of livestock units carried in 1950.

$L.\ U_1$ is the number of livestock units carried in 1965.

Figure shows the regression line for the relationship, the equation for which is

$$YC = a + b\ X \pm e$$

Where a=1.1209.

$b=1.9076$

$e=$the residual.

Gravity Model

1. The gravity model adopted was given by the formula[37]

$g=\Sigma\frac{P}{d}$, where P is density of hops in any parish; d is distance away. When the value for g has been ascertained for the parishes in and around the centre of hop cultivation, the parish with highest value was then designated as the 'centre of gravity' of hop cultivation.

Application of a Multiple Regression Model

The following model was examained decennially for the years 1910-1960 using the Chicago data.[38]

$$V'_i=a+b_1\,C'_i+b_2\,R'_i+b_3\,M'_i+b_4\,E'_i+b_5\,P'_i+b_6\,N'_i+e$$

Where:

$V'_i=$Logarithm of front-foot land value at the ith place;

$C'_i=$Logarithm of distance (in hundreds of feet) of the ith place from the central Business District;

$R'_i=$Logarithm of distance (in hundreds of feet) of the ith place from the nearest regional shopping centre.

$M'_i=$Logarithm of distance (in hundreds of feet) of the ith place from Lake Michigan.

$E'_i=$Logarithm of distance (in hundreds of feet) of the ith place from the nearest elevated subway system;

$P'_i=$Logarithm of population density at the ith place (by census tract);

$N'_i=$the per cent non-white at the ith place (by census tract);

and $a, b, \ldots\ldots b_6$ are empirically derived constants, i=1, 2...... 484, and e is the error term.

The application of the above hypothesized model at each one of the time periods yields some interesting results.

Mathematical Determination of the (MTP) Minimum Travel Point[39]

I. (X_1,Y_1), (X_2, Y_2), (X_3, Y_3), and $(X_1\ Y)$ be the coordinates of A, B, C and P respectively, and let a, b and c be the weights to be moved along the lines AP, BP and CP respectively. The problem is to minimize the function

$$S = ar + bs + ct$$

With $r = [(x - x_1)^2 + (y - y_1)^2]^{\frac{1}{2}}$

$$S = [x - x_2)^2 + (y - y_2)^2]^{\frac{1}{2}}$$

$$t = [(x - x_3)^2 + (y - y_3)^2]^{\frac{1}{2}}$$

Thus S must be minimized as a function of X and Y by putting

$$\frac{\delta S}{\delta X} = 0$$

$$\frac{\delta S}{\delta Y} = 0$$

This yields

$$0 = a\frac{x - x_1}{r} + b\frac{x - x_2}{S} + c\frac{x - x_3}{t} \qquad \text{...(1)}$$

$$0 = a\frac{y - y_1}{r} + b\frac{y - y_2}{S} + c\frac{y - y_3}{t} \qquad \text{...(2)}$$

But, as may be seen from Fig. 13.3 $a\frac{x - x_1}{r}$,

$$a\frac{y - y_1}{r},\ b\frac{x - x_2}{S},\ b\frac{y - y_2}{S},\ c\frac{x - x_3}{t},\ c\frac{y - y_3}{t}$$

are, respectively the x and y components of three forces of the absolute strengths a, b and c which point towards P from A, B and C, respectively. If they are all three inserted so that they issue from P, the relation remains undisturbed. This proves the correctness of the mechanical procedure[40].

II. Equations (1) and (2) express that the three for vectors of lengths a, b, and c form a triangle. As shown in Fig. 13.3, the supplementary angles of this triangle are (α-β), (β-y) and ($2\pi - \alpha + y$), respectively. This leads to the geometrical construction of P.

III. (a) First we construct the triangle out of a, b and c, thus obtaining the angles (α-B), (β-y) and ($2\pi - \alpha + y$); respectively

(b) We notice from Figure 13.3 that $\measuredangle\ BPA = \alpha - \beta$. Similarly, of course, $\measuredangle CPB = \beta - y$, and $\measuredangle CPA = 2\pi - \alpha + y$.

(c) By a familiar theorem of elementary geometry the locus of all the points which subtend a fixed angle to two fixed points forms a circle. This may be easily constructed and P obtained as the point of intersection of these three circles.

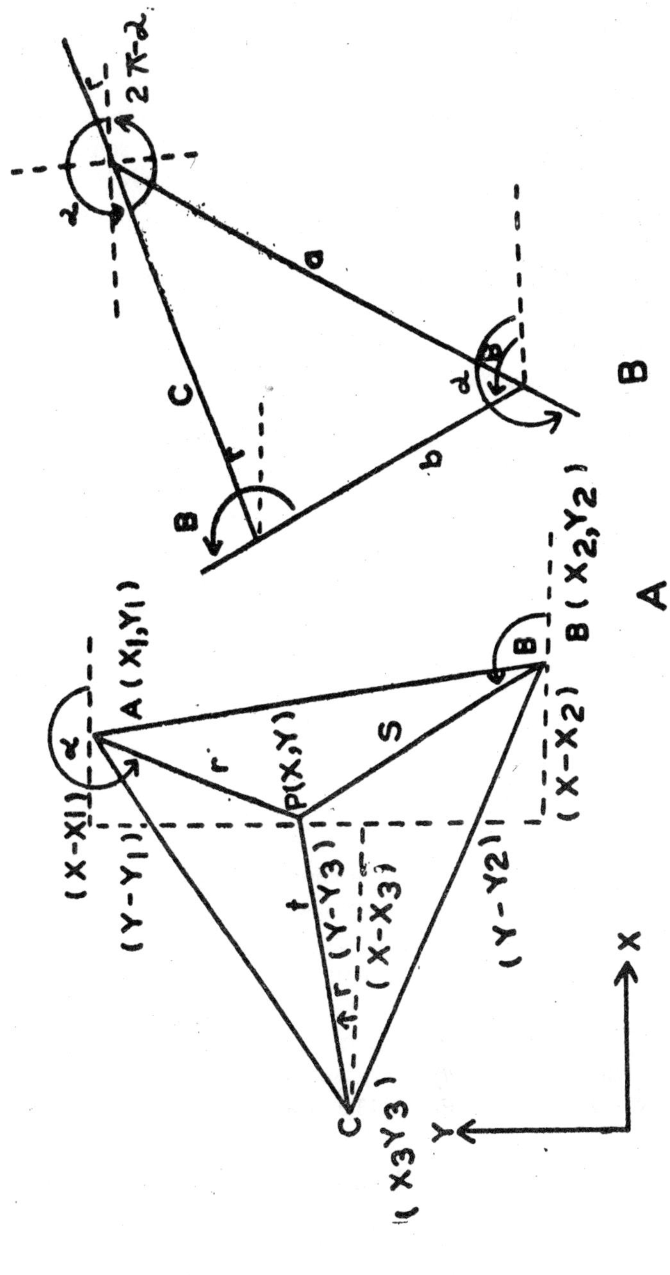

FIG. 13.3 : Mathematical Determination of the M.T.P.

Component Standardization

If we are concerned with mapping a complex phenomenon like the '*development level*' of a set of regions, we may have to use a number of measurements rather than any simple one. Thus Ginsburg (1961) in his *Atlas in Economic Development* uses some ninety separate indices in an attempt to prove the elusive concept of development. Any method of reducing such measurements to a single index must overcome the problem of comparability. Some measurements may range from 1 to 10, others may range from 0.98 to 1.02 and clearly some need to reduce these ranges to a single comparative level.

Ranking the regions in order from the highest to the lowest (rank 1, 2, 3,......n) on any given measure provides such a common yardstick; adding rank of regions on all measures provides an overall index. The problem with ranking, however, is that it, throws away so much of the information gained in the measurement process; 10.7, 6.2, 6.1, and 0.004 become simply 1, 2, 3, and 4 despite the great differences in the original measurements. To retain the relative differences of regions along the measurements, Berry (1961-B) has suggested the use of standard scores or values. Such values may be derived from the formula

$$w = \frac{(x_i - \bar{x})}{Sx}$$

Where x_i is the value for the ith region, $\bar{x}$ is the arithmetic mean, and Sx is the standard deviation of the measurement in question. Thus for any observation, the w-score is expressed as a deviation from the mean value in units of the standard deviation.

One recent example of the use of rank sums in building up a combined picture of regional variation has come from a study of the '*Economic Health*' of New York state by Thompson, Sufrin, Gould, and Buck (1962, P.S.) Nine separate indicators of the economic activity of the state were taken, including items like per capita income and per cent employment, and the fiftyeight countries were ranked in order on each of the nine indicators. For each county its rank on each of the indicators was noted, the nine ranks summed, and the resulting total re-ranked to give a combined index. As the authors point out, the method involves some curious additions; thus New York city has a high rank (4th) among the counties in economic levels, but a low rank (46th) in economic trends. Its overall position was given by this method as twentyeighth. They concluded, therefore, that although the method gave reasonably accurate pictures of different levels of development within the state, they could find no basic mathematical logic in the method and went on to replace it by a more reliable method, factor analysis.

Factors of Regionalisation

The first factor of Regionalisation (R) is built up by the following method taking into cansideration the above three factors.[41]

$$R = \sqrt{DW(W-A)}$$

Where D=Density of population

W=Proportion of all workers

$W-A$=Non-Agriculture in total population.

R=Regionalisation.

This factor of regionalisation is used for forming economic regions. The unit having higher value of R is treated as economically better. The results obtained for four such divisions of Hooghly district of West Bengal are as follows.

Name of Sub-division	$\sqrt{D}$ *(per km.)*	*W*	*W.A*	*R*
1. Sadar	21.59	0.331	0.091	0.6502
2. Chandernagore	26.41	0.291	0.088	0.6762
3. Serampore	28.42	0.290	0.146	1.2031
4. Arambagh	21.64	0.269	0.066	0.3843

Rural population is considered for finding the Value of R.

Mapping technique is a powerful tool for having a visual impression. The value of R can be used for mapping and regions can be formed by grouping the values of R. There are various methods of grouping. Here the use of geometric scale is made to classify the values of R. The basic classification is given below:

Group I 0^3 to 1^3 i.e. 0 to 1

Group II 1^3 to 2^3 i.e. 1 to 8

Group III 2^3 to 3^3 i.e. 8 to 27

etc.

From this, effective classification can be obtained by multiplying the class limits by suitable constant number. In this particular case the constant F is obtained by the following method.

$$F=\frac{\text{Maximum } R-\text{Minimum } R}{N^3}$$

Where N=Number of required classes.

The power in basic classification and the power of N is taken as 3 because R is obtained by multiplication of three factors; then

Group 1=Minimum $R+0^3 F$ to Minimum $R+1^3 F$
Group 2=Minimum $R+1^3 F$ to Minimum $R+2^3 F$
Group 3=Minimum $R+(N-1)^3$ to Minimum $R+N^3 F$
Let the value of R in a Table is
$R=2.9202$: Minimum $R=0.1215$
$N=8$ then $F=0.0055$ and 8 classes would be as follows:

Group 1 0.1215 to 0.1270
Group 2 0.1270 to 0.1652
Group 3 0.1652 to 0.2691
Group 4 0.2691 to 0.4713
Group 5 0.4713 to 0.8047
Group 6 0.8047 to 1.3021
Group 7 1.3021 to 1.9963
Group 8 1.9963 to 2.9202

It may be mentioned here that the above method of classification would be useful only when higher or lower values of R are not too high or too low. In that case separate the groups and then apply the above technique.

The Second Factor of Regionalization

In the first factor of regionalization only the population characteristics are considered. No economic factors such as salary and wages of workers, value of population, etc. are taken into consideration. In fact such type of data are not available for smaller areas. They are generally taken into consideration while building up state income-estimates. It is, therefore, felt that the association of State income figure with the factor of regionalization may give better result. So attempt is made towards this direction and second factor of regionalization R' is evolved as per following method:

$$R'_i = \frac{R_i \frac{1}{2} T}{\Sigma R_i \frac{1}{3} P_i}$$

Where R_i=First factor of regionalization for i^{th} unit. T= state income; P_i=Population of ith unit. As a test case the values of R and R' are computed for districts of Bihar.

District	Value of R	Value of R'	Per capita income 1955-56
1. Patna	1.0700	174	168
2. Gaya	0.5480	140	140
3. Shahabad	0.6487	147	177

Source : "Resource Development Regions and Divisions of India", 1965.

After formation of regions on the basis of either R or R' the same can be further delineated on the basis of predominant categories of workers. For determining the predominant types of workers here the lower limit method of crop combination regions have been used. The original method is explained here in brief. First lower limit is fixed by the following method.

$$\text{Lower Limit} = \frac{100}{3N}$$

Where N = Number of categories of workers in which at least 1 per cent workers are engaged. Then the categories in which the percentage of workers engaged is more than the lower limit are treated as predominant. So for example, if the distribution of 1000 workers into 9 categories is as follows:

400; 300; 50; 90; 10; 5; 2; 3; 140.

Then N=6; Lower limit = 5.6 per cent and predominant categories of workers would be 1, 2, 4, and 9.

The lower limit can also be determined by any of the following ways:

$$\text{Lower limit} = \frac{100}{2N} \text{ or } \frac{100}{N} \text{ or } \frac{\text{Percentage of 1st ranking category}}{N} \text{ etc.}$$

It may be mentioned here that it is not necessary to calculate percentages. The lower limit can be fixed for absolute figures by putting total number of workers in place of 100 or putting number of workers in the 1st ranking category instead of its percentage.

Inter-Regional Differences of Income

A world-wide review of inter-regional difference of income, a real tour de force, has been prepared by Williamson.[42] He develops an index of regional divergences of income, weighted according to the population of each region as follows:

$$V_w = \sqrt{\frac{(y_i - \bar{y})^2 f_i / n}{\bar{y}}}$$

Where f_i is the population of region i.
n is national population.
y_i is income per head in region i.
$\bar{y}$ is national average income per head.
V_w is weighted measure of inequality.

There is a strong general tendency for regional differences to be at their least in the advanced countries, and at their greatest in the poorest countries.

Eliminating the Bias of Scale

All the variables selected for regionalisation, measured on a continuous scale, may be converted to discrete variables. Each observation is assigned a value on a discrete scale on the basis of its relative position in the series. The method of ranking, popular among regional scientists for its simplicity and computational case, falls in the category. This scale transformation equalises length, variance and coefficient of variation of each variable.[43]

$$\text{Mean} = \frac{(n-1)}{2},$$

$$\text{Length} = [n\,(n-1)\,(2n-1)\,]^{\frac{1}{2}}$$

$$\text{Variance} = \frac{(n_2-1)}{12}$$

There is an additional loss of information as unit difference in ranks do not reflect an equal difference in the variable values. This loss, however will be less significant, greater the number of variables because of the Law of Large Numbers.

Weights to Maximise Sum of the Squares of Correlations

The application of factor analysis is hailed as a landmark in 'objective' regionalisation (Geer, 1976). This method enables to determine a vector known as the first principal component (linearly dependent on the variables) having the maximum sum of squared correlations with the constituent variables. The eigen vector (F) corresponding to the maximum eigen values of the correlation matrix (R) gives the required factor loadings (Anderson, 1958). The composite indicator of the ith geographical unit (y_i) can be obtained by linearly combining the standardised variable values X_{ij} ($j = 1, \ldots\ldots, m$) where n is the number of variables. The weight for the jth variable F_j will be the jth element in the vector F.

$$y_i = \sum_{j=1}^{m} x_{ij} F_j$$

Often the variables are divided into sub-groups in such a way that in a sub-group they have a high inter-correlation while cannonial correlation between pair of sub-groups is low on an average. The first principal component of different sub-groups may be treated as a new set of variables and the method of factor analysis may, again be applied to construct the final index. This method reduces the computational load and alleviates the necessity of combining two or more orthogonal components in order to explain a greater percentage of variation having doubtful theoretical interpretation.

REFERENCES

1. B.Y. Levin, *The Origin of the Earth and Planets*, Moscow (Year is not given) p. 15.
2. Gunar, Myrdal, *Asian Drama*, Penguin Books, Middlesex; 1968, Vol. III, p. 2177.
3. G., Myrdal, Ibid., p. 2166.
4. M.R.C. Coulson, "The Distribution of Population Age Structure in Kansas City", *Annals of the Association of American Geographers*, Vol. 58, No. 1, March 1968, pp. 155-176 (p. 172).

5. H.W. Richardran, "Emperical Aspects of Regional Growth in United States", *The Annals of Regional Science*. Vol. No. 2, 1974, pp. 8-23.
6. A.A Mints & T.I. Petryakova, "The Use of Territory as a geographical Problem", *Soviet Geography*, Vol. XV, No. 2, 1977, pp. 61-73.
7. N. Kadmon, "The Mapping Synthetic Statistical Values", *Abstract of Papers*, 21st IGC, 1968, Calcutta, p. 375.
8. J.W. Pawling, "An Arithmetic Method for Determining elevation Changes on Aerial Photographs", *The Professional Geographer*, Vol. XII, No. 5, 1960, pp. 10-12.
9. J.T. Andrews & K. Shimiza, "Three-Dimensional Vector Technique for Analysing Till Fabrics: Discussion and Fortran Program", *Geographical Bulletin*, Vol. 8, No. 2, 1966, pp. 155-156.
10. A.W. Gilg, "Diffusion of Fowl-Pest Disease", *Transactions af the British Council of Geographer*, Vol. 59, 1973, p. 91.
11. S. Dasgupta, "Some Measures of Generalization on Thematic Maps", *Geographical Review of India*, Vol. XXVI, No. 2, 1964, pp. 73-78.
12. J.K. Wright, "Some Measures of Distribution", *Annals of the Association of American Geographers*, Vol. XXVII, 1937, pp. 177-211.
13. S. Dasgupta, "Characteristics of Thematic Scales on Maps", 60th Session of the Indian Science Congress, *Section of Geology and Geography*, pp. 282-283.
14. G.R. McBoyle, & D. Steiner *Geografiska Annals*; Vol. 54A, No. 1, 1972, pp. 12-27.
15. W. Zenker, 1888, Dia Verteilung der Warm auf der Erdolerflache, Berlin, *Quoted from Geografiska Annaler*, Vol. 54A, 1972.
16. W. Gorczyuski, Geografiska Annaler, No. 2, 1920, pp. 324-337.
17. O.U. Ghansson, "Die Haupt Charakteristika des Jahrlichen Temperatur-gauges: Gorsands Beiter", *Geographysik*, Vol. 33, 1931, pp. 406-428.
18. V. Conrad, "Useful Formulas of Continentality and their Limits of Validity", *Trans A.G.U.*, Vol. 27, 1946, pp. 663-664.
19. N.N. Ivanov, "Belts of Continentality on the Globe,', *Izvestia Geographiia*, Vol. 91; pp. 410-423.
20. U.P. V. Haufler, *Regional Differentiation*, Sborink, Universita Karlova, Praha 1970, pp. 44-45.
21. J.A. Sheor, "A General Measure of Diversity", *The Professional Geographer*, Vol. XVII, No. 2, 1965, pp. 14-17.
22. J.A. Howard, "Stereoscopic Profiling of Land Units from Aerial Photographs", pp. 260-61.
23. D.C. Armond, "Point Scales in Geography", *Soviet Geography*, Vol. XIV, No. 8, 1973, pp. 491-505.
24. R.D. Hecock & J.F. Rooney, "Towards a Geography of Consumption", *The Professional Geographer*, Vol. XX, No. 5, 1968, p. 392.
25. E.W. Miller, "Changing Patterns in the Mineral Economy of the United States: 1939-54", *The Professional Geographer*, Vol. XIII, No. 3, 1961, p. 3.
26. J.J. Cirrito, "Radar Photogrammetry: A Cartographic Process", *The Professional Geographer*, Vol. XV, No. 1, 1963, p. 24.

27. L. Pernarowski, "Mathematical Method of Regionalization and its Application to Poland", Abstract of Papers, 21st I.G.C. Calcutta 1968, p. 134.

28. V.L.S. Prakasa Rao, "Macro-Urban Analysis Geographers Contribution", *Bombay Geographical Magazine*, Vol. 10, No. 1, 1962, p. 14.

29. M.N. Pal, "Quantitative Delimitation of Regions", *Bombay Geographical Magazine*, Vol. VIII, IX, No. 1, 1961, pp. 71-72.

30. A.A. Mints & V.S. Preobrazhenskiy, "The Function of Place and Its Changes", *Soviet Geography*, Vol. XIII, No. 5, pp. 267-282.

31. A.D. Sorensen, "A Method for Measuring the Spatial Association Between Point Patterns", *The Professional Geographer*, Vol. XXVI, No. 2, 1974, pp. 172-176.

32. E. Raisz, *Principles of Cartography*, McGraw Hill Book, Tokyo, 1962, p. 56.

33. E. Raisz, ibid., p. 151.

34. M.J. Cooper, "Container Shipping and Regional Economic Growth", *New Zealand Geographer*, Vol. 30, No, 1, 1974, pp. 55-56.

35. P. Toywe & T. Newby, *Techniques in Human Geography*, Macmillan, 1971, p. 110.

36. G.T. McDonald, "Recent Pasture Development on the North Tablelands of New South Wales", *The Australian Geographer*, Vol. X, No. 51, 1968, p. 389.

37. R.W. Smith & E.J. Taeffe, and L.J. King, *Readings in Economic Geography* Rand McNally and Company, Chicago, 1969, p. 80.

38. Ibid., p. 98.

39. Ibid., p. 156.

40. This solution was developed by Dr. Gustav Bergman of the Deptt. of Philosophy, State University of Iowa.

41. A.G. Thawab, "Delineation of Area into Economic Regions by Population Characteristics", *Geographical Review of India*, Vol. 36, No. 3, 1974, pp. 188-195.

42. Willamson, *Economic Development and Cultural Change*, Chicago, Part 2, 1965, p. 19.

43. Amitabh Kandu, "Construction of Indices for Regionalisation—An Enquiry into the Methods of Analysis", *Geographical Review of India*, Vol. 37, No. 1, 1975, pp. 21-29.

FOURTEEN

R. B. Mandal

LAND UTILISATION

THE CHIEF elements of land utilization are the availability of land for cultivation, pasture, forests, water bodies, etc. and the determinants are the intensity of rainfall, wind speed, hydrological and soil conditions, temperature and the amount of evaporation, etc. This is because, the human society is mostly related with the agricultural potentiality of land, its classification, capability, land consolidation and land tenure; hence the localization of crops, their intensity, combination, diversification, concentration and changing patterns of crop land use are the dominant aspects of land utilization. The measurement of agricultural efficiency, deficiency, dynamicity, nutrient indices, disease concentration and the problem of human employment in agricultural land use predominate the scene of quantitative methods in the geography of land utilization.

The use of statistical approach in most of the research papers necessitates the review of such statistical formulas which are useful for students and research scholars in bringing the land use studies up to the mark in comparison with the other branches of geography and to note the advancement of the subject in relation to statistics as a base of all science subjects.

Measures of Land Potentialities

In India the food production is not sufficient to meet the requirements of the growing population. It is a sorry state of affairs that the rural inhabitants may suffer from a deteriorating standard of food nutrition and they may be under nourished due to the occurrence of flood and famine at intervals. Hence the formulas derived by the various geographers[1] are considered here in short to measure the overall potentiality of the land.

1. Agricultural Density (Feranczi, 1938 and Trewartha, 1953) $= \dfrac{\text{Total Agricultural Production}}{\text{Total Cultivated Area}}$

2. Nutritional Density $= \dfrac{\text{Total Rural Population}}{\text{Total Cropped Area}}$

3. Economic Density (Singh, J., 1967) $= \dfrac{\text{Total Rural Population}}{\text{Total Net Sown Area}}$

4. Intensity of Population Pressure $= \dfrac{\text{Caloric Density}}{\text{Optimum carrying Capacity}} \times 100$

5. Caloric Density $= \dfrac{\text{Total Rural Population}}{\text{Total Area under Food Crops}}$

6. Man-Soil Density (Kuperus, 1938) or Physiological Density $= \dfrac{\text{Total Population}}{\text{Total Cultivated Area}}$

7. Weighted Average Caloric Production Available for Ingestion $= \dfrac{\text{Total Caloric Output available for Ingestion}}{\text{Total Area under Food Crops}}$

These formulas are also applicable especially in measuring the industrial potentialities and transportational potentialities of an area.

Techniques of Land Classification

Land use classification is the systematic arrangement of various classes of land on the basis of certain similar characteristics, mainly to identify and understand their fundamental utilities, intelligently and effectively in satisfying the needs of human society. Thus the classification of land must be clearly presented with the help of productivity index and the yield and quality of crops grown under physically defined systems of management or according to Stone index based upon soil profile, soil texture and other physical factors combined to control the use capabilities of particular soil and its productivity under favourable environmental conditions.

Unit Area Method

In this method parameters relating to physical and human environment are tabulated and numbered and the numbers for each unit area of land are expressed in a fractional code, while digits in the denominator represent physical conditions and digits in the numerator stand for human environment. The three parts of Roman numericals, short fraction and long fraction are as follows:[2]

$$V \frac{2}{3} \quad \frac{1G^{*}\,234}{2331123}$$

The Roman numericals from I to V show the severity or absence of problems in the area e.g. I—Problems insignificant; II—Problems not critical; III—moderately critical; IV—very critical; V—so critical that a radical change in land use is desirable. The short fraction is a summary of long fraction. In numerator (1-5) i.e. excellent, good, medium, poor and very poor, while seven digits of denominator relates to 1 for slope, 2 for drainage, 3 for erosion, 4 for stoniness, 5 for rock exposure, 6 for soil depth and 7 for soil fertility. Each of these seven digits is further classified into five categories except storiness and soil depth which have four categories each.

Storie Index or Inductive Method

This method is based on the ratings of the soil profile. Three factors are used i.e. general character of the profile including texture, surface texture and soil modifying conditions such as drainage, acidity and erosion. These three variables are expressed as percentage of the optimal conditions for plant growth are multiplied together to obtain the rating and the product is expressed as the percentage of the maximum. A soil may have excellent profile if it is related at 100 for the first factor, excellent texture if rated 100 for the second factor, but due to acidity and alkalinity it is rated as 40 for the third factor. The results of these three variables give a percentage rating of the soil of 40 per cent as despite other favourable conditions the alkalinity makes the soil unproductive. After addition the rating will be 80 per cent. Mathematically the results expressed is as follows :

$$\text{Case } I = \frac{100 \times 100 \times 40}{100 \times 100 \times 100} = 40 \text{ per cent}$$

$$\text{Case } II = \frac{100 + 100 + 40}{100 + 100 + 100} = \frac{240}{300} = \frac{8}{10} = 80 \text{ per cent}$$

In spite of undecided optimum conditions, the greater degree of comparability can be achieved through this method.

Deductive Method

This method has been taken from 'U.S. Soil Survey' productivity ratings which illustrate the deductive type of classification which are as follows :

Soil Tye	Corn	Wheat	Rye	Alfalfa	Sugar	Productivity Grade
	A—B	*A—B*	*A—B*	*A—B*	*A—B*	*A—B*
Miami Loam	70-90	70-100	80-100	70-90	60-70	3-1

Source : Jacks, G.U. Land classification for Land Use Planning", Imperial Bureau of Soil Science; Technical Communication No. 43, 1964, p. 69.

Under *A* and *B* in each column figures refer to the percentage of the standard yield of crop obtainable (*A*) with common practice of farming, while (*B*) the best current practices. The 'productivity grade' can be calculated by a simple percentage weighting of the crops local importance. The values 70(*A*) and 90(*B*) can be calculated as follows :

$$A = \frac{35 \times 100}{50} = 70$$

$$B = \frac{45 \times 100}{50} = 90$$

The values of other columns can be estimated in the same way. The productivity grade is an arbitrary procedure which indicate the general productivity of the soil for the commonly produced crops of the region.

Measures of Land Capability

The measures of land capability is "a field investigation of soil properties, slope, degree of soil erosion and changing land use patterns which form the basis for future planning of soil and water conservation".[3] Thus the term 'land capability' is broad and relative which shows that certain land can produce better crops with fewer conservation practices than others.

Method 1

According to V.R. Singh land capability classes should be determined with the help of physical and economic factors i.e. physiography, soil fertility, power input, land use type, agricultural intensity and crop superiority. In this method[4] "the first step is to obtain for each variable, data (*D*) as percentage of the total area of each village or other unit area (Table 14.1). The percentages are then ranked (*R*) in descending order and rank coefficients (*RC*) are obtained by dividing the total of the rank values (*TR*) by the number of *R* observations. Rank coefficients (*RC*) of all variables are summed to give a composite index (*CI*). On the basis of the composite indices (*CI*), land well suited for agriculture can be distinguished from that not so suited. These two characteristics can be grouped together and possible combinations identified"

Method 2

R.B. Mandal[5] while studying the land utilization of Bihar, measured the land capability index with the help of two sets of variables i.e; (a) positive, e.g. soil fertility, amount of rainfall necessary for crop raising and better irrigation facilities and (b) negative variables which reduce the potentialities of the land e.g. higher relief features, flood, drought, forest. rugged topography, waterlogging and soil erosion. After giving proper weighting to these variables capability index can be derived with the help of formula given below :

$$LCI = \frac{\left(\dfrac{\sum\limits_{i=1}^{n} R_1 + S_2 + I_3 + \dots\dots + Q_n}{\Sigma\, PFV} \times 100 \right)}{\left(\dfrac{\sum\limits_{i=1}^{n} R_{n_1} + S_{e2} + W_3 + \dots\dots + Q_n}{\Sigma\, NFV} \times 100 \right)}$$

Where *LCI* is land capability index; *PFV* is positive factor value; *NFV* is negative factor value; *R, S, I* are rainfall, soil fertility and

TABLE 14.1 : **Land Use Base**

No. of	Non-cultivable			Cultivable Waste			Arable Land			TR	RC
Village	D	%	R	D	%	R	D	%	R		
1	6.58	4.98	IV	23.75	16.02	III	69.66	21.79	III	10	3.33
2	4.62	3.50	V	10.98	17.40	VI	84.39	26.41	I	12	4.00
3	2.78	2.10	VI	15.65	10.55	IV	81.56	25.51	II	12	4.00
4	87.63	66.38	I	11.77	7.81	V	0.60	0.19	VI	12	4.00
5	18.95	14.35	II	60.42	40.94	I	20.62	6.45	V	8	2.66
6	11.51	8.69	III	25.65	17.30	II	62.82	19.65	IV	9	3.00
GT	132.07	100		148.22	100		319.65	100			

Source : V.R. Singh, "Land Capability : A Method of Classification", *The Professional Geographer*, 1970.

irrigation facility respectively as positive variables and R_n, S_e, W, F, D, H and T are ruggedness number, soil erosion, waterlogging, forest density, drought, higher degree of slope and flood respectively as limiting factors of land

TABLE 14.2 : **Weighing and Evaluation of Land Capability**

Variables	Weighing values			
	25	15	10	5
1. Rainfall	Above 150 cm.	149 to 130 cm.	129 to 119 cm.	Below 119 cm.
2. Soil Types	Younger Alluvium	Older Alluvium	Sandy	Stony waste
3. Irrigation	Canals. Tubewells.	Wells	Tank	Wells & Ditch.
4. Slope	Above 3°	2°--3°	1°--2°	Below 1°
5. Flood	High flood	Partly flooded	Only in rainy season.	No flood
6. Drought	Even Water is scarce	Medium drought	Low drought	No drought
7. Forest Density	Above 20%	15% to 20%	5% to 15%	Below 5%
8. Ruggedness Number*	Above 1.00	0.15 to 1.00	0.05 to 0.15	Below 0.05
9. Waterlogging	Whole year	6 months	3 months	No Waterlogging
10. Soil Erosion	High in Plateau region	Medium in scarp region	Low in upland plain	Very low in plain land

Source : Compiled by the author.

*Ruggedness number (R_n) has been calculated by using the formula :

$R_n = D_d \times \frac{RR}{5280}$, where D_d is drainage and RR is the relative relief.

TABLE 14.3 : **Summary of Land Capability Index in Relation to Net Sown Area in Bihar**

Regions/Districts	Positive Index	Negative Index	Final L.C. Index	Percentage of NSA to Total Area (1971)
1. Saran	66.66	34.28	1.94	73
2. East Champaran	45.66	37.20	1.25	53
3. West Champaran	53.33	54.28	0.98	45
4. Muzaffarpur	53.33	34.28	1.56	74
5. Darbhanga	59.99	34.28	1.56	57
6. Saharsa	73.33	51.42	1.42	40
7. Purnea	59.99	54.28	1.10	49
8. North Monghyr	59.99	40.00	1.49	50
9. South Monghyr	26.66	77.77	0.34	45
10. North Bhagalpur	73.33	42.45	1.70	64
11. South Bhagalpur	26.66	61.14	0.43	55
12. North Gaya	53.33	42.45	1.25	55
13. South Gaya	26.66	72.00	0.37	47
14. Bhojpur	66.66	34.28	1.44	70
15. Rohtas	46.66	77.77	0.60	45
16. Patna	80.00	54.28	1.47	78
17. Santal Parganas	20.00	88.57	0.22	40
18. Dhanbad	33.33	74.85	0.44	38
19. Hazaribagh	33.33	88.57	0.38	19
20. Palamau	20.00	88.57	0.22	10
21. Ranchi	33.33	88.57	0.38	38
22. Singhbhum	26.66	88.57	0.30	18

Note : NSA is Net Sown Area.

Source : Calculated by the author.

Table 14.3 shows the relationship between land capability index in relation to net sown area. A product moment coefficient of correlation has also been estimated which comes to $r = 0.67$. This alone signifies the relationship of the capability index and the net sown area available in the region. For these two components of land use the regression value comes to $Y=0.022$, $X=0.10$. In this way, it can be said that the intensity of net sown area is the real indicator of the potentialities of land capability in a region.

Method 3

Soil capability classes are ranked on the merit of land.[6] Class one land, has no significant limitation i.e. severity in the past. Thus it is supposed to be 100 per cent suitable for agricultural uses.

The method for calculating the indices in an area of 100 acres with class one land (a) measuring 40 acres and class two (b) 60 acres, is discussed here as an example. As the class one land measures 40 acres, the capability index per acre of this land is 100. Therefore, total capability value would be the sum of the total area multiplied by the capability index per acre of land. At the same time the class two land covers some 60 acres and capability index per acre of this land is 83.33. The total capability of this hypothetical area of 100 acres would be as shown below.

(a) $40 \times 100 = 4{,}000.00$

(b) $60 \times 83.33 = 4{,}999.80$

Therefore, the total capability Index = 8,999.80. Weighted Average capability has been worked out by dividing the total capability indices of different classes of land by their total area.

Therefore,

$$\text{Weighted Average Capability} = \frac{\text{Area} \times \text{Capability Value}}{\text{Total Area}}$$

$$\text{In short, } WAC = \frac{(A_a \times C_a) + (A_b \times C_b)}{TA_a + TA_b}$$

$$= \frac{(40 \times 100) + (60 \times 83.33)}{40 + 60}$$

$$= \frac{4000 + 4{,}999.80}{100} + \frac{8{,}999.80}{100}$$

and $WAC = 89.99$.

Such weighted average capability index of 89.99 indicates a capability states of above class two type of land. With the help of this method the average capability indices can be worked out for each region.

Index of Land Consolidation

The holding management is only advantageous when the farmstead is in the centre of a circular holding. This is expressed by the formula :

$$S = \frac{r}{2}$$

When S is index of land consolidation, r is radius from the centre of a circular holding.

The square shape of farm is equally advantageous. There are several formulas for computing the optimum shape of farm. One of them is Kosicki's formula

$$U = \frac{L_r}{L_i}$$

Where U is the index of the configuration of the holding, L_r is the actual average distance between farm land and farmstead, L_i the ideal distance calculated on a holding of the same area but in the shape of a circle or square. Moszczanski's formula

$$U = \frac{O.S}{P}$$

U is the index of the configuration of the holding, O the circumference of the farm, S average distance of the farmstead and P is the farm area. The average distance is calculated with the assistance of a square network.

The above indices may be supplemented with one more very essential index[7] depicting the degree of land consolidation.

$$K = \frac{\sum_{i=1}^{n} a_i}{\sum_{i=1}^{n} a_i} = \frac{a_1 + a_2 + a_3 \ldots + a_n}{a_1 + a_2 + a_3 \ldots + a_n}$$

Where $a_1, a_2, a_3, \ldots\ldots, a_n$ are symbols of various plots of different size.

If the index value is *1*, the holding is composed of a single plot. Indices of holding composed of several plots are included between *0* and *1*. In practice indices are multiplied by 100 to obtain percentages of consolidation. It is useful in showing changes which have occurred in a given period of time. It is a relative value independent of the size of holding. The index of consolidation may be supplemented with the dispersion of plots and it is calculated with the help of following formula:

$$R = 100 - 100K = 100(1-K)$$

Distribution of Dividends over Farm Land

A formula has been devised for justifiable distribution of dividends on the basis of participants contribution to the farm in the form of land, labour or capital.

The application of this formula will show that the income of each family or cultivator within the farm will be much more than his present income from the individualistic operation of land holding. The formula is as follows:[8]

$$D = \frac{T-E}{M \times 100}(L + I + C)$$

Where D = Dividend to be obtained by each participant.
T = Total earning of the farm.
E = Total Expenditure of the farm on various heads.
M = Total number of member participants in the farm.
L = P.C. dividend of each participant for his land.
I = P.C. dividend of each participant for his labour.
C = P.C. dividend of each participant for his capital share in the farm.

If each participant would have possessed equal amount of land in the farm, offered labour of equal value and contributed equal share of capital, then it would have been easier to find out the dividend per head to be distributed equally. There will be variance in respect of the shares of each participants land, labour and capital. For the fact, the above formula can be modified in the following way for the maintenance of justice from all sides in the distribution of dividend.

$$D = \frac{T - E}{M \times 100}\left(\frac{L \times L_s}{L_{avg}} + \frac{I \times I_s}{L_{avg}} + \frac{C \times C_s}{C_{avg}}\right)$$

OR

$$D = \left(\frac{I \times Is + I \times Is + C \times Cs}{Is + Is + Cs}\right)\frac{T - E}{M \times 100}$$

Where L_s = Amount of land of a participant member.

L_{avg} = Average amount of land of the participants in the farm.

I_s = Amount of labour offered by a participant in terms of time.

I_{avg} = Average amount of labour offered by the participants in terms of time.

C_s = Amount of Capital share of a participant.

C_{avg} = Average amount of capital share of the participants.

According to Das, M. N., "In Galia village of Kamrup district in Assam the total land is 850 bighas. In case of a joint farming the total income from this farm would have been about Rs. 5 lakhs. If Rs. 1 lakh is spent in various activities of the farm including the activities of social welfare of the villages, then Rs. 4 lakhs will remain to be distributed among the participants of the farm. If all the participants offer equal amount of labour in all the working days of the year if the average land per head in the farm is 17 bighas and if the average amount of capital share per head is Rs. 800 then the annual dividend of the above farmer may be calculated as follows :

$$D = \text{Rs. } \frac{500{,}000 - 100{,}000}{50 \times 100}$$

$$= \text{Rs. } \left(\frac{30 \times 15}{17} + \frac{50 \times 365}{365} + \frac{20 \times 500}{800}\right)$$

$$= Rs.\ 7117.00$$

Therefore, the above farmer will get annual income of Rs. (7,117.00 – 2,000.00) = Rs. 5000.00 more than what he would get from the same amount of land, labour and capital, if he would operate his land holding individually".

Changing Patterns of Crop Land Use

The patterns of crop land use in each of the sample years are studied in two ways:[9] by ranking of crops according to their intensity and by deriving crop associations on the basis of Weaver's formula.

$\sqrt{\Sigma \frac{d^2}{n}}$ as applied in its modified form by Doi (Σd^2).

A comparison of these two aspects in different years gives a true assessment of the evolution of land use patterns in a region.

Agricultural Dynamicity

(1) The calculation of Index values of total volume of change by the following equation: [10]

$$I_c = \frac{V}{A_n} \times 100$$

Where I_c is the Index of total volume of change, V is the value of total volume of change and A_n is the proportion of acreage under crops to gross cropped area.

(2) The compilation of index of dynamicity on the basis of the average index values of total volume of change for each pargana is calculated by the following equation.

$$ID = \frac{AVU}{AVR} \times 100$$

Where ID is the index of dynamicity,
AVU is the average index value of the unit, and
AVR is the average index value of the Region.

Location Quotient for Agricultural Employment

The location quotient measures the degree to which a specific region contains more or less than its share of employment in any particular industry briefly indicating a ratio of ratios. To calculate the ratio of employment in agriculture to total population the formula is :[11]

$$\frac{\text{Employment in agriculture}}{\text{Total Population}} = \frac{7,382,009}{20,155,600} = 36.6\%$$

The ratios of these two variables of each district have been calculated and then by dividing these district ratios with the State percentage of 36.6 the location quotient for each district was found out.

Agricultural Deficiency Regions

To measure the agricultural deficiency regions of India, output per unit area has been taken as the basic tools for consideration. In this particular case per hectare yield of twentyfive crops of the year 1963-64 has been taken into account. All the twentyfive crops have been grouped into cereals, pulses, oilseeds and money crops. While per hactare yields of pulses, cereals and oilseeds have been computed in tonnes, for money crops all the yields have been converted into values (Rs.) as per the ruling price for that particular year.

Mean figures for India as a whole have been found out for cereals, pulses, oilseeds and money crops. Thus standard deviation formula has been applied for cereals, pulses, oilseeds and money crops. After that the standard score for each district of India has been computed by adopting the well known formula.[12]

Standard Value $Zic = \frac{c_i - c}{\sigma c}$ for cereals

$$Zip = \frac{p_i - p}{\sigma p} \text{ for pulses.}$$

$$Zio = \frac{o_i - o}{\sigma o} \text{ for oil seeds.}$$

$$Zim = \frac{M_i - M}{\sigma M} \text{ for money crops,}$$

Where for example $\sigma\, C$ is equal to $\sqrt{\frac{\Sigma(C_i - \bar{C})^2}{n}}$

The standard for Zic, Zip, Zio and Zim are calculated. To measure the volume of production Zi values have been multiplied by the acreage figure i.e. $Zic \times AC$, $Zip \times Ap$, $Zio \times AO$ and $Zim \times ZM$.

These values are arrayed in descending order for plus values and ascending order for minus values, separately. Quartiles are found for all the figures with plus and minus values which are above and below the country's average respectively. For this below (—) values have been taken into account.

At the end to determine the total aggregate efficiency of the component areal unit the following formula has been adopted in case of all districts to find out the weighted average of Z'S.

$$Z = \frac{(Zic \times A_c) + (Zip \times A_p) + (Zio \times A_o) + (Zim \times AM)}{AC + AP + AO + AM}$$

All the Z minus values indicate the below country's average while + are equal to the average efficiency of India.

All the values are mapped and deficiency regions so far as cereals, pulses, oilseeds and money crops and also aggregate deficiency are mapped.

Efficiency Measurement

The following statistic is suggested as a way of making a comparative evaluation of the efficiency with which various counties utilize their total food resource environments.[13]

$$\frac{A'}{A}\ (y_1\, n_1\, a_1 + y_2\, n_2\, a_2 + \ldots\ldots y_n\, h_n\, a_n),$$

Where: A is the total cultivated area,
A' is the adjusted food producing area
y is yield relative for each food.
n is the nutrition relative for each food and a is the area relative for each food.

The ratio A'/A indicates the change in a county's cultivated area through the use of non-cultivated land, non-land food resources, food imports or exports, etc. The quantity inside the parentheses combines the diverse components that affect food availability.

The computations indicate that Japan uses its food resource environment ten times more efficiently than the United States. The statistic may also be adopted for use as a measure of population pressure and economizing inputs into agriculture for food production.

Index of Disease Concentration

For each disease a concentration index can be calculated for any areal unit and the formula is as follows.[14]

$$C = \frac{d}{S} \times 100$$

Where C = index of concentration for disease X in a district,

d = percentage of persons suffering from disease X in a district to the total number of disease sufferers from the district, and

S = percentage of persons suffering from disease X in all the districts to the total number of persons suffering from all diseases of the State.

If the index value derived from the above formula is greater than 100 the district has a share greater than it would have. If the distribution were uniform in the entire State, it points to the fact that the district has a concentration of the particular disease. All the index values that show concentration are then arranged in a descending order and divided into five equal parts to show very high, high, medium, low and very low degree of concentration.

Localization of Crops

The locational coefficient show the measure of concentration of any entity and has been used by industrial economists to demarcate areas of industrial concentrations and has also been used to indicate localisation of crops. Formula is as[15]

$$LC = \frac{P}{A}$$

Where P is the percentage of a particular crop of the district in the total crop production of State; and A is the percentage of area of the district to the total area of the State. If the locational coefficient for a district is (1) one means that the district is having just the production of a particular crop which it should have.

Crop Intensity

The procedure adopted for arriving at a intensity of cropping per year is discussed below:

Let A_{ijk} and C_{ijk} be respectively the cropped areas and the area left fallow in ith year, jth holding and kth season denoted by p_{ijk} was obtained as[16]

$$P_{ijk} = \frac{A_{ijk}}{A_{ijk}+C_{ijk}} \quad ...(1)$$

and crop intensity in the ith year and jth holding by

$$P_{ij} = \sum_{K=1}^{2} \frac{A_{ijk}}{A_{ijk}+C_{ijk}} \quad ... (2)$$

Intensity of cropping in jth holding for each period was worked out by averaging out the corresponding pij's over the year to which the period pertained. This method takes area under the perennial crops as double cropped. The numerations of expression (2) equals gross cropped area in the ith year, jth holding and the denomination equals net area sown plus cultivable area left fallow in both the crop, seasons. Applying this method, intensity of cropping in different holding size classes was obtained.

Mechanization in the Employment of Human Labour

The following models were used to study the various stages of mechanization in the employment of human labour

$$L_{ij} = D_i + e_{ij} \quad ... (1)$$

$$L_{ij} = D_i + Bx_{ij} + e_{ij} \quad ... (2)$$

Where L_{ij} represents the employment of human labour (family/casual/hired/total) per hectare on the jth sample form of the ith farm category and X_{ij} is the farm size. And D_i is the mean,

i = 0 for bullock farms.
= 1 for bullock + tubewell/oil engine
= 2 for bullock + tubewell and other farm machinery
= 3 for tractor + tubewell and other farm machinery

Both the models were used. However, model (2) which eliminated the linear effects of farm size did not show any improvement in the result, as the correlation coefficient between farm size and the employment human labour per hectare turned out to be non-significant. This indicates that by transforming the variables to per hectare basis, the farm size effect is reduced to a large extent. Therefore, the results presented in the discussion refer to analysis of variance only.

In the above models, the differences $D_i—D_o$ (i = 1, 2, 3,) and

D_i—D_k (*ijk*, *i*, *k* = 0, 1, 2, 3,) represent the effects of mechanization stages on the employment. These differences were also subjected to test of significance by calculating the critical differences.

Percentage change in the employment of human labour was worked out as $\frac{D_i - D_o}{D_o} \times 100$, Share of the *j*th farm operations to the total change in employment was calculated as[17]

$$\frac{D_{ij} - D_{io}}{D_i - D_o} \times 100$$

Where i = 1, 2, 3, (represent per hectare).
j = 1, 2, 3, (farm operations).

Impact of Hydrology on Land Utilization

An elaborate discussion on correlation of different hydrological character, and land use items of sub-divisions have been after a study of the precipitation, drainage, floods and droughts, river sub-basins and ground water properties and extraction in the region. Each sub-division has a pattern ≠ difference or a difference in intensity of mòno pattern form which has been determined by a deviation from a standard set for the region. The standard is obtained by applying the formula[18]

$$S = \frac{(a_1 + a_2 + a_3 + \ldots) \times 100}{t}$$

Where S represents the standard, a_1, a_2, a_3 ... areas under different items at t the total area of the region.

Measurement of Spatial Discontinuity—An Index in Land Tenure

The basic index outlined here may be referred to as the relative index of land. Percentization, P_i and is given by the equation.[19]

$$P_i = \frac{\frac{1}{S} D_t}{100}$$

Here D_t is the aggregate distance between land fragments: the actual total distance covered by the operator in a single round that takes him to all his parcels, and S represents the hectarage of each parcel. Instead of total land area, average parcel size is used for each land fragment is in effect a separate firm or establishment especially where large distances are involved (where there is no fragmentation P_i=0).

Index of Water Requirements to Cultivated Area

Potential evapotranspiration as a basis can be used for reckoning the need of water in cubic metre per square kilometre of the cultivated area, after taking into account the actual rainfall. It has been done as:[20].

Water requirements per sq. kilometre of the cultivated area

$$= \frac{\text{Potential Evapotranspiration} - \text{rainfall}}{100} \times 10^6 \text{ cubic metres.}$$

Where 100 is used for the conversion of cms into metres, 10^6 to change square metres to square kilometres.

Nutrient·Indices

This has been calculated for different soil nutrients i.e. phosphorus, potash, organic carbon based on the percentage of soil samples tested, which fall in the low, medium and high categories for a particular nutrient. The calculation is based on the following formula:

$$\frac{\text{Low} + 2 \times \text{Medium} + 3 \times \text{High}}{100}.$$

The obtained figure is compared with the following three given indices for low, medium and high which will coincide in any one of the indices.[21]

Index of Crop Diversification

In a study of the regional pattern of agricultural land use in India, Bhatia has evolved a formula for determining crop diversification.[22]

$$\text{Index of Crop Diversification} = \frac{\text{Percent of cultivated Area under } X \text{ crop}}{\text{Number of } X \text{ crops}}$$

Where X crops are those crops which individually occupy ten per cent or more of cultivated area in a district. The lower the value of Index, the higher is degree of crop diversification. The index value, however, cannot be less than ten.

A Method for Measuring and Comparing Agricultural Land Use

It has been developed by P.M. Enggass and tested the method in Las Torres of Seville Province.[23]

TABLE 14.4 : **Determination of the Index of Cultivated Area**

Crops	Sown Area in Hectares
Green Beans	84.00
Early Potatoes	84.00
Sugar Beats	67.20
Wheat	78.40
Barley	67.20
Alfalfa	5.60
Olive	9.30
Hurto	44.80
Sown Area First Crop	440.50
Maize	89.75
Sweet Potato	44.80
Late Potato	15.82
Green Alfalfa	5.60
Hurto	44.80
Sown Area Second Crop	200.27
Total Sown Both crops	640.77=A
Total Area of Project	512.13
Less than Productive land	41.61
Total Cultivable Area	470.52=B

The reforms project of the 'Institute National Decolonization factor=1.24......(unknown)

$$\text{Index}=\frac{A\times 100}{B\times 1.34}=\frac{640.77\times 100}{470.52\times 1.34}$$

$$=\frac{64077}{630.497}=101.63$$

TABLE 14.5 **Determination of the Index of Yield**

Crops	Production (00's Kg/ha		Theoretically Sown area	Actually Sown
	A	B	C	D
	Total obtained	Previous Project Goal		
Wheat	25	20	1.25	1
Barley	20	19	1.05	1
Maize	58	24	2.42	2
Beans	10	15	0.67	1
Early Potato	150	117	1.28	1
Clover	400	333	1.20	1
Cotton	18	14	1.24	1
Sugar Beats	380	232	1.64	1
Total	1061	774	10.80	9

$$\text{Index of Yield } (I_y) = \frac{C \times 100}{D}$$

$$= \frac{1080}{9} = 120$$

Measurement of Crop Concentration

In order to determine the regional concentration of crops in Uttar Pradesh, an index, the location quotient has been calculated:[24]

$$\text{Index for determining concentration of crops} = \frac{\text{Area of crop } X \text{ in an areal unit}}{\text{Area of all crops in the areal unit}} \div \frac{\text{Area of crop } X \text{ in the entire region}}{\text{Area of all crops in the entire region}}$$

Public Agriculture and the Chemistry of Food Production

Public agriculture, as developed by S.S. Bose is a new discipline in Human Geography which highlights certain possibilities and impossibilities in the matter of food production. The circulation of each of the thirteen fertilizer elements, *N, P, K, Ca, Mg, S* .... in nature are governed by the equation:[25]

$(K_1 - K_2)P = G + A$

Where $K_1 = K_2$ is a constant depending on the food habits, conservancy practices etc. of the community.

P=the population.

G=Soils annual quota of fertilizer elements from natural sources, and A the quantity of artificial fertilizers.

If $G+A$ falls short of $(K_1 - K_2)P$, the use of high yielding varieties cannot, in view of the operation of the law of conservation of Mass Element-wise, Lielcig's law of the Minimum and the law of Diminishing Returns, helps in producing the food required by the country.

Methods of Crop Combination

In order to have a more intensive structure of agricultural production the study of crop combination is necessary.

Method 1 : Weaver[26], the pioneer of quantitative idea on crop combination has considered a decreasing order arrangement of the percentage of different crops such as :

$A=50$ and $B=50$

In the same way $A=33$, $B=33$, $C=33$ and $A=52$, B=28, $C=12$, $D=8$.

In these arrangements the combinations are AB, ABC as unambiguous ones but the last combination is $ABCD$ and as the decreasing order percentage; it is an ambiguous combination.

In this way, Weaver's standard theoretical distributions are as follows :

100.0% of GCA for monoculture i.e. Single combination
50.0% of GCA for each crop i.e. two crop combination
33.3% of GCA for each crop i.e. three crop combination
25.0% of GCA for each crop i.e. four crop combination

Where GCA is gross cropped Area. The deviations (d) plus or minus of the actual percentage from any one theoretical values for every set of crops are squared and summed. Each sum is divided by the number of crops in the set. The square root of this result will give us a measure of crop combination from the theoretical distribution. The formula for Standard Deviation is as follows :

$$\sigma = \sqrt{\frac{\Sigma(d^2)}{n}}$$

Where σ = Standard Deviation.
d^2 = deviation from mean square.
n = number of crops.
Σ = Summation value of d^2 in an areal unit.

Weaver considered all the crops having at least one per cent acreage of the total cropped land for the analysis.

The main objections against this method is as follows :

(a) The time involved for all these calculations is too much.
(b) The crop combination cannot be determined till all the nth crops are considered.
(c) Sometimes quite a large number of crops are included in the crop combination.

Method 2

In 1957 Doi has presented his quantitative method of crop combination through his paper entitled *Industrial Structure of Japanese Perfecture*[27]. His modified formula is that of Weaver is as follows:

$$\Sigma d^2$$

The combination of crops is as follows :

1. Monoculture = $(600—35)^2$ = 4225.00
2. Crop combination = $(50—35)^2 + (50—24)^2$ = 901.00
3. Combination= $(33.33—35)^2 + (33.33—24)^2 + (33.33—15)^2$ = 423.70 and so on.

In the above calculation, as the variance for 3-crop combination is (423.70) smallest, the region is designated with a combination of three crops, i.e., wheat, Barley and Paddy.

In deriving crop combination for any unit on Doi's method, it was found that the results are more realistic in comparison with those worked out by any other methods. The calculation are also very simple.

Method 3

In 1965 Rafiullah introduced a new deviation formula which is quite suitable for delineating the primary crop combination. The formula is as follows[28] :

$$d = \sqrt{\frac{\Sigma D^2P - \Sigma D^2n}{N^2}}$$

Where d is the deviation,

DP is the Positive difference

D_n is the negative difference from the mean value of the theoretical curve value of the combination.

N is the number of crops in the combination.

As only relative rank of the deviation is needed the root sign may be ignored to save calculations and the formula may be used as :

$$d = \frac{\Sigma D^2P - \Sigma D^2n}{N^2}$$

Just like Weaver's method, the difference of actual values is calculated from the middle value of the theoretical standard and the maximum positive deviation determines the number of combinations.

Method 4 : Maximum Distance Method

Deviating from Weaver's idea of crop combination N.P. Ayyar has tried to simplify such investigation and his grouping method is as follows :

All cereals (7) .... *C*
All Pulses (3) .... *P*
All Oil Seeds (6) .... *O*
All Other Food crops .... *F*
All other non-food crops................ *N*

The figures in brackets show the number of crops in each group while *C, P, O, F, N* indicate the symbol for each group. Thus the standard values may be obtained which are as follows :

C 100.00%
C—P .... 70.0%—30.0%
C—P—*O* .... 43.7%—18.8%—37.5%
C—P—*O*—*F*.................. 31.8%—14.3%—28.6%—23.8%
C—P—*O*—*F*—*N* .... 26.9%—11.6%—23.1%—19.2%—19.2%.

The deviations of the actual percentages from the above values would be found out, squared and summation (Σ) will give σ^2 (standard deviation) as required by the crop group combination. Now the method of difference, method of summation and lower-limit method have been used here in words of A.G. Athawals.[29]

Method 5 as Method of difference

In case of greater difference between first and second ranking crop than second crop is not important and this will be a case of mono-culture. If this difference is little and the difference between second and third ranking crops is much, then the third crop is not important and it is a case of biculture and so on. The solution of obtaining high or low values are as follows :

(a) With higher difference between the percentage of 1st and 2nd ranking crops i.e. $\frac{100}{1} - \frac{100}{2} = 50$. In this case first ranking crop is important and the combination would be mono-culture.

(b) In relation to theoretical difference if the difference between second and third ranking crop is higher i.e. $\frac{100}{2} - \frac{100}{3} = 16.67$ and this is the case of bi-culture.

(c) When the difference between the percentages of nth and $(n+1)$ the ranking crops is more than the difference between the theoretical values i.e. $\frac{100}{n} - \frac{100}{n+1} = \frac{100}{n(n+1)}$ than the crop combination would consist of first n ranking crops.

By inserting different values to n in $\frac{100}{n(n+1)}$ a table has been prepared. Then by taking actual differences and referring those to table values the crop-combination can be determined immediately.

n	1	2	3	4	5	6	7	8	9
$\frac{100}{n(n+1)}$	50.00	16.67	8 33	5.00	3.33	2.33	1.79	1.39	1.11

Example 2 : Let values of X_1, X_2 and X_3 be 70, 25 and 5 respectively. The difference between 1st and 2nd ranking crop is 45 which is less than the corresponding table value for $n=1$. The second difference is 20 which is more than table value for $n=2$. Hence this is a case of bi-culture crop combination.

Example 3 : Let value of X_1, X_2, X_3, X_4 and X_5 be 54, 24, 13, 5 and 2 respectively. The various difference are 30, 11, 8,,3 and 2 respectively none of which is more than the corresponding table value. When such interminate results are obtaining, it is suggested that the case may be treated as all crop combination.

Method 6 : Method of Summation

In this method instead of observing the differences, progressive sums are taken. The actual procedure is explained below :

(a) If the percentage of first ranking crop is more than the average of theoretical value for 1st and 2nd crops i.e. more than $\frac{1}{2}\left(\frac{100}{1}+\frac{100}{2}\right)=75$ then it is a case of monoculture.

(b) If the sum of percentage of first-two ranking crops is more than twice the average of theoretical values for 2nd and 3rd ranking crops i.e. more $2\times\frac{1}{2}\left(\frac{100}{2}+\frac{100}{3}\right)=83.33$, thus it is a case of biculture.

(c) If the sum of percentages of first n ranking crops is more than n times the average of theoretical values for nth and $(n+1)$ the crops i.e. more than $n\times\frac{1}{2}\left(\frac{100}{n}+\frac{100}{n+1}\right)=\left(\frac{2n+1}{n+2}\right)100$ thus it is a case of n crop combination.

By preparing table for $\left(\frac{2n+1}{2n+2}\right)100$ by giving different values to n it is possible to determine the crop combination just by taking progressive sums and referring to table values.

TABLE 14.6 : **Determine the Crop Combination**

n	1	2	3	4	5	6	7	8	9
$\frac{(2n+1)}{(2n+2)}\times 100$	75	83.33	87.50	90	91.67	92.86	93.75	94.44	95

Example 4 : Let values of X_1, X_2 and X_3 be 70, 25 and 5 respectively First cumulative sums is 70 which is less than table value for $n=1$ and hence crop combination is not monoculture. The second cumulative sum (70 + 25) is more than the table value for $n=2$ so it is a case of biculture crop combination.

Method 7 : Lower Limit Method

The above method of taking difference or progressive sums is discontinued after getting the result. If the value of X_1, X_2, X_3 and X_4 be 76, 12, 10 and 2 respectively then by method 1 and 2 the result is monoculture crop combination even though the percentage acreage of second and 3rd ranking crops are quite high. Of course, by any method such marginal cases are bound to occur. In lower limit method an attempt is made to associate the value of N to determine the crop combination. This is achieved in the following way which gives the lower limit for percentage values.

$$P = \frac{100}{3N}$$

The crops having percentage acreage more than P will be included in the crop combination. By giving different values to N a table of P values can be had :

TABLE 14.7 : **Different Values to N**

N	1	2	3	4	5	6	7	8	9
P	33.33	16.67	11.11	9.33	6.67	5.56	4.76	4.17	3.70

Example 5 : Let X_1, X_2, X_3 and X_4 be 76, 12, 10 and 2 respectively. The value of P for $N = 4$ is 8.33.

So all the crops having percentage acreage more than 8.33 will be included in the crop combination. There are 3 crops above this limit. So it is a three crop combination.

As we have seen above the result is monoculture crop combination by method 1 and 2. It is interesting to note that by Weaver's method also it comes to monoculture. So the result obtained by method 3 is more appealing. Also the saving of time is considerable. In addition to the advantage gained in time factor

TABLE 14.8 : **Place of Crop Combination**

District	1	2	3	4	5	6	7	8	9	10	11	12	Result			
													Method	Method	Method 3	
													1	2	$\frac{100}{3n}$	$\frac{100}{2n}$
District	22.5	18.5	15.9	10.9	7.4	7.2	6.9	4.2	2.6	1.6	1.1	1.1	7	9	8	8
Chittoor	31.1	24.9	12.0	10.1	6.7	5.4	4.2	3.7	1.4	—	—	—	8	7	8	5
Cuddapah	20.2	19.9	16.2	14.0	7.1	6.4	6.1	4.1	3.9	1.2	—	—	4	9	9	7
East Godavari	60.5	8.4	8.0	4.5	4.2	4.1	3.8	1.6	1.6	1.3	—	—	1	8	7	3
Guntur	22.5	19.6	14.7	9.6	8.7	7.9	5.3	5.2	4.3	1.0	—	—	6	9	9	8

one of the greatest advantage of this method is its flexibility. If broad regions are required then instead of $100/3N$ one can take $100/2N$ or $100/N$, or any multiple of N in the denominator to decide the lower limit.

In all the methods discussed above the percentages are first to be calculated. The part of calculations also take a lot of time and it can be saved by modification in Method 3 which is as follows :

$$A = \frac{G}{3N}$$

Where G = Gross cropped area

N = number of crops having more than or equal to $G/100$ i.e. 1%.

A is the lower limit in terms of acreage. Then all crops having acreage more than A will find place in crop-combination. An example is as follows :

When $G = 1260$ and the crop acreages be 800, 200, 160, 50, 25, 10, 10, 10 and 5 than $G/100 = 12.60$.

Hence $N = 5$

and lower limit $A = G/3N = 70$.

There are three crops above 70 and hence combination would be triculture.

This idea of lower-limit method is best due to its simplicity and flexibility, but the diversification of crops may arise mapping problem.[30]

Method 8

Betal Himansuranjan, while studying the problems of agricultural growth in India sets out Weaver's method of crop combination after taking mean and standard deviation of different crops for the States of India. The calculation for this method is given here for rice crop in Bihar State.

Then the calculation of Standard Deviation is as follows :

$$\sigma^2 = \sqrt{\frac{\Sigma (d^2)}{n}}$$

$$= \sqrt{\frac{4553}{17}}$$

$$= \sqrt{268}$$

$$\sigma^2 = 16.37$$

TABLE 14.9 : **Rice Crop in Bihar**

Districts	% acreage under rice crop	Deviation from Mean	Deviation Squared
1. Patna	39	—11	121
2. Gaya	53	+3	9
3. Shahabad	49	—1	1
4. Saran	30	—20	400
5. Champaran	50	000	000
6. Muzaffarpur	42	—8	16
7. Darbhanga	48	—2	4
8. Monghyr	21	—29	841
9. Bhagalpur	40	—10	100
10. Saharsa	44	—6	36
11. Purnea	62	+12	144
12. Santal Parganas	69	+19	361
13. Hazaribagh	56	+6	36
14. Ranchi	62	+12	144
15. Palamau	30	—20	400
16. Dhanbad	78	+28	784
17. Singhbhum	84	+34	1156
Total (17)	857		4553

(Mean) X = 50%

Then the range is as follows :

Rank III — Mean + 1 = 50 + 16.37 = 66.37
Rank II — Mean + 2 = 50 + 66.37 = 82.74
Rank I — Mean + 3 = 50 + 82.74 = 99.14

Then for different districts the following ranks for the rice crop are determined.

Singhbhum	—	I
Dhanbad	—	II
Santal Parganas	—	II
Gaya	—	III
Champaran	—	III
Purnea	—	III
Hazaribagh	—	III
Ranchi	—	III

In this way, the worksheet for the major crops are computed to facilitate the crop combination. The results are in Table 14.10.

TABLE 14.10 : **Facilitate the Crop Combination**

Sl. No.	Major Crops	Symbols	Mean (X)	Standard Deviation σ^2	Ranks of Distinctiveness. $\frac{X+3}{1}$	$\frac{X+2}{2}$	$\frac{X+2}{3}$
1.	Rice	R	50	16.37	99.14—82.74	82.74—66.37	66.37—50.00
2.	Sugarcane	S	1.38	1.53	5.97— 4.44	4.44— 2.91	2.91— 1.38
3.	Marua	Ma	1.86	2.06	8.04— 5.98	5.98— 3.92	3.92— 1.86
4.	Wheat	W	10.38	7.66	33.36—25.70	25.70—18.04	18.04—10.38
5.	Arhar	A	1.52	1.10	4.82— 3.72	3.72— 2.62	2.62— 1.52
6.	Maize	M	10.23	6.87	30.84—23.97	23.97—17.10	17.10—10.23
7.	Gram	G	2.32	1.70	7.42— 5.72	5.72— 4.02	4.02— 2.32
8.	Barley	B	2.24	1.90	7.94— 6.04	6.04— 4.14	4.14— 2.24

Source : Calculated by the author from Season and Crop Report, 1970-71

On the basis of the scale of distinctiveness the crop combinations can be computed for the different districts of Bihar State.

TABLE 14.11 : **Distinctiveness of the Crop Combination, Bihar**

Districts	Crop Combination		
	1st Rank	2nd Rank	3rd Rank
1. Patna	...	...	R, S, W, G
2. Gaya	...	...	R, S, W, G
3. Shahabad	...	W	G
4. Saran	S, A, B,	M	W
5. Champaran	S	B	R, W
6. Muzaffarpur	...	B	S, W, A, M
7. Darbhanga	...	Ma	S, W, A, M
8. Monghyr	M, G	W	A, B
9. Bhagalpur	...	...	M, G, B
10. Saharsa	...	Ma	W, M
11. Purnea	...	...	R, W
12. Santal Pargana	...	R	S
13. Hazaribagh	Ma	A	R, M
14. Ranchi	Me	...	R, A, G
15. Palamau	...	B	M
16. Dhanbad	...	R, G	Ma, A
17. Singhbhum	R	...	...

Source : Compiled by the author.

With the above two tables of crop combination it has been revealed that higher percentage of crops have come under third rank and lower percentage of other crops on the first rank. This is the main defect of this method of crop combination.

Method 9

A new method has been envisaged here for crop combination on regional level. The formula is as follows.[31]

$$\text{C.C.} = \begin{cases} \dfrac{HP}{1} - 1 = \text{I ranking crop.} \\ \dfrac{HP}{2} - 2 = \text{II ranking crop.} \\ \dfrac{HP}{3} - 3 = \text{III ranking crop.} \\ \dfrac{HP}{4} - 4 = \text{IV ranking crop.} \end{cases}$$

Lower Limit for crop diversification.

Where C.C. is crop combination.

H.P. is highest percentage of a crop in the study area. 1, 2, 3, 4 are denominators for crops of a particular rank.

And —1, —2, —3, —4 are diversity index for crops of low percentage.

Computation of crop combination on the basis of above formula is as follows :

Singhbhum District

Rice	= 84%		Sugarcane	= 0.25%
Arhar	= 1.25%		Marua	= 0.16%
Maize	= 3%		Wheat	= 0.40%
Gram	= 2%			

84/1—1 = 83% I ranking crop.

84/2—2 = 42—2 = 40% II ranking crop.

84/3—3 = 28—3 = 25% III ranking crop.

84/4—4 = 21—4 = 17% IV ranking crop.

Rice (I ranking crop).

A case of monoculture.

An other example is also presented here for multiple crop combination.

Monghyr District

Rice	— 21%		Barley	— 2.24%
Wheat	— 22%		Arhar	— 2.35%
Maize	— 28%		Sugarcane	— 0.70%
Gram	— 7.28%			

28/1—1 = 27% I ranking crop.

28/2—2 = 14—2 = 12% II ranking crop.

28/3—3 = 9.33—3 = 6. 33% III ranking crop.

28/4—4 = 7—4 = 3% IV ranking crop.

I	II —	II	III	Multiple crop
Maize	Wheat	Rice	Gram	combination

TABLE 14.12 : Ranking of crops in Bihar (1970-71)

Sl. No.	District	Ranks of Distinctiveness according to the Percentage of crops			
		I	II	III	IV
1.	Patna	38 Rice	38-17.5	17.5-10 Wheat	10-5.75 Maize
2.	Gaya	52 Rice	52-24.50	24.50-14.66 Wheat	14.55-9.25
3.	Shahabad	48 Rice	48-22.50	22.50-13.33 Wheat	13.33-8.25
4.	Saran	29 Rice	29-13 Maize, Wheat	13-7 Barley	7-3.50 Sugarcane, Arhar
5.	Champaran	49 Rice	49-23	23-13.66	13.66-8.50
6.	Muzaffarpur	41 Rice	41-19	19-11 Maize, Wheat	11-6.50
7.	Darbhanga	47 Rice	47-22	22-13 Wheat, Maize	13-8
8.	Saharsa	43 Rice	43-20	20-11.66 Wheat, Maize	11.66-7
9.	Purnea	61 Rice	61-29	29-17.66	17.66-11.50
10.	Monghyr	27 Maize	27-12 Wheat, Rice	12-6.33	6.33-3
11.	Bhagalpur	39 Rice	39-18 Maize	18-10.33 Wheat	10.33-6
12.	Palamau	29 Rice	29-13	13-7 Maize	7-3.50 Barley, Wheat
13.	Santal Parganas	68 Rice	68-32.50	32.50-20	20-13.50
14.	Hazaribagh	55 Rice	55-26	26-15.66 Maize	15.66-10
15.	Ranchi	61 Rice	61-29	29-17.66	17.66-11.50
16.	Dhanbad	77 Rice	77-37	37-23	23-15.50
17.	Singhbhum	83 Rice	83-40	40-25	25-17

Source : Compiled by the author.

The above method of crop combination has advantage over several other methods in two respects. Firstly, it takes less time in calculations and secondly it maintains the integrity of higher percentage of crops grown in a region. For instance, in Betal's method of crop combination through Standard Deviation, lower percentage value emerges as 1st ranking crop over highest percentage of other crops in a region. But this new method relegates such lacuna so far as the methodological efficiency is concerned. The other lacuna in Betal's method is that lesser percentage of crops emerges for combination with the dominant crops, but such association may not persist for a longer period, because less important crop in a region never represents the agricultural situation and ecological condition of a region.

Measures of Agricultural Efficiency

The term agricultural efficiency connotes the productivity of a particular unit of land in foodgrains. It is a scientific tool to study the inherent fertility, productivity and capability of the land, so that its misuse, under use, over use and unusability can be checked by planning for future use.

Method 1 : Quite a large number of techniques have been used to measure the agricultural efficiency of an area. In 1939 Kendall[32] has used the ranking coefficient of yield of main crops per unit area to assess their efficiency. The formula applied is as follows:

$$\frac{r_1+r_2+r_3+\ldots+r_n}{n}$$

Where r is ranking of yield of individual crops and n is number of crops. This method was applied by Stamp and Shafi (1960). In this method the ranking of crop's yield is measured with the help of per hectare production without considering the output of other food grains.

Method 2 : In 1938 Ganguli[33] has measured the agricultural efficiency of the Ganga Valley in India after multiplying the percentages of crop's share with percentages of crop's yield in an areal unit and later averaging them into one. These are the stages of an equation to reach up to final formula :

$$I_{yn} = \frac{Y_i}{Y_r} \times 100$$

$$E_n = \frac{(I_{yn}\ C_n)}{100}$$

$$E_i = \frac{E_{i1} + E_{i2} \times E_{i3} + \ldots . . E_{in}}{n}$$

Where I_{yn}=% yield of crop n; y_i=yield of individual crops in an areal unit; y_r=yield of individual crop in the total area; E_{in}= efficiency of crop n in an areal unit; C_n=crop land share in %; E_i=agricultural efficiency of all crops in an areal unit.

Method 3 : Spare[34] and Despande have devised an equation after modifying Kendall's approach. They multiplied the ranking of crops with that of crop land share divided by the total of crop land share. The formula is as follows:

$$E_i = \frac{r_1\ c_1 + r_2\ c_2 + r_3\ c_3 + \ldots + r_n\ c_n}{c_1 + c_2 + c_3 + \ldots + c_n}$$

Where E_i is agricultural efficiency, r is ranking of yield of individual crop,

C_n is crop land share in per cent.

Method 4 : Bhatia[35] has used the Ganguli's percentage yield of crops (I_{yn}) instead of yield ranking of individual crops (r_n) which has been formulated as follows :

$$E_i = \frac{I_{v1}\ C_1 + I_{v2}\ C_2 + I_{v3}\ C_3 + \ldots\ldots + I_{yn}\ C_n}{C_1 + C_2 + C_3 + \ldots\ldots\ldots\ldots + C_n}$$

Where I_{yn} is % yield of crop n;, E_i is agricultural efficiency of all crops in an areal unit: C_n is crop land share in per cent.

Method 5 : Gupta, P.S. (1968)[36] while studying the agricultural efficiency of Cannanore district in South India presented a method for measuring the agricultural efficiency, after considering cereals and pulses, because they provide as much as two-thirds of the calorie intake of average Indian district. He measured the agricultural efficiency by considering yield efficiency of crops weighted by their proportionate share of land. The formula for measuring yield efficiency is as follows :

$$Er_1 = \frac{yr_1}{YR} \times 100$$

Where Er_1 is yield efficiency,

Yr_1 indicates the yield of crop

r_1 per hectare in any unit area in calories, and

YR is the yield of the same crop for the country as a whole in calories.

The yield efficiency of each crop is to be weighted by its proportionate share of land to the total area under all food crops in any areal unit. The agricultural efficiency in food production of a particular area may be measured with the help of the formula as follows :

$$Ia_1 = \frac{yr_1+cr_1+yr_2+cr_2+yr_3+cr_3+ \ldots\ldots +yr_n+cr_n}{Cr_1+Cr_2+Cr_3+ \ldots\ldots +Cr_n}$$

Where Ia_1 indicates agricultural efficiency of the area a_1; while Yr_1, Yr_2, Yr_3, etc. and Cr_1, Cr_2, Cr_3, etc. indicates the proportionate share of land under crops r_1, r_2, r_3, etc. to total area under all food crops. These methods are praise worthy attempts to measure the agricultural efficiency but the purpose is to give an alternative method of agricultural efficiency measurement, so that easily available data in season and crop report and simple calculation may open a new horizon with simple quantitative methods to study such a difficult aspect of agricultural geography and land use planning.

Method 6 : Agricultural efficiency in Bihar has been assessed by R. B. Mandal in terms of food grains[37] which comprise rice, wheat, barley, maize, ragi, jowar, bajra, gram, arhar, mansoor and khesari. Per hectare production of all these crops in different districts and their summation values multiplied by 100 and divided by the total of all the districts in the region gives a value of location coefficient and that value serves here as agriculture efficiency. The formula is as follows[38]

$$AE = \frac{\Sigma D_1\ (X_1 + X_2 + X_3 + \ldots\ldots + X_n)}{\Sigma\ R\ (X_1 + X_2 + X_3 + \ldots + X_n)} \times 100$$

Where AE denotes agriculture efficiency, D_1 district in particular, X_1, X_2, X_3, are crops grown in the region and ΣR is the regional total of all crops.

Method 7 : An analysis of the productive efficiency of various crops in different districts of region or State is based on the relative average yield index (*RyI*) and relative spread index (*RSI*) or the crops[39]

$$RyI = \frac{\text{Mean yield for the crop in the district}}{\text{Mean of the State yield}} \times 100$$

$$RSI = \frac{\begin{array}{c}\text{\% area of the crop in relation to the}\\ \text{total cultivated area of the district}\end{array}}{\begin{array}{c}\text{\% area of the crop in relation to the}\\ \text{total cultivated area of the State}\end{array}}$$

The indices may be grouped into categories. It has been assumed that average relative yield below 90% be taken as an index of relatively inefficient areas.

Method 8 : Productivity rating of crops can also be measured with the help of soil rating formula, which is known as Storie index method. The formula is as follows[40]

$$\text{Productivity Rating} = \frac{A}{100} + \frac{B}{100} + \frac{C}{100} + \frac{D}{100} \times 100$$

Where *A*, *B*, *C*, and *D* are production of different crops in a region in percentage in comparison with the total crops grown in the region.

Zoning Mechanism in Von Thunen Model

The German economist Von Thunen has considered a settlement as consuming centre, absolutely lying in an isotropic plain surface. In the surrounding area of that centre equal rates of transportation cost, wages to the labourers and productivity of soils were available. Ultimately he proposed concentric zones of production around that settlement, where the bulky and perishable products being produced close to the settlement while less bulky and less perishable products in outer margin of the settlements that may be either towns or rural settlements.

Von Thunen regarded the distribution of locational rent as the key factor sorting the uniform area of his isolated state into distinct land use zones. The location is given by[41]

$L = Y(P—C) - YD(F)$
L = the locational rent (in Rs./Km^2)
Y = the crop yield (in tone/Km^2).
P = the market price of the crop (in Rs./tonne)
C = the production cost of the crop (in Rs./tonne),
D = the distance to the central market (in Km), and
F = the transportation rate (in Rs./tonnes Km)

The competition of two crops (i and j) for the same area depends on their yield (Y) and relative profitability ($P—C$). When the condition

$$I < \frac{Y(P—C)_i}{Y(P—C)_j} < \frac{Y_i}{Y_j}$$

Obtains for crops i and j, they form two distinct spatial zones; crop i dominates a circular area adjacent to the city, and crop j occupies a ring-shaped zone immediately outside it. The symbol means "is less than". Any other relation of the terms in the question above results in the two crops being reversed so that j occupies the inner ring, or one crop dominating all available land to the complete exclusion of the other, or both crops being grown side by side with no spatially differentiated zoning.

Cereal Sufficiency and Deficiency Regions

The cereal sufficiency and deficiency have been computed for the year 1967-68 for which relevant data are available. The lowest level of areal unit for discussion are different districts of India. In all seven major cereals have been taken into account. These areas under cultivation and production are the bases of calculation. For 1968, population figures have been estimated by following the well-known formulae

$$\log \frac{P_2}{P_1} = n \log (1 + r)$$

$$\log (1 + r) = \frac{\log \frac{P_2}{P_1}}{n}$$

Where P_1 = is the number of people at the initial date.
P_2 = is the number of people at the later date.
r = is the annual rate of growth.
n = is the exact number of years between P_1 and P_2

From the total production, wastage of 10% and seed requirement of 6.8% have been deducted and the rest 83.2% of the total production is available for consumption. At the district level, total consumption units have been computed and the consumption per annum has been estimated at the rate of 16.62 ounces per capita per day or at the annual rate of 171.7 kg per consumption unit. For this, the following formulae has been adopted.

$$S_p = (0.832F/0.760) - C$$

Where S_p = Per consumption unit of the total population

p = Total population in 1968.

C = Average normal consumption of food grains per consumption unit per year (171.7 kg.)

F = Total cereal production of the district in question.

For each district the consumption factor is calculated based on effective food consuming population estimated as follows

$$p_c = \text{Population } (0.42 \times 0.533 + 0.58 \times 0.925).$$

Where 0. 42 indicates 42% of population below the age of 15, and 0. 533 unit is the capacity of food-consumption in that age-group on an average, while the rest (*i.e.* 0.58 or 58%) has an average consumption capacity of 0.925 unit each. The results have been mapped and it is found out that most efficient districts are not necessarily cereal sufficient districts. In other words, there does not exist any correlation between productive efficiency and sufficiency. Rather, less productive areas are food sufficiency areas since the population pressure in them are less due to hostile terrain.

At the end it can be said that the above list of different methods applied by different scholars is not exhaustive. The important statistical methods in the field of land utilization and agricultural geography are those related with the agricultural efficiency, land capability, land classification, crop combination, measurement of land potentialities, land use changes, crop intensity, land use zoning and several others of land tenure, hydrology, location of crops, etc. As the present scientific world is moving towards interdisciplinary approach, for this fact the method as discussed above can also be used in other branches of geography, sociology, economics and in the field of agricultural land utilization.

REFERENCES

1. Jasbir Singh, "Optimum Carrying Capacity of Land, Caloric Density and Intensity of Population Pressure Changes in Punjab, 1951-61", *The National Geographical Journal of India*, Vol. XVII, Part 1, 1971, p. 40.
2. N. A. Siddi, "Land Classification for Agricultural Planning : A Study in Methodology", *The Gographer*, Vol. XVIII, 1971, pp. 25-38.
3. H. S. Sharma, *Land Capability Classification of the Lower Chambal Valley*, Symposium on Land Use in Developing Countries, 21st IGC, Aligarh, 1972, pp. 173-79.
4. V. R. Singh, "Land Capability : A Method of Classification", *The Professional Geographer*, Vol. XXII, No. 1, 1970, pp. 40-41.
5. R. B. Mandal, *Land Use Studies in Geography*, Associated Book Agency, Patna, 1978.
6. Maheshwari Prasad, "Factor Analysis: A Study of Soil Capability and Land Utilization", *Indian Geographical Studies*, No, 3, 1974, pp. 4-5.
7. Jozof Janceszewaki, "Index of Land Consolidation" Abstract of Papers: 21st I.G.C., 1968, pp. 144-45.
8. M.M. Das, "Plea for Joint Agricultural Farming: A Case Study in Rural Economic Geography", *Journal of the North-East Indian Geographical Society*, Vol. V, No. 1 & 2, 1973, pp. 46-56.
9. M.F. Siddiqu, "Changing Patterns of Crop Land Use in Bundelkhand, India", 21st I. G. C., 1968, Abstract of Papers, p. 181.
10. R.R. Tripathi, "Agricultural Dynamicity and Efficiency Levels : A Case Study of the Upper Ganga-Gomati Doab", *The Indian Gographical Journal*, Vol. XLVII, No. 1 & 2, 1972, pp. 12-19.
11. N. L. Gupta, and S. L. Hiran, "Agricultural Regions of Rajasthan", *The Deccan Geographer*, Vol. XI, Nos. 1 & 2, 1973, p. 61.
12. B. N. Sinha, "Agricultural Deficiency Regions in India", Proc. 57th Ind. Sci. Congress, Part III, Abstract, p. 227.
13. L. Zoblar, "A Proposed Food Resource Efficiency Statement for the United States and Japan", Abstract : *Annals of the Association of American Geographers*, Vol. 49, 1959, p. 222.
14. M. F. Siddiqu, "Concentration of Deficiency Diseases in Uttar Pradesh", *The Geographer*, Vol. XVIII, 1971, p. 94.
15. Sundrama Sastry, "Location of Crops", *The Indian Geographical Journal*, Vol. XVII, 1942, pp. 110-114.
16. P. N. Soni, "Cropping Pattern and Crop intensity in various Size Class of Farmers in some I.A.D.P. Districts", Agricultural Situation in India, October, 1974, p. 484.
17. H. S. Bal, "Impact of Mechanization on Farm Labour Employment", Agricultural Situation in India, Sept. 1974, pp. 385-386.
18. S.L. Ambastha, "Hydrology and its Impact on the Land utilization of Magadh, Bihar", First Indian Geographical Congress, New Delhi, 1972, (Bibliography Hydrology).

19. M. Uzolgleozurike, "Land Tenure Relations, Social Relations and the Analysis of Spatial Discontinuity", *Area*; Vol. 6, No. 2, p. 133.
20. Jasbir Singh, "Seasonality and Regionalism in the Agricultural Meteorology of Hissar District, Haryana", *The Geographical Review of India*, Vol. XXXIII, No. 3, 1971, p. 172.
21. S. K. Sadhukhan, "A Quantitative Analysis of an Agro-Irrigation Face of West Bengal", *The Geographical Review of India*, Vol. 36, No. 3, 1974, pp. 196-203.
22. Harpal Singh, "Crop Diversification in the Malwa of Punjab", *The Indian Geographical Journal*, Vol. XXXVIII, No. 1, 1963, pp. 75-78.
23. P.M. Enggass, "A Method for Measuring and Comparing Agricultural Land Use : An Andalusian Experiment", *The Professional Geographer*, Vol. XIX, No. 3, 1967, pp. 125-126.
24. Majid Hussain, "Patterns of Crop Concentration in U.P.V. *Geographical Review of India*, Vol. XXXII, No. 3, 1970, p. 170.
25. S.S. Bose, "Human Geography, Public Agriculture and the Chemistry of Food Production", Proc. 55th Ind. Sc. Cong., Part III, Abstract, p. 260.
26. J.C. Weaver, "Crop Combination Regions in the Middle West", *Geographical Review*, 1954, Vol. 44, pp. 175-200.
27. K. Doi, "The Industrial Structure of Japanese Perfecture", *Proceedings of the I.G.U. Regional Conference in Japan*, 1957 (1959), pp. 310-316.
28. S.M. Rafiullah, "A New Approach of Functional Classification of Towns", *The Geographer*, Vol. XII, 1965, pp. 40-53.
29. A.G. Athawala, "Some New Methods of Crop Combination", *Geographical Review of India*, 1966, Vol. XXVIII, No. 3, pp. 28-34.
30. H. Betal, "*Crop Combination Regions of India*", Geographical Aspect of Indian Agriculture", Department of Geography, Calcutta University 1976, pp. HRB 7.
31. R.B. Mandal, *Land Use Studies in Geography*, Associated Book Agency, Patna, 1978.
32. M.G. Kendall, "The Geographical Distribution of Crop Productivity in England", *Journal of the Royal Statistical Society*, Vol. 102 (New Series), 1939, pp. 21-62.
33. B.N. Ganguli, *Trends of Agriculture and Population in the Ganges Valley* London, 1938, pp. 93-94.
34. S.G. Sapre, and V.D. Despande, "Inter-District variations in Agricultural Efficiency in Maharashtra State", *Indian Journal of Agricultural Economics*, Vol. 19, No. 1, 1964, pp. 242-252.
35. S.S. Bhatia, A New Measurement of Agricultural Efficiency in Uttar Pradesh, India, *Economic Geography*, Vol. 43, No. 3, 1967, pp. 244-260.
36. P.S. Gupta, "Agricultural Efficiency in Food Production in India", Abstract of Papers, 21st International Geographical Congress, 1968, p. 166.
37. R. Chaudhury, et. al., Soils of India, ICAR, New Delhi, 1963.
38. R.B. Mandal, *Land Use Studies in Geography*, Associated Book Agency, Patna, 1978.

39. Ali Mohammad, "Spatial and Temporal Variations in Levels and Growth of Agricultural Output", Paper presented in a syposium at NEHU, Shillong, 1977.
40. G. Azzi, *Agricultural Ecology*, Constable & Co. London, 1955.
41. E. S. Dunn, *The Location of Agricultural Production*, University of Florida Press, Gainesville 1954, Chaptered.

FIFTEEN

R. B. Mandal

POPULATION GEOGRAPHY

IN RECENT decades the overwhelming growth of population has created much of the problems of human society than it solves as the base of all resources. The enumeration of population and its evaluations require the use of statistical methods. The effects of physical and cultural factors are highly helpful in analysing the distribution of population. In this regard, in any study of population, some variables must affect the result and that should be interpreted quantitatively. Considering statistical use, population geography has a wider scope, which covers both rural and urban population. The population movement, migrations, concentration, dispersion, growth rate, fertility-ratio and mortality whatever may be the subject matter of this branch of geography, statistics have been widely used as these are tools in the hands of geographers, in performing their studies, investigation, teaching and future estimation of population. Besides this, the same statistical method can also be used in any branch of geography, e.g., spatial measurement of standard distance.

Population Growth and Variability

The actual population is the population as enumerated at each census year. The exponential population is characterized by the property of constant growth rate at all decades but equal to the actual population at 1901 and 1951. When P and P_1 are the 1901 and 1951 actual population respectively, the exponential population at 1911, 1921, 1931 and 1941 are P_r^1, P_r^2, P_r^3 and P_r^4. Where r being equal to $5\sqrt{P_1/P}$.

The term general growth is directly proportional to the net **variation (in 5 decades)**[1].

$$\text{It may be noted that } \left(\frac{G + 1{,}000}{1{,}000}\right)^{5}$$

$$= \frac{N + 1{,}000}{1000}$$

Where G is general growth and N is net variation per thousand population. Here the general growth is used instead of the other, since this is readily comparable with each decades actual growth rate.

Index of Population Change

Index of change is the measure for studying intercensal changes in population. It can be calculated with the help of following formula :[2]

$$IC = \frac{P_1}{P_2}$$

Where P_1 = population in 1971, and P_2 = population in 1961. Such index can be compared with National and State rates and it can be plotted with choropleth to identify regional patterns.

An another index of change in percentage of population size is considered here, which involves population of two time periods of a region. The formula is as follows :[3]

$$r = \frac{(P_2 - P_1)/t}{(P_2 + P_1)/2} \times 100$$

where, r is the annual growth rate
P_1 is the population of the region in one time.
P_2 is the population of the region in another time.
t is the time interval P_2 and P_1.

The formula expresses change in numbers on an annual basis $(P_2 - P_1)/t$ as a per cent of the average population size $(P_2+P_1)/2$ over the period of time (t).

Exponential Population Growth

Exponential models of population growth describe a simplified situation in which growth (or a decline) is unchecked and the rate of change is constant. We express this simply as[4]

$$\frac{dN}{dt} = rN$$

Where N = the number of people,

r = the rate of natural increase (a constant), and

$\frac{d}{dt}$ = the rate of change per unit of time.

The expression states that the amount of growth is related to the size of the population, the larger a population is, the faster it grows.

To simplify the computation, we can rewrite this as

$$N_t = N_o^{ert}$$

Where N_t = the number of people at time t,

N_o = the number of people at time o, and

e = 2.71828.

The constant e is the base of Napierian, or natural, logarithms and is the sum of the infinite series

$$1 + \frac{1}{1} + \frac{1}{2 \times 1} + \frac{1}{3 \times 2 \times 1} + \frac{1}{4 \times 3 \times 2 \times 1} + \ldots . + n.$$

In case, we start with 1,000 people (N_o) and assume a growth rate of 1 per cent per annum (r=0.01), then we can show, by substituting these values in the equation, that after 70 years (t=70) the original population will have doubled (N_t=2,000). In another 70 years the population will have doubled again. Exponential models show the critical importance of small changes in the rate of natural increase. If we halve this rate (r=0.005), the population doubles only every 140 years. Despite their simple structure, exponential models are a useful way of describing recent phases of human population growth.

Logistic Population Growth

When a population is allowed to develop in an optimal environment of unlimited size, its growth follows an exponential curve. If we now introduce a fixed carrying capacity, or saturation level (k), the potential, for biological growth, or the biotic potential, will be modified by environmental pressure.

We can introduce this environmental pressure into the exponential growth model,

$\frac{dN}{dt} = rN$, by subtracting

$\frac{k - N}{k}$ from rN, we then have

$\frac{dN}{dN} = rN - \frac{k - N}{k}$, where N = the number of individuals in the population, k = the maximal number of individuals allowed by the carrying capacity, r = the rate of growth per individual and $\frac{d}{dt}$ = the rate of change per unit of time. Both N and k also can be expressed as population densities. This modified growth curve is termed logistic growth curve and has a characteristic S-shape.

Growth of Urban Population

The decadal growth of urban population can be calculated on the basis of this formula :[5]

$$\text{Decadal Growth Rate} = t\sqrt{\frac{P_i}{P_o}} - 1 \times 100$$

Where t is the number of years interval,
P_i is the last census population,
P_o is the previous census population.

The two main components may be census returns or population estimates, and so the rate is subject to the general qualifications for such data. It is a useful rate, and may be helpful in assessing the accuracy of vital and migration statistics.

Potential Natural Increase Rate

Natural increase is the difference between the number of births and deaths, and the natural the crude birth and death rates. The formula for measuring natural increase rates is :[6]

$$\frac{(B \times E) - (D \times A)}{PL}$$

Where B is the number of births, E the expectation of life at birth, D the number of deaths, A the average age of deceased persons, and PL the life potential of the total population (i.e. the total number of years that the population has yet to live).

Declining Trend of Population Growth Rate

The declining trend of population growth rate with time is fully explained by the second degree parabolic curve given by the equation :

$$Y = 95.391 - 39.952x + 6.434x^2$$

The estimated values of the growth rates $(\hat{y})$ are very close to observed values which are tabulated below :

TABLE 15.1 : **Observed and Estimated Growth Rates of the City Population in Bihar**

Period	Growth Rate in Thousand Persons Per Annum	Estimated Growth Rate
1931-41	95.29	95.391
1941-51	62.18	61.873
1951-61	40.93	41.235
1961-71	33.57	33.468
1971-81	—	38.575

This also indicates the possibility that the growth rate during the next decade and the decade after may go up.

Annual Rate of Population Increase

Apart from the absolute increase or decrease per annum, one of the west common measures of growth in the annual rate of increase. The U.N. Demographic year-books use the following formula :[7]

$$\left(\sqrt[t]{\frac{P_1}{P_o} - 1} \right) \times 100$$

Where P_o is the population at the beginning of the period, P_1 is the population at the end of the period, and t is the number of years. The two main components may be census returns or population estimates, and so the rate is subject to the general qualifications for such data. It is a useful rate, and may be helpful in assessing the accuracy of vital and migration statistics.

Population Growth and Variability

The actual population is the population as enumerated at each census year. The exponential population is characterized by the property of constant growth rate at all decades but equal to the actual population at 1901 and 1951. When P and P_1 are the 1901 and 1951 actual population respectively, the exponential population at 1911, 21, 31 and 41 is P_{r1}, P_{r2}, P_{r3} and P_{r4}, r being equal to $5\sqrt{P_1/P}$

The term general growth is directly proportional to the net variation (in 5 decades).[8]

$$\text{Note that} \left(\frac{G+1000}{1000}\right)^5 = \frac{N+1000}{1000}$$

Where G is general growth and N is net variation

N is net variation per thousand population. Here the general growth is used instead of the other, since this is readily comparable with each decade's actual growth.

Estimate of Total Fertility

To estimate total fertility by the end of reproductive period is very necessary. Such extrapolation can clearly be a good deal more precise for one ten years younger. In fact, a formula has been found which, tested on past data, gives a good extrapolation for any cohort which has reached the age of 30.

The fact that births are recorded separately for the different sequences makes such extrapolation a good deal more precise than if we had merely dealt with birth as a whole. Many attempts have been made to find a mathematical expression which would fit (with some) curves and serve to extrapolate then when in complete—a very urgent task. The solution was discovered by Martin[9], who showed that the Gompertz equation gives a close fit. The method of fitting is as follows :

$X =$ Exact age of mother. X_0, X_1 and X_2, are three selected values of X such that $X_2 - X_1 = X_1 - X_0$.

$Y =$ Cumulated number of births (Per 1000 women) prior to specified age—y_0, y_1, y_2 are the three volumes of y corresponding to X_0, X_1, X_2.

$Z = X - X_0$

$n = X_2 - X_1 = X_1 - X_0$.

The Gompertz formula is :

$Y = K_a b^z$ (where K is the asymptotic value of y).

or $\log y = \log K + (\log a)\, b^z$

$$\text{Where } b^n = \frac{\log y_2 - \log y_1}{\log y_1 - \log y_0}$$

$$\log a = \frac{\log y_1 - \log y_0}{b^n - 1}$$

$\log K = \log y_o - \log a$

values of a, b and k obtained for diagrams were as follows :

U.S.A. 1928 Cohort

	1st births	2nd births	3rd births	
a	0.281	0.215	0.179	The deviation of the first births for a few years in U.S.A. proves the rule except the beginning of the period.
b	0.739	0.778	0.823	
k	892	824	665	

Some Measures of Fertility

The general formula for crude rates and specific rates are given below. The fertility rates are also calculated on the same basis.

$$\text{Crude Rate} = \frac{\text{Number of vital events which occurred among the population of a given geographic area during a given period}}{\text{Mid period total population of the given geographic area during the same period.}} \times 1000$$

$$\text{Specific Rate} = \frac{\text{Number of vital events which occurred among a specific group of the population of a given geographic area during a given period}}{\text{Mid period population of the specified group in the given geographic area during the same period.}} \times 1000$$

Crude Birth Rate (*CBR*)

$$\frac{\text{Number of live births which occurred among the population of a given geographic area during a given year}}{\text{Mid year total population of the given geographic area during the same year}} \times 1000$$

General Fertility Rate (*GFR*)

$$\frac{\text{Number of live births which occurred among the population of a given geographic area during the given year}}{\text{Mid year female population in the reproductive ages (usually 15-44) in the given geographic area during the same year}} \times 1000$$

Age-Specific Fertility Rate (*ASFR*)

$$\frac{\text{Number of live births to mothers of a specified age group of the population of a given geographic area during a given year}}{\text{Mid year female population of the specified age group in the given geographic area during the same year.}} \times 1000$$

Age-Specific Marital Fertility Rates (*ASMFR*)

$$\frac{\text{Number of (legitimate) live births to mothers of a specified age group of the population of a given geographic area during a given year}}{\text{Mid year currently married female population of the specified age group in the given geographic area during the same year}} \times 1000$$

Both *ASFR and ASMFR* can also be computed by age of father.

Other Specific Fertility Rate

In a similar way, we can also calculate fertility rates specific for 'duration of marriage', 'age and duration', 'occupation' etc. The only difference is with regard to 'parity-specific fertility rates'.

Parity-Specific Fertility Rates

$$\frac{\text{Number of live births of } n^{\text{th}} \text{ order which occurred among the population of a given geographic area during a given year}}{\text{Mid year female population of } (n-1) \text{ the parity in the given geographic area during the same year}} \times 1000$$

We can calculate these rates by age group of women and then they are known as age parity specific fertility rates. Similarly 'parity-duration of marriage specific' and 'age-parity-duration specific' fertility rates can also be computed.

Total Fertility Rate (TFR)

This is the sum of age-specific fertility rates of women by single year of age from 15 to 44 (or 15 to 49) and is expressed 'per woman', i.e., the sum is divided by 1000. If the ASFR's age given for 5 year age intervals the sum is to be multiplied by 5. This index given the average expected number of children born *per woman* as she passes through the reproductive ages under the given fertility schedule.

Gross Reproduction Rate (GRR)

This is the sum of the age-specific fertility rates of women *for female births only* by single year of age from 15 to 44 (15 to 49) and is expressed 'per woman', i.e., the sum is divided by 1000.

GRR varies fron TFR only to the extent that in the computation of GRR only female births in TFR.

Thus GRR = TFR × proportion of female births to total births

$$= TFR \times \frac{100}{100 + Sex\ ratio\ at\ birth}$$

$$= TFR \times \frac{100}{205} \text{ if we assume the sex ratio at birth to be 105.}$$

$$\text{Sex ratio at births} = \frac{\text{Number of male births}}{\text{Number of female births}} \times 100$$

GRR may be defined as the ratio of female births expected in two successive generations under the given fertility schedule assuming no mortality.

Net Reproduction Rate (NRR)

NRR is defined as the ratio of female births in two successive generations under the given schedules of fertility as well as mortality.

Thus NRR= $\sum_{15}^{44} bx \frac{LX}{10}$ where bx is the rate of female births per person year at each age X.

If the data are in 5 year age intervals.

$$NRR = \Sigma\, bx \frac{SLX}{10}$$

Where $bx = \frac{ASFR}{1000} \times \frac{100}{205}$ and $\frac{SLX}{10}$ can be obtained from an appropriate Life Table.

TABLE 15.2 : **Appropriate Life**

Age group	Number of women	Number of ASFR	3/2×1000
15-19	w_1	b_1	f_1
20-24	w_2	b_2	f_2
25-29	w_3	b_3	f_3
30-34	w_4	b_4	f_4
35-39	w_5	b_5	f_5
40-44	w_6	b_6	f_6
Total	$\Sigma w_i = w$	$\Sigma b_i = b$	

GFR, ASFR and TFR

(i) $ASFR\, f_i = \frac{b_i}{w_i} \times 1000$

(ii) $GFR = \frac{b}{w} \times 1000 = \frac{w_1 f_1 + w_2 + \ldots\ldots + w_6 f_6}{w_1 + w_2 + \ldots\ldots\ldots + w_6}$

In other words, *GFR* is the weighted average of the *ASFR*'s with number of females in each age group as the weights.

(iii) $TFR = \frac{f_1 + f_2 + \ldots\ldots\ldots\ldots + f_6}{1000} \times 5$

Thus in calculating *TFR*, equal weight is given to all ages, viz., one woman at each age.

Weights of Age Specific Fertility Rates

The U.N. population Division has recommended the weights 1, 7, 7, 6, 4, 1, which correspond to the average pattern of age specific fertility rates, for the five year age groups 15-19, 20-24, 25-29, 30-34, 35-39 and 40-44 respectively. These have been obtained from a study of data for 52 countries of which 15 were having comparatively high fertility and 32 were with low fertility. The following observations made in Manual III may be noted. "Since it was considered convenient that these synthetic rates should be of the same order of magnitude as crude birth rate per 1000 of the total population, the weights were so chosen, not only that they would be roughly proportionate to the average of the age-specific birth rates, but also that the sum of their products with the corresponding numbers of women in the various age groups would ordinarily be of the same order of magnitude as the total population.

Sex Age Adjusted Birth Rate (SAABR)

This is a simple device for eliminating the effects of changes in age-sex structure of the population in the measurement of fertility for trend, comparability and estimation of future births.

$$SAABR = \frac{\text{Number of live births in the year}}{1Xw_1+7Xw_2+7Xw_3+6Xw_4+4Xw_5+1Xw_6} \times 1000$$

Where, w_1, w_2, w_3, w_4, w_5 and w_6 are the number of women in age groups 15-19, 20-24, 40-44 in the middle of the year.

The numerator may be obtained from vital registration if the same is sufficiently complete, otherwise the same can be estimated indirectly by reverse survival method or by using an estimated value of *CBR* from any other method.

e.g.

TABLE 15.3 : **Female Age Group**

India

Age Group	Females 1961 ('00000)	U.N. Weights	Actual Weights
15-19	17,284	1	3.6
20-24	19,134	7	6.7
25-29	18,050	7	6.1
30-34	14,855	6	4.7
35-39	11,860	4	3.2
40-44	10,774	1	1.7

If the *CBR* is 41.7 per 1000 in India, the estimated number of births in 1961 = 18,304,000.

$$= \frac{18304000}{1\times17284+7\times19134+7\times18050\times14855+4\times11860+1\times10774}$$

$= 43.1$

Child-Woman Ratio (CWR)

CWR may be defined in any one of the following four ways :

(i) $\frac{C(0-4)}{w(15-44)}\times1000$, (ii) $\frac{C(0-4)}{w(15-49)}\times1000$

(iii) $\frac{C(5-9)}{w(20-49)}\times1000$, (iv) $\frac{C(S-9)}{w(5-9)}\times1000$

Where C and w respectively denote the number of children and women in the corresponding age groups. This is a simple measure of fertility useful for comparing the fertility performances of different groups. It can be easily computed from the census age-sex distribution, and is found to be of utmost use when other data on fertility are lacking. It serves only as a relative measure.

Average Parity

The average parity of women can be calculated for each age group if the distribution of women by age and parity is available. The average parity of women by age provides a useful measure for comparing the parts fertility performance of women in different groups. Of particular importance is the parity of women who have passed their reproductive ages, that is, aged 45 and above, which is also known as the average completed family size.

Parity Progression Ratio

Parity progression ratio for parity n can be defined as the proportion of women who having achieved n^{th} parity, proceed to achieve at least $(n+1)^{th}$ parity.

It is normally computed for women of completed fertility, but may also be computed for women of a specified age or cohort.

Parity Progression Ratio

$$= \frac{\text{Number of women of } (n+1)^{th} \text{ and higher parity in a group of women}}{\text{Number of women of } n^{th} \text{ and higher parity in the same group}}$$

Standard Mortality Ratio

By comparing observed and expected numbers of infants deaths, a standardised ratio was obtained as follows[10].

$$SMR = \frac{\text{Observed number of infant deaths in sub-group}}{\text{Expected number of infant deaths in sub-group}} \times 100$$

The value of *SMR* exceeded 100 in case of increased mortality in the particular sub-group of population and was less than 100 in the opposite case. A simple test of significance with 5 per cent level was applied throughout the study. A similar procedure was followed for the analysis of neo-natal and post-neo-natal mortality.

TABLE 15.4 : **Greater Bombay– Infant Mortality, 1962-68**

Causes of Death	Death Rate Per 100,000 live born in the year			
	1965	1966	1967	1968
Smallpox	160.44	21.0	113.6	5.1
Measles	137. 3	26.4	89.4	14.0
Dysentery	1145. 8	1119.0	1069.6	1029.4
Disease of respiratory system	2907.2	2132.6	2725.9	2001.5

Source : After: Rusicka, L.T. and Kanitkar, T.

Job Availability Index (Job 1) :

According to Stouffer it has been assumed that an outflow point has a zone around it in which the potentially mobile population moves in search of jobs. Soon the zone was formed a circle having a radius of families. Within this circle the mobile workers would be attracted to the job available at various inflow points which may be equated to the various urban (service) centres situated within the circle of mobility.

The basis for framing a specific index is the assumption that the number of jobs available at different inflow points is directly proportional to the total population living there. Thus the job availability position at different inflow points with respect to the overall job availability within the whole zone of mobility can be worked out. The job availability index, expressed on a percentage basis will be as follows[11].

$$\text{Job 1} = \frac{\text{Population of the inflow Point}}{\text{Total Population of the Service Centres.}} \times 100$$

An illustration would make the point clear. If a particular zone of mobility having three inflow points (urban centres) of 6,000; 24,000 and 270,000 persons respectively (i.e. a total population of three lakhs), the job availability indices for the various inflow points will be 2%, 8% and 9% respectively.

The following points emerge from the above study

(1) Diurnal movement of population (M) is due to the movement of a part of the potentially mobile population (PMP).

(2) Diurnal movement of population is directly proportionate to the availability of jobs (at a particular inflow point) in relation to the overall job availability within the total zone of mobility (i.e., job 1).

(3) Diurnal movement of population is also directly proportionate to the accessibility factors related to the distance between the inflow and outflow points (i.e. $ACC1$.)

The final mobility equation is as follows

$$M = PMP \times Job\ 1 \times ACC1.$$

Where M is the total movement of daily wage workers and PMP is the potentially mobile population. Since the indices have been expressed as percentages the equation could also be worked out as below

$$M = \frac{PMP \times Job1 \times ACC1}{100 \times 100}$$

Gross and Net Reproduction Rates

It is the sum of the age-specific fertility rates of women for female births only by single year of age from 15 to 49 and is expressed per woman.

The gross reproduction rate is equal to[12]

$$\sum_{i=15}^{49} f_i \frac{BF}{B}$$

Where f_i denotes the age specific fertility rate, BF female births and B total live births. The net reproduction is equal to

$$\sum_{i=15}^{49} f_i L_i \frac{BF}{B}$$

Where L_i is the probability of female population surviving up to age i in a life table (below)

TABLE 15.5 : **Net Reproduction Rate**

Caste	Gross Reproduction Rate	Net Reproduction Rate
Brahmin	3.4	1.8
Jat	3.4	1.8
Gujjar	3.4	1.8
Functional Castes	3.4	1.8
Bhangi	3.5	1.9
All Castes	3.4	1.8

Source: After Agarwala, S.N.

Standardised Mortality Rate

The most satisfactory method of comparing by a single figure the mortality in each local area with that of the country as a whole is by means of the Standardised Mortality Rate (SMR) which allows for peculiarities of age structure in local populations. Whether the S.M.R. for a particular locality differs significantly from the national mean may be examined by χ^2 test as follows[13]

$$\chi^2 = \frac{Ot\,(O-E)^2}{EOt + E^2}$$

Where O = Observed deaths in locality

Ot = Observed deaths in total population.

E = expected deaths in locality if age specific rates for total population prevailed.

Shift in Fertility

To measure shift in fertility over a period of time, all the females have been divided into two-marriage-cohorts, pre-1930 and post-1930. In calculating age-specific fertility rate we have avoided the problem of truncated distribution, that is, of incomplete fertility performance in some age-groups, by taking it as quotient of the females above that age and the number of children born to them during that age-interval. For example if $\sum_{i=15}^{19} f_i$ denotes the age specific fertility rates of a group of women during the age interval 15-19, then

$$\sum_{i=15}^{19} f_i \quad \frac{\sum_{g=20}^{45\text{-}49} \sum_{i=15}^{19} k_{ig}}{\sum_{g=20}^{45\text{-}49} F_g}$$

Where k_{ig} denotes the total number of births during the ith age interval to the women of the gth age. F_g represents the women of the gth age, Table 15.6 gives the age-specific and cumulative fertility rates of pre-1930 and post-1930 marriage cohorts.

TABLE 15.6 : **Age Specific and Cumulative Fertility**

Age at Maternity	Pre-1930 Marriage Cohort Fertility Rates		Post-1930 Marriage Cohort Fertility Rates	
	Age specific	Cumulative	Age specific	Cumulative
Below 15	0.07	0.07	0.04	0.04
15-19	0.86	2.93	1.92	1.02
20-24	1.65	2.58	1.89	2.91
25-29	1.69	4.27	1.99	4.90
30-34	1.62	5.89	1.69	6.59
35-39	1.14	7.03	1.29	7.88
40-44	0.65	7.68	—	—
45 above	—	—	—	—

Source: After Agarwala, S. N., p. 95.

Age Standard Average Family Size

If f_i is the age-specific fertility rate of the ith age period, then the standardised average size of family, F_s, can be obtained by

$$F_s = \frac{\sum_{i=15}^{49} f_i d_i}{\sum_{i=15}^{49} d_i}$$

Where d_i is the number of females in the ith age group in the Standard Population. The female age distribution by age of all the villages has been taken as the standard population, the age standardised average size of family has been obtained by multiplying the age specific fertility rates of different castes to the relevant female population by the age of standard population.

TABLE 15.7 : **Age of the Standard Population**

Caste	Standard Average Family Size
Brahmin	4.2
Jat	4.2
Gujjar	4.2
Functional Caste	4.3
Bhangi	4.4
All Caste	4.3

Source : After Agarwala, S. N., *A Demographic Study Six Urbanizing Villages*, Asia Publishing House; 1970; Bombay: p. 101.

Measures of Births and Deaths

Information on births and deaths was collected along with the fertility inquiry. Our problem was to find out the total number of births and deaths which may have occurred in the villages. To put it differently, the assumption implied that [14]

$$\frac{B}{b} = \frac{b'}{b''} \quad \text{or } b'' = \frac{b.b'}{B}$$

Where

B = numbers of events recorded by both chowkidars and the investigators.

b = number of events correctly reported by the village chowkidars.

b' = number of events not reported by village chowkidars but correctly reported by our investigators.

b'' ≐ estimated number of events missed by the chowkidars.

Then

$b + b'$ = estimated total number of events that occurred during the specific period

$(b + b'')\ (b' + B)$ = estimated number of events missed by the investigators.

To put it differently the assumption implies that[15]

$$\frac{X}{T} = \frac{A + X}{T} \cdot \frac{B + X}{T} \cdot \frac{C + X}{T}$$

or, $2P\,(A + B + C)\,X^2 + [P^2\,(AB + AC)]\ X - ABC = 0$,

Where

P = Total number of events recorded by all the agencies.

X = Number of events not recorded by all the three agencies.

T = Total number of events or $(P + X)$.

Y = Number of events commonly reported by all the three agencies.

A = Number of events not recorded by agency a, but recorded by some other agency.

B = Number of events not recorded by agency b, but recorded by some other agency.

C = Number of events not recorded by agency C, but recorded by some other agency.

Isoentropy of Demographic Partition

The information theory, a recent development has formulated this concept. Thus Brillouin defines information as[16]

$$I = K l_n P_o$$

Where I is the information, K is the logarithmic constant and P_o is the number of equally probable outcomes. The average expected information of a set of probabilities associated with the components of a system can be derived on the basis of the Shanon formulation as[17]

$$H = \sum_{i=1}^{N} X_i \, l_n \, X_i$$

Where : $\Sigma X_i = l$

Pelto [18] introduced the term relative entropy, which defines as the ratio of the actual entropy to the maximum entropy

$$100\, H_r = \frac{-100 \sum_{i=1}^{N} P_i \, l_n \, P_i}{H_m}$$

Where P_i=the proportion of the ith component in an N component system and H_m = the maximum possible entropy.

$$= -\Sigma \frac{1}{N} \, l_n \, \frac{1}{N}$$

$$= l_n \, N$$

The application of the relative entropy is particularly attempted to assess the degree of demographic mixing of different categories of workers in the working population of the settlements and the formula is as follows[19]

$$H_r = \frac{100 \sum_{i=1}^{N} P_i \, l_n \, P_i}{H_m}$$

$$H_r = \frac{100 \times 1.864}{2.187} = 85\%$$

Table 15.8 is given below to show the example of relative entropy.

TABLE 15.8 : **Kaonali**

Category	i	ii	iii	iv	v	vi	vii	viii	ix	Total
Workers	102	45	193	31	175	29	40	17	259	891
Ratio	.11	.05	.21	.04	.19	.03	.05	.02	.29	0.99
Entropy	.243	.150	.328	.129	.313	.106	.150	.084	.339	1.864

Maximum entropy of a component = 2.187.

Source : Har Prasad, 1971.

The Movement of Population

The movement of population over a considerable period of time is continuous. In intercensal intervals it exhibits considerable irregularity, but it is smoothed out in forty or fifty years averages. Demographers have found regularity of movement such that they are able for various purposes to regard population increase as obeying a law mathematically formulable. The equations formulating the law can be and often are abused. They are none the less serviceable in competent hands. Such equations are affirmative of the view that in a given country and at a given epoch there is a normal population density for a homogeneous population.

One of the most interesting of these equations is the so called logistic of Verhulst. It assumes (i) a tendency to increase at a geometrical rate, and (ii) a restrictive influence varying to a first approximation with the square of the population. This is expressed by the differential equation.[20]

$$\frac{1}{P} \cdot \frac{dp}{dt} = r - up$$

and the finite equation $P = \dfrac{r}{n + ke^{-rt}}$

Where P is the size of population at a given time, r its natural rate of increase when growth is unrestricted, t the time reckoned in appropriate units from an appropriate initial date, n a natural constant, k a constant of integration, and e the basis of Napierian logarithms.

Dislocation of Population within Metropolitan Areas

It needs some measure of the amount of dislocation or displacement which transpired between 1951 and 1960. Most segregation indices, concentration indices, and forms of the Lorenz curve fail to observe the location of the phenomena within the region. It is apparent that any measure must make some assumptions about the "expected" population at the later period.

If the rate of change in each of the subregions is equal to that for the region as a whole, then a subregion "computed" population for 1960 can be obtained from :

$$C_{i2} = \frac{\sum_{i=1}^{n} X_{i2} - \sum_{i=1}^{n} X_{i_1}}{\sum_{i=1}^{n} X_{i_1}} X_{i1} \qquad i = 1, \ldots\ldots n$$

Where X_{i_1} is the 1451 population in the ith subregion and X_{i2} is the 1960 population in the ith subregion. C_{i2} is the "computed population of the ith subregion in 1960. There are n subregions in each metropolitan area and m population (racial or ethnic) groups.

By calculating the difference between each subregion's actual 1960 population and the "computed" population summing by 2, it is possible to detain the net absolute dislocation (δ) for each population group[21]

$$\Delta = \frac{\sum_{i=1}^{n} X_{i2} - C_{i2}}{2}$$

The values for each group in the five major metropolitan areas are detailed in Table 15.9.

TABLE 15.9 : **Net Absolute Dislocation of Population (δ) in Major Metropolitan Areas of South Africa, 1951-1960**

	White	Coloured	Asiatic	Bentu	Total
Pretoria	21,238	1249	208	48551	68415
Johansberg	10,638	8250	1465	190444	222068
Durban	7,358	241	3825	28191	32672
Cape Town	24,352	28725	331	22423	56928
Bloemfontein	3,100	339	0	2364	6121

Perhaps more meaningful is a relative measure or index of dislocation (δ) which weights the net absolute dislocation by the total 1960 population. From Table 15.10 it can be observed that relative dislocation

$$\delta = \frac{\Delta}{\sum_{i=1}^{n} X_{i2}}$$

TABLE 15.10 : **Index of Dislocation of Population (δ) in Major Metropolitan Areas of South Africa, 1951-1960**

	Total	White	Coloured	Asiatic	Bantu
Pretoria	.326	.216	.346	.053	.480
Johansberg	.405	.055	.289	.108	.612
Durban	.100	.076	.020	.033	.276
Cape Town	.159	.175	.157	.073	.690
Bloemfontein	.087	.101	.109	.000	.067

Migration Rate

The study of differences in the rate of migration between various demographic, economic and social groups of population is termed differential migration. Migration rates are generally estimated with the help of following formulas :

$$\text{Out migration rates} = \frac{O}{P} \times 1{,}000$$

$$\text{In migration rates} = \frac{I}{P} \times 1{,}000$$

$$\text{Net migration rates} = \frac{I-O}{P} \times 1{,}000$$

$$\text{Gross migration rates} = \frac{I+O}{P} \times 1{,}000$$

Where O, I and P are the number of out-migrants, in-migrants and mid year population of the area. The sources of migration data are frontier traffic data, sample surveys of the census and population registration.

Internal migrations are measured by two methods (a) direct and (b) indirect. In direct method the number of migratory population are generally counted under boundary of a territory in a particular migration interval. This method becomes more easier with the help of national registration system. This type of data is also obtained with the help of questionnaire method in the census about the person's residence one, five or ten years ago. With the comparison of the place of enumeration with the place of residence $\times$ years ago, it is possible to identify the persons who have been movers and those who are migrants. In case it is assumed that each migrant moved directly to his present residence from his residence x years ago, it is possible to tabulate streams of in- and out-migration for particular place. Detailed study of migration is possible with the help of census data.

Indirect methods of measuring internal migrations area[22].
(a) Vital statistics method, (b) survival ratio method, and (c) place of birth method.

The Vital Statistics Method

With the help of the equation of population growth the estimate of net migration is possible such as :

$p_t - p_o =$ population growth

$= B_{o-t} - D_{o-t} + IM_{o-t} - OM_{o-t} = B - D + NM$

$\therefore NM = (p_t - p_o) - (B - D)_{o-t}$

With the help of the availability of reliable data for births and deaths, it is possible to estimate net migration in between two successive census counts only. Although this procedure is simple but owing to certain limitations, the idea of adjustment in the result is highly favourable.

The Survival Ratio Method

The areal unit where the vital statistics for birth and death are not available for the application of vital statistics method, in such a situation survival ratio method is used.

In case a cohort of age x survive at the first census and if the survivors among them would be of age $x+t$ at the second census taken after t years. In case of no migration, the number of persons in the cohort would be slowly diminished by the occurrence of death during the intercensal period. The total migration for this cohort will be the difference between the population at the terminal census data and the expected number of survivors.

$$Mx+t = \overset{t}{\underset{s}{P}}x+t - (\overset{o}{\underset{s}{P}}x - D)$$

Where $\overset{t}{\underset{s}{P}}x+t$ is the population of a given State at time t of those aged $x+t$ and $\overset{o}{\underset{s}{P}}x$ is the population of the State aged x at time o, and D denote deaths among members of the cohort during intercensal period.

Quite often, the number of deaths are not known. In such a situation the number of survivors $(\overset{o}{\underset{s}{P}}x - D)$ can be estimated by applying a survival ratio or proportion of population that would survive at the date of terminal census assuming no migration during the intercensal period. The term $(\overset{o}{\underset{s}{P}}x - D)$ is replaced by $\overset{o}{\underset{s}{P}}x + sx$ where $\underset{x}{ts}$ is the survival ratio. In case the population of the cohort at the second census is greater than expected number of survivors, then the difference is a measure of net emigration. In lieu of individual ages the groups of ages may be used, in such a situation the survival ratio wlll be meant for the extra age group. The total estimate of intercensal net migration is given by

$$\Sigma M_x + t = \Sigma \overset{t}{P}_x + t - \Sigma \overset{o}{P}_s x - \overset{s}{t}x$$

But in reality this method is not suitable for the estimate of net migration for those born during the intercensal period. For this purpose life table survival ratio or census survival ratio may be taken into consideration.

Life Table Survival Ratio

This is the number of persons in the stationary population of a given age who would survive to a specified older age. The life table survival ratio for single year age and age group are given by the formula

$$\overset{s}{t}x = \frac{Lx+t}{Lx} \quad \text{(Single year age)}$$

$$\overset{s}{t}x+n = \frac{\overset{L}{nx\,t}}{\underset{nx}{L}} \quad \text{(age group)}$$

Where $\overset{s}{t}x$ is the survival ratio at age x, time t. $\overset{L}{n}$ is L survivors in n years.

Census Survival Ratio

These are of two types : (a) forwarded census survival ratio, and (b) reverse survival ratio. In forwarded census survival ratio the numerator is the number of persons in age-sex group of a closed population at a given census and the denominator is the number of ten years younger in preceding census e.g. :

$$\frac{\text{Males aged 10-14 in 1971, India}}{\text{Males aged 0-4 in 1961, India}}$$

On the other hand, reverse survival method is just reciprocal of forward census survival ratio. Hence, the true amount of migration of a State is :

$$\Sigma Mx+t = \sum \overset{t}{\underset{s}{P}} x+t - \sum \frac{\overset{t}{P_c} x+t}{\underset{c}{\overset{o}{P}} x} \overset{o}{\underset{s}{P}} x$$

But such estimate is dependent upon the accuracy of enumeration in the census.

Estimating Migration from place of Birth Data

Generally in census statistics we observe the tabulated data in the form given below :

State of Birth	State of Enumeration	
	West Bengal	Tamil Nadu
West Bengal	5555	50
Tamil Nadu	66	6666

From the above table it has been drawn that :

(a) The number of persons born in each State and living there at the time of census is called non-migrant.

(b) The number of persons living in each State and born in any other State.

(c) The number of persons born in each State and living in other State.

Estimation of Migration

In case we suppose that P_{xy} denote the number of born in State x and enumerated in state y and P_{yx} the number born in State y and enumerated in x then the net gain or loss in migration is given by

$$P_{yx} - P_{xy}$$

In case $i = \sum_{y} P_{yx}$ stand for the number of persons enumerated in a State x and born elsewhere in the country, and $O = \sum_{y} P_{xy}$ the number of persons born in the State x and enumerated elsewhere in the country in such a situation I is referred as life-time in migrants and O as life time out-migrants. The difference between $I - O = M$ referred to as life net migration.

$I_2 - I_1$ = net internal change during the decade to the State among persons born outside the State.

$O_1 - O_2$ = net internal change during the decade to the State among persons born in the State.

$I_2 - I_1 + (O_1 - O_2)$ = net migrants during the decade to the State.

In these formulas, we have not taken into consideration the cases of return migration and number of deaths occur during the intercensal period. Hence correction is needed for such analysis.

Models of Migration

In order to explain the mechanisms and processes of migration it is essential to explain such complexities in a simple way through mathematical precision models. Models refer to "an integrated system of hypothesis, theories or postulates useful for prediction, experimentation and analysis at the aggregate level". Models may be deterministic, stochastic or probabilistic in nature.

Inverse Distance Law of Migration

First of all G.K. Zipf has postulated that the volume of migration is inversely proportional to the distance travelled by the migrants. This has been mathematically expressed as :

$$N_{ij} \propto \frac{1}{D_{ij}}$$

Where N_{ij} is the number of migrants from town i to j and D_{ij} is the intervening distance between the two towns.

The Pareto type simple deterministic model which shows relationship between migration and distance is given by the formula :

$$M = aD^{-b}$$

Where M is the out-migration rate and D is the distance moved. a and b are constants. Empirical results showed that the value of b ranges from 0.4 to 3.0. Hagerstrand noted a decline in the value of b over time. This indicates that migrations and counter-migrations were confined to shorter distance formerly than now. Further, the value of b was found to be smaller for towns than for rural settlements indicating a more extensive field for those living in urban places than those living in rural settlements. It contradicts Ravenstein's conclusion that the "natives of towns are less migratory than those living in the rural areas". The weakness of this model lies in the fact that as $D \rightarrow 0$, $M \rightarrow \infty$ which indicates that there is a tendency to overestimate short distance migration. But this limitation could be avoided by the use of normal type of relationship given by:

$$M = K_e^{-D^2}$$

On the other hand, Kulldorff has suggested log-normal type of relationship with the help of a formula as :

$$M = \frac{1}{\sqrt{2\pi}\ \sigma D}\ \bar{e}\ \frac{(\log D - M)^2}{2\sigma^2}$$

This gives a more precise result in comparison with the above two.

In 1940, S. Stouffer proposed his theory of intervening opportunities, in which he suggested that 'the number of persons going a given distance is directly proportional to the number of opportunities at that distance and inversely proportional to the number of intervening opportunities'. The formula is as follows:

$$N_{ij} \propto \frac{O_j}{O_{ij}}$$

Where N_{ij} is the number of migrants moving from town i to j, oj

is the number of opportunities at j, and oij is the number of opportunities between i and j. Opportunity is defined by Stouffer in terms of vacant bouses but it has also been expressed in terms of employment opportunities.

Gravity models has been also applied in migration analysis. It implies that variations in the size of destination centres will influence the strength of their attraction, as the volume of migration decreases with increasing distance from the source, hence a nearby city may be expected to have a stronger attraction than a distant one of equal size. This may be expressed as :

$$N_{ij} \propto \frac{P_i \times P_j}{D^a{}_{ij}}$$

Where N_{ij} is the number of migrants moving from town i to j, p_i is the population of town i, p_j is the population of town j, D_{ij} is the distance between towns i and j, and a is an exponent of distance. The above formula can be expressed in a number of ways so that a closer approximation of the reality is represented :

$$N_{ij} \propto \frac{x\, P_i \times YP_j}{D^a{}_{ij}}$$

Where x and y might be values of 0.5 and 1.0 respectively to take into account the different levels of attraction of the two towns, low at i and high at j.

Johnson (1952) suggested that migration from a centre might be like to the emission of a ray of light. Light is gradually absorbed by the medium in proportion to flow per unit of distance. In the same way migration of population thought to be gradually absorbed by the areas in which it moves. This idea may be written as a formula :

$Mx = Kx^{-1}\, c\, —aX$, where Mx is the percentage of the in-migration to a centre from a zone at distance x; K is a constant, and a is the absorption coefficient.

Index of Migration Flow

The index of migration flow (mle) is given by the following equation :[23]

$$mle = \frac{(t+Q) - t\,M_i}{\sum_{i=1}^{m} (t+Q) - t\,M_i/M}$$

$$\text{where } (t+Q) - t\,M_i = \frac{t+QP_i - tP_i - (t+Q) - tN_i}{P_i}$$

and $t+QP_i$ = total population of region i in the year $t+Q$.
tP_i = total population of the region i in the base year t.
$(t+Q)-tN_i$ = natural increase of population in the region i during the period t to $t+Q$, and m is the total number of regions.

Inductive Models of Population Migration

A somewhat different type of approach has been used in the analysis of Scandinevian population migration. Kant (1946) in a study of in-migration to the Estonian town of Tartu suggested a pareto-type formula :

$M = a^{D-b}$, where M is the number of migrants reduced to a standard population, D is the distance, and a and b are constants. This value for Asly town of Sweden for the period 1860-69 comes to $M = 2.1 \times 10^6\,(D-3.0)$ for out-migration.

The Concentration of New Commonwealth Immigrants

To measure the changes in the degree of concentration shown by the entire immigrant population in the borough as a whole between 1961-66, index of concentration may be computed with the help of the formula as :[24]

$$Ic = \frac{\sum_{i=1}^{n} x_i^2}{100}$$

Where x_i^2 is the square percentage of the coloured population in the *ith* area and where n is the number of areal sub-units. In this case that value is 74. The theoretical limits to this computation would be 100 where all the immigrants were concentrated into one sub-area and 1.35 where every sub-area would have an equal number of immigrants in other units.

Migratory Zones

The method involves use of an equation that defines the position of a dividing line, between the zones of influence of two adjoining cities :[25]

$$R \times B = \frac{RAB}{1 + \sqrt{\frac{HA}{HB}}}$$

where $R \times B$ is the position of the dividing line between the zones of influences of two adjoining cities ; RAB is the distance between two cities : HA and HB are the population size of the A and B cities. The dividing line joins those points on roads linking A and B where the probability of a trip to either city is identical and equal to 0.5 ($R \times A = 0.5$; $R \times B = 0.5$). We measured RAB along railroads and motor roads to have the measurements reflect the transport accessibility of areas.

Migration Preference Index

This index is defined as "the ratio of the actual to the expected number of migrants in a stream when the expected number is directly proportionate to both the population at origin and the population at destination".[26] The expected number of migrants in each stream is obtained by distributing the expected total number of out-migrants from a given province to all destination among the other provinces in proportion to their population. The preference index is given by the following formula :

$$PI = \frac{M_{od}\ (\Sigma P_i - P_o)}{m\ P_o\ p_d} \times 100$$

where M_{od} = actual number of migrant from o to D

P_o = population at origin o,

P_d = population at destination D

P_i = national population

m = proportion of inter-area migrants in the national population.

Economic Potential of Internal Migration

Tachi and Oyama have shown that in the case of Japan there is a very close association between migration and economic disparity among the various regions. They measured economic disparity by "economic potential of internal migration" (j) expressed by the following formula :[27]

$$J = \frac{P - P_i}{P_i} = \frac{Y_i}{P_i} - \frac{P}{Y} - 1$$

where P = denotes the total population of the country
Y = denotes the per capita national income
P_i = denotes the population of region "i"
P_i = denotes the expected population of region "i" under assumption of theoretical distribution of population
Y_i = denotes the per capita income of region i.

Marriage Migration

Libbee derived a formula to predict in effect the probability that any straight line or fixed length, D taken at random and having at least one end within a circle of radius R will lie completely within that circle. The apparent simplicity of the problem is deceiving and in the end an approximate formulation was obtained for the probability value G. Only a graphic solution could be found for the next step, the calculation of D, given G and $R.G$ was then replaced by the ratio of females born outside the village, and R by the value $\sqrt{A/\pi}$. A being the district area, the corresponding value D read from the derived probability tables, was taken as the estimated median marriage distance.

If in circular district x with radius r, everyone of an evenly spread population of points (or women) migrates a distance d, what percentage of the population (G) will find their new homes within the district ? At a distance d or more away from the circumference of the district, 100 per cent of the population will terminate within the circle. That area is represented by A. In the outer ring, labled B, only a portion of the migration will be within the district. The formula for G can be written as :

$$G = \frac{\text{area } A + PAV\,(\text{area } B)}{\text{area } x}$$

or

$$G = \frac{\pi(r-d)^2 - PAV\,[\pi\, r^2 - \pi\,(r-d)^2]}{\pi\, r^2}$$

The problem now is to find PAV. This solves the sine of angle ϕ and as an indicator of the chance a person living on the circumference of the district has of migrating within the district (the sine of the angle must be divided by 2). This figure (POB) is then averaged with the probability that a person on the inner circumference of area B will migrate within the circle (POA). This is 100 per cent when $r \geqslant d$, but when d becomes greater than r, the probability drops to 0. This creates an odd-looking and illogical break in the results, caused by the assumption that every female migrates the marriage distances instead of having the distance distributed around the medium. Sacrificing mathematical rigour for intentive appeal. POA was set equal to 100 per cent in all cases. The likely effect is to slightly over predict G when $d>r$. Find PAV:

POA = Probability of indistrict migration at $(r-d)$ or the centre of the circle when $d<r$.

POB = Probability of indistrict migration at district circumference, equal to sin $\phi/2$.

PAV = Mediam probability of indistrict migration from any points in area B. The value between POA and POB which falls on the circle dividing area B into two donuts of equal area is taken as the last estimate, for according to the assumption of uniform population distribution, half the population would be in each segment.

Let circular district x be centred at the origin with radius r, and the marriage circle be centred at the intersection of the x-axis and the circumference of district x. The formula for district x is the general formula for a circle :

$$X^2 + Y^2 = r^2$$

and the formula for the marriage circle is :

$$(x - r)^2 + y^2 = d^2$$

Solving simultaneously for x : $x^2 + y^2 = r^2$

$$\frac{-(x - r)^2 - y^2 = -d^2}{x^2 - (x - r)^2 = r^2 - d^2}$$

or

$$x^2 - x^2 + 2xr - r^2 = r^2 - d^2$$

or

$$2xr = -d^2 + 2r^2$$

Prog (3) : $x = \dfrac{2r^2 - d^2}{2r}$

and

Prog (4) : $Z = r - x$

and since

$$Z^2 + y^2 = d^2$$

Then $Y^2 = d^2 - z^2$

or

Prog (9) : $Y = (d^2 - z^2)^{\frac{1}{2}}$

and

Prog (10) : $\sin Q = Y/d$

and

Prog (11) : $POB = \sin Q/2$

The radius (RAV) of the circle dividing area B into two equal parts was found by equating the inner and outer segments of area B :

$$\pi(RAV)^2 - \pi(r - d)^2 = \pi r^2 - \pi(RAV)^2$$

Cancelling the π's and re-arranging

$$2(RAV)^2 = (r-d)^2 + r^2$$

$$\text{Prog 4 : } RAV = \left[\frac{(r-d)+r^2}{2}\right]^{\frac{1}{2}}$$

Since POA 1, the marriage probability at RAV is :

$$\text{Prog (12) : } PAV\ POB = \left[\frac{(R - RAV)}{D} \times (1 - POB)\right]$$

and G equals, after cancelling π and reducing

$$\text{Prog (13) : } G = \frac{(r - d)^2 + PAV\,(2\,rd - d^2)}{r^2}$$

P_x, P_y and P_z are dummy variables inserted into the program to prevent impossible solutions. Prog [1] and [2] assigns values to d and r. Prog. [14], [15] and [16], reduce and round off the answers".[28]

Effectiveness of Migrations

The index of effectiveness of migrants is as below :

$$M_e = \frac{I_m - E_m}{I_m + E_m},$$ Where M_e is the index of effectiveness ot migration, I_m number of immigrants to the area investigated and

E_m number of immigrants from the area investigated. The index of effectiveness of migrations (M_e) in theory can take the values :

$$-1 \leqslant 0 \leqslant +1$$

From the above formula the following possibilities can occur :

(1) $M_e = -1$, when $I_m = 0$, and $E_m \neq 0$;
where $\neq$ is a sign of not equal to
(2) $M_e = +1$, when $I_m \neq 0$, and $E_m = 0$
(3) $M_e = 0$, when $I_m = 0$, and $E_m = 0$;
(4) $M_e = 0$, when $I_m = E_m$;
(5) $-1 < M_e < 0$ when $I_m < E_m$, and,
(6) $+1 > M_e > 0$ when $I_m > E_m$.

The value $M_e=0$ only occurs when immigration is numerically equal to emigration ($I_n=E_m$). To avoid the use of fractions the index may be multiplied by 100 or 1000. The absolute value of the index of effectiveness of migration is of decisive importance for the study of the problem, because the symbol ($\pm$) only indicates the character of the balance (increase or decrease).[29].

Fields of Migration

Applebawn and Cohen (1961, p. 81) described the shape of the trading areas of outlying shopping centres as parabolically elongated away from the central business district. Similarly Park (1929) was able to show that the circulation fields for daily newspapers in part of the central United States (South Dakota) were asymetrically elongated away from the areas of intense competition.

In the migration fields Stouffer (1940), a sociologist, has argued that we should not in any case expect fields to be circular because there is no necessary deterministic relation between migration and geographical distance. His hypothesis may be stated in the form.[30]

$$M_{ij}=(N_j/N_{ij})K$$

where M_{ij} is the expected interaction between location i and location j, where N_j is the number of opportunities at location j, N_{ij} is the number of opportunities in the intervening area between i and j, and K is a constant. This intervening opportunity model states simply that the amount of interaction over a given geographical distance is directly proportional to the opportunities at that distance but inversely proportional to the number of intervening opportunities. This model has been used by Isbell (1944) and Folger (1953) in studies of internal migration in Sweden, and the Tennessee valley in the eastern United States. In both cases the intervening opportunity model gave better prediction than the simple gravity type model.

Fragmented Fields of Migration

Consist of a contiguous inner area with one or more outliers of high interaction. Hagerstrand (1957, pp. 126-54) has drawn together a number of regional examples of this type of field in migration movements.

Hagerstrand finds three common elements in all these cases : (1) The importance of the information chain represented by individual contacts, (ii) the division of migrants into active and passive elements with the latter following the definite channels made by the farmer, and (iii) the random element in the initial choice of areas. While it was not possible to build a comprehensive feedback model to accommodate all three elements, Hagerstrand has suggested a short term model. This is given by the expression

$$M_{ij} = (V_j - I_j)/P_j.K$$

where M_{ij} is the number of migrants from location i to j, V_j are the vacancies at j. I is the information level about these vacancies existing at the source i, P_j is the population of the destination j and K is a constant. The volume of migration is related to the vacancy density of the destination and the level of information about it. Here distance is introduced indirectly through the information level since more information is likely to be available about very near places than very distant places. Hagerstrand has tested his model on Swedish migration data using Monte Carlo methods to simulate the random selection of areas and their relative growth. Some indication of the considerable relative success of the feedback model is shown in Table 15.1 where its predictions are compared with a simple intervening opportunity level.

TABLE 15.11 : **Comparison of Intervening Opportunity and Feedback Models in Predicting Migration**

Distance zone (kms)	0.19	20.39	40.59	60.79	80-99
Observed Migration	86	132	42	18	12
Predicted Migration Using Model					
Intervening opportunity	105	89	39	29	29
Feedback	89	131	41	17	13

Source : Hagerstrand, 1957, pp. 125, 153.
Dadesjo, Sweden, 1946-50.
(Estimates within 10 per cent of observed migration)

Methods for Estimating the Amount of Net Rural-Urban Migration

In a fixed territorial limit of urban areas the number of increase in the urban population attributable to net migration can be estimated on the basis of the formula :

$$(1)\ N_m = P_{u2} - P_{u_1} - B_u + D_u$$

where, N_m is net rural-urban migration, P_{u1} and P_{u_2} represent the size of the urban population at the start and the end of the period, and B_u and D_u represent the number of births and deaths occurring in the urban population (place of occurrence) during the period.[31] (This is a derivation of the formula which expresses change in Population size : $P_2 = P_1\ (B - D) + (I - O)$, with I and O representing number of persons who enter and leave the territory occupied by the population).

Such formulas yield the balance between the number of people who enter and leave the urban territory during the period, but they do not necessarily reflect rural-urban movement involving a change in the place of residence.

An overestimation stemming from foreign immigration can be avoided if urban residents are tabulated in census reports by country of birth. In areas of the availability of such data the formula for net rural-urban migration is :

$$(2)\ N_m = P_{u2} - P_{u1} - B_u + D_u - P_u f_2 + P_u f_1$$

Where P_uf_1 is the number of foreign-born persons in the urban territory at the start of the period and P_uf_2 the number of foreign born in the urban territory at the end of the period.

The international migration also taken into account with the help of data which can be used to determine the number of immigrants arriving in the country during the period and their destination in terms of rural or urban. Such information may be represented by the formula :

$$(3) \quad N_m = P_{u2} - P_{u1} - B_u + D_u - F_{iu}$$

where F_{iu} is the number of immigrants destined to settle in urban territory.

In case only the number of immigrants during the period have to be determined (i.e., not their destination in terms of rural or urban) the appropriate formula is :

$$(4) \quad N_m = P_{u2} - P_{u1} - B_u + D_u - F_i\,(P_{u1}/P_1)$$

Where F_i is the number of foreign immigrants, and P_1 represents the total population of the country at the start of the period. Its application involves the distribution of immigrants among the urban and rural populations in proportion to the size of the two populations at the start of the period. It varies from place to place and time to time.

In case the nativity of the urban residents can be established at the beginning of the period, the adjustment formula for immigration is :

$$(5) \quad N_m = P_{u2} - P_{u1} - B_u + D_u - F_i\,(P_uf_1/P_1).$$

According to this formula the foreign immigrants who arrive during the period settle in urban and rural areas in much the same proportion as those who arrived before the period.

The overseas departures of origin are such that (rural or urban) emigrants can determine a reliable estimate of net rural-urban migration by these formulas [32]:

$$(6) \quad N_m = P_{u2} - P_{u1} - B_u + D_u - P_uf_2 + P_uf_1 + U_e.$$

$$(7) \quad N_m = P_{u2} - P_{u1} - B_u + D_u - F_{iu} + U_e.$$

$$(8) \quad N_m = P_{u2} - P_{u1} - B_u + D_u - F_i\,(P_{u1}/P_1) = U_e.$$

$$(9) \quad N_m = P_{u2} = P_{u1} - B_u - D_u - F_i\,(P_uf_1/P_1 = U_e$$

with U_e representing the number of urban residents who have emigrated during the period.

It takes into account F_{iu} which leads to an overestimation of net rural-urban migration, but the reverse is true when U_e is ignored. Thus errors in estimation can be substantially reduced if the balance between the two, F_{iu} and U_e, can be determined. In case the two numbers are not known the balance can be estimated indirectly. This is accomplished by first determining the balance between immigrants and emigrants for the nation as a whole (i.e. regardless of their origin or destination) during the period on the basis of the formula :

$$(10) \quad I_e = P_2 - P_1 - B + D$$

where I_e is the excess of immigrants over emigrants and other letters relate to the nation as a whole. For the given I_e the appropriate formula for estimating net rural-urban migration is :

$$(11) \quad N_m = P_{u2} - P_{u1} - B_u + D_u - I_e \,(P_{u1}/P_1)$$

According to this formula the net loss or gain of population through international migration is shared by the rural and urban components in proportion to their size at the beginning of the period.

The assumption is that the rural and urban population had the same rates of natural increase during the period. This helps in estimating the size of urban population at the end of the period. There would be no net rural-urban migration by applying the rate of growth of the total population to the urban component. The difference between the estimated size and the actual size of the urban population is an estimate of the amount of net rural-urban migration :

$$(12) \quad N_m = P_{u2} - P_{u_1} \left(\frac{P_2 - P_1}{P_1}\right) - P_{u_1}$$

The above formula favours P_1 for higher rates of fertility for rural than urban population. Formula 12 produces an estimate of the minimum amount of net rural-urban migration.

Even in the absence of vital statistics some adjustment for differential fertility can be made, provided that the ages of the urban population are tabulated in census reports. This adjustment is reported through the formula as given below :

$$(13) \quad N_m = P_{u2} - P_{u1} \left[\left(\frac{P_2 - P_1}{P_1}\right) - \left(\frac{T_c \,.\, U_c}{P_1 \, P_{u1}}\right)\right] - P_{u1}$$

where T_c and U_c represent the number of children in the total population and the urban population at the end of the period who, as judged by their reported ages, were born during the period. The above formula rests on two assumptions : (a) There is no major rural-urban difference between the mortality of children or people of any age groups, and (b) there is no net rural-urban migration of families with children in these age groups.

Fertility can be used in attempting to adjust rural-urban differentials of migration when age by sex is tabulated in census reports at the beginning and end of the period :

$$(14)\ N_m = P_{u_2} - P_{u1}\left[\frac{P_2 - P_1}{P_2} - \left(\frac{T_{c1}+T_{c2}}{T_{w_1}+T_{w2}} - \frac{U_{c1}+U_{c2}}{U_{w1}+U_{w2}}\right)\right] - P_{u1}$$

In the above formula c is the number of children 0-4 and w is the number of women 15-49. This formula is subject to all assumptions that apply to formula 13. It should also be noted that the age group 0-4 is one of the most unreliable as far as accuracy of census enumeration is concerned, but this is compensated in case where the same enumeration bias applies to both the urban and the total population.

Simple Allocation Model of Migration

Here a simple allocation model of United States inter state migration is presented and later modified to include the concept of opportunity cost.

The model viewed as 'traditional' is the following [33]:

(1) $A_{ij} = f\,(d_{ij},\ P_j,\ U_j,\ W_j$ random errors)

where A_{ij} = the number of persons, five years of age and older, residing in state j on April 1, 1980, who resided in state i on April 1, 1975, divided by the total number of persons, five years of age and older, who resided in state i on April 1, 1975 and in another state on April 1, 1980.

d_{ij} = 1975 highways mileage between the principal city of state i and the principal city of state i

P_j = Population of state *j*, 5 years and older, in 1980.

U_j = Per cent of the civilian labour force unemployed in state *j* during oppt. the first week of April, 1980.

W_j = Median 1974 money income of males living in state *j* in 1980.

We have selected the allocation of 1975-1980 U.S. inter-state migrants as our dependent variable. Our model then attempts to explain the number of migrants from state *i* to state *j* as a proportion of the total out migrants from state *i*, as only explanatory variables are relevant in an allocation model. The origin proxy variables partially account for cyclic costs for a potential migrant but are irrelevant for actual migrants in the allocation for the aggregate migration model.

Function of Migration Potential

This is based on a spatial interaction model which is developed from the assumption that the spatial aspect of human behaviour may be analysed like physical matter treated by the early set of Newtonian equations.

Similarly population is considered as a mass located at a point in space. Interaction (I_{ij}) between two populations (P_i, P_j) is directly proportional to the product of the two populations (P_i, P_j) and inversely proportional to the distance intervening between the points of two populations D_i-J) or to that distance taken to same exponent ($D_{xi}-J$).[34] The spatial interaction is a probabilistic concept which treats only sets, or sub sets, of population. The probability (P) for interaction by a single particle of a mass, or individual of a population is not at issue, but is regarded as equal to the mass probability to interact for all individuals of a population located at a point in space. Often, P is not known, except as a derived function of an interaction model.

The simple gravity model is an analogue of the gravitational energy law of Newton. Mathematically, in its basic form, interaction is defined as :

$$I_{ij} = \frac{P_i \ P_i}{D_{xi-j}} \quad \dots (1)$$

Potential (${}_iS_j$), as Stewart defines the term, is essentially the influence of a population located at one point on a population located at another point. This may be represented as potential produced at point i by a population located at point j is defined as the population at point j divided by the intervening distance (D_i—J).

This formulation is analogous to Newton's gravitational potential law. In its basic form :

$$ {}_iS_j = P_i/D_{xi}-j \qquad \text{... (2)} $$

Expanding the basic models, to estimate interaction by groups of individuals among all points within an area, or to estimate potential exerted at all other relevant points by a population located at point j, the equations are :

$$ I_i\, n-j, \qquad n=\Sigma(P_i,\, nP_{jn}/D_{xi},\, n-i,\, n) \qquad \text{... (3)} $$

and

$$ i,\, {}_nS_j = \Sigma\,(P_i/D_{xi},\, n-j) \qquad \text{... (4)} $$

Adopting (4), potential exerted upon a population at point i by populations located at all other points (j, n) is given by :

$$ {}_iS_j,\, n = \Sigma\, P_i\, n/D_{xi},\ j,\, n) \qquad \text{... (5)} $$

In case spatial interaction models differentiated from gravity and potential models the ramification involves a more restrictive gravity and potential models. These are regarded as an example of logically affected spatial interaction of masses of population at varying locations.

Index of Primacy

Browning and Gibles have proposed an index of primacy.[35]

$I = \dfrac{P_1}{P_2}$, where P_1 is the population of the first city and P_2 the population of the second. A high degree of primacy among the towns of Bihar usually means low conformity to the rank-size rule.

Potential of Population

The concept and term potential of population was introduced by a non-geographer-astrophysicist J.Q. Stewart. The use of population potential has been further developed by W. Warntz and other American geographers.

J.Q. Stewart calculates population potential at point j of the area studied according to the formula :

$\frac{E_j P_i}{D_{ij}}$, where P_i = population number in place i, D_{ij} = distance between this place i and the place j investigated.

The calculations proper— it was necessary to carry out tens of thousands of operations, though rather simple ones—were carried out by a common type computer according to the formula :

$$P_i = \frac{L_i}{A_j} + \sum \frac{L_j}{D_{ij}}$$

Where P_i = Population potential of the areal element i, L_i = Population number in point i, A_i = index of reduction in point i, L_i = Population number in any other point j, D_ij = distance between points i and j.

Distances are then calculated by the computer using the formula :

$$D_{ij} = \sqrt{(x_i - x_j)^2 + (Y_i - Y_j)^2}$$

which is a Pythagoras theorem. The results were typed by the computer in place of the position of point i. Thus it was necessary to incorporate into the map and to interpolate the isolines of population potential.

The formula we use is formally identical and we only introduce the value L_{io} instead of $L_i - L_{io}$ being equal to h_{pi}, where P_i = area of district, h = average density of population in the *CSSR*. The same expression is valid for index j; the other expressions remain the same.

One single corresponding calculation of the ideal field is sufficient, since

$$P_i = \frac{L_{i_o}}{A_i} + \sum \frac{L_{io}}{D_{ij}}$$

$$= h\left[\frac{P_i}{A_i} + \sum \frac{P_i}{D_{ij}}\right]$$

The caloulation can thus take the areas of districts into account, the resulting potential is then an h multiple, when the density h_1 is used, we can get over to the case for density h_2 by multiplying the result by the coefficient $\frac{h_2}{h_1}$.

Central Tendency in Areal Distribution

Warntz and Neft (1960) have shown how three familiar statistical measures of central tendency, the mean, median and mode, may be translated into areal dimensions. The mean centre of a distribution they define as the point where[36]

$$\int R^2 \cdot G\,(dA)$$

is minimized, while the median centre is defined as the point where

$$\int R \cdot G\,(dA)$$

is minimized. In both expressions G is the density of population over a very small part of area dA and R is the distance from each part to the mean or medium point in question. In the same way, mode can be defined in terms of the position of high point on density surface.

Measures of Range or Deviation

Similar to the measures of central tendency in statistical descriptions of frequency distribution are the measures of range or deviation Stewart and Warntz (1958) have drawn 'dynamical radius' of a population. The formula is as follows :

$$\sqrt{\frac{\{\Sigma\,(p_d{}^2)\}}{P}}$$

Where P equals the position of small area, d_i is the distance from the mean centre, and P is the total population.

Potential Model and the Spatial Distribution of Population

Rural Density (DR) (in person per km^2) is found to vary as the square of the potential (V) (in persons per km)

We have [37]:

$$DR = KV^2 \qquad \dots (1)$$

where K is a constant which has been determined for Vidarbha as

$$\frac{\Sigma\,(DR/V_2)^{2}}{N} \qquad \dots (2)$$

which comes to 0.005067. The expected rural density was calculated from equation (1) with respect to the potential square, so that the expected value from the actual one can immediately be known.

Apart from calculating the expected rural population density the expected rural population can also be calculated for each tehsil by using the equation (Stewart, 1987) :

$$P_r = K_1 V^2 A \qquad \ldots (3)$$

P_2 is the rural population of a given small area A having a potential V_2. The overall ratio of

$$K_1 = \frac{\Sigma P_r}{\Sigma V_2} \qquad \ldots (4)$$

which gives the value of constant when multiplied into each V_2 compares well with the observed values. The 3 per cent error between the actual and the expected rural population shows the regularity of the agreement. The equation thus describes an equilibrium of rural population in relation to total population.

Occupancy Rate

A measure of the degree of overcrowding of people in any area, calculated as :

$$= \frac{\text{No of Population}}{\text{No. of Habitable Rooms}}$$

In general, living conditions are comfortable up to an occupancy rate of about 1.0, and overcrowding occurs above a rate of about 1.5. (All rooms other than bathroom and kitchen are known as habitable room).

Centrographic and Potential Measures of Distribution

The mean centre of a distribution is the exact equivalent of the arithmetic mean of conventional statistics. It is the balancing point or 'centre of gravity' of the distribution. Several alternative methods are available for the computation of the mean centre of a distribution, but the most convenient system depends on the use of grid-to-ordinates. A square grid is superimposed on a map showing the base areal units for which data is available. On the assumption of even distribution within each base area, the geographical centre of each unit is determined and expressed in terms of the two co-ordinates of the grid, X and Y. Each X and Y measurement is weighted by the population of the base unit concerned. The sum of these terms for the whole city, divided by the total city population, gives weighted mean positions along each of the co-ordinates. The mean centre of the distribution may be found at the intersection of the mean of the X's and the mean of the Y's (formula 4).

Formula : 4 $= \bar{X} \; \frac{\sum_i (X_i \; P_i)}{\sum_i P_i}$, where X_i = Vertical co-ordinate of the ith area

P_i = Population of the ith area

$$\bar{\Delta} = \bar{Y} = \frac{\sum (Y_i \; P_i)}{\sum_i P_i}$$

Y_i = Horizontal co-ordinate of the ith area

Two measures are available to describe the scatter of a distribution around its centre. Both may be computed with reference to any given point, e.g. the centre of Central Business District or the point of highest land value. The simple measure, the Mean Distance Deviation (*MD*), is an expression of the average distance separating the units of the distribution from the given centre. The formula for computing the mean distance deviation is given in formula 5, where d_i represents the linear distance between the geographical centre of the ith base unit and any given point O, and P^1 is the population of the ith unit,

Formula 5 : $$MD\Delta = \frac{\sum_i (P_i \; d_i)}{\sum_i P_i}$$

The second measure of dispersion, the Standard Distance Deviation (*S*), is strictly analogous to the standard deviation of conventional statistics. If a system of grid co-ordinates has been used to compute the mean centre of population it is a simple matter to extend the tabulation in order to derive the standard distance by either version of formula 6 :

Formula 6:

(a) $$S\Delta = \sqrt{\frac{\sum (P_i \, [X_i - \bar{X}^2])}{\sum P_i} + \frac{\sum (P_i \, [Y_i - \bar{Y}^2])}{\sum P_i}}$$

(b) $$S\Delta = \sqrt{\frac{\sum (P_i \, X_i^2}{\sum P_i} - \bar{X}^2 + \frac{\sum P_i Y_i^2}{\sum P_i} \; \bar{Y}^{-2}}$$

Formula 6 gives both the standard distance *S*, and the two component 'latitude' and 'longitude' deviations, S_x and S_y, which comprise it. The standard distance itself is invariant under the rotation of co-ordinate axes, but this is not true of its *x* and *y* components.

For this reason no truly satisfactory means has yet been devised for plotting the standard distance on a map, although a possible solution to this problem, involving respectively the maximization and minimization of the x and y deviations has recently been proposed (Bachi, 1963).

An interesting theorem connects the standard deviation to the average root-mean-square deviation of a total frequency distribution. The proof of the theorem can be readily generalized to include areal distributions and it may then be shown that the General Standard Distance Deviation ($S\Delta$) is equal to the standard distance deviation times the square root of two (formula 7). Thus, the amount of scatter revealed in the distances separating all the units of a distribution can be shown to be related to the amount of scatter of those units around the mean centre of the distribution.

Formula 7 : $S = S\Delta = S\Delta\sqrt{2}$

Closely related to centrographic measures of distribution are the concepts variously known as population potential, market potential, and workplace potential. For a full discussion of the population potential concept the reader is referred to an article by Stewart and Warntz (1958B). The population potential (V) at a point may be considered as a measure of the influence of total population on that point. According to its strict definition, the potential at point O, is obtained by measuring the distance from that point of each individual in the population inhabiting the universe of territory under study, computing the reciprocals of each distance, and summing the reciprocals. In practice a simpler procedure is followed. The universe of territory is subdivided into a manageable number of sub-units, the population of each unit is ascertained and, on the assumption that this population is concentrated at a single central point within each unit, the distance is measured between this centre and point O. The population at point O is then given by formula 8.

Formula 8 : $V_o = \sum_{1}^{k} \frac{P_i}{d_i}$ Where d_i is the distance of the ith areal unit from point O

Conversion of population potential into market potential is simply attained by the use of an income or expenditure multiplier. Work-place potential (Duncan and Duncan, 1960) is defined, anologously to population potential, as the sum over all workplaces in the city of the reciprocals of the distances separating each workplace from any given point. The measure is interpreted as an index of the accessibility of a point to workplaces, on the assumption that accessibility declines as distance increases. Exploratory work using the workplace potential in the analysis of intra-urban residential structure is part of the Chicago Urban Analysis Project. The population potential measure has primarily been used in inter-urban studies of economic activity, but as a measure of accessibility and of 'sociological intensity' offers several possibilities for inter-urban research.[38]

Economic Density of Population

The notion of economic density of population was proposed in 1934 by the French demographer Simon. The formula is as follows [39]:

$\Delta = 100\frac{\delta}{\alpha}$, where δ is the index of population size, and α the general index of production for the same year. Later refinements attempted to relate population to its production capacity, first by incorporating into the formula the index of productivity α'

$$\Delta' = \frac{100\delta}{\frac{(100\alpha)}{\alpha'}}$$

Then by replacing the index of production α by an index of latent productivity as :

$$D = \frac{100\ \delta}{\frac{100\ a}{\alpha'}}$$

$$\therefore \frac{D}{\Delta'} = \frac{\alpha}{a},$$

$$\Delta' = \frac{a}{\alpha}D$$

Simon believed that he had discovered an index capable of determining the optimum density of population in relation to a base period. If Δ' exceeded D and a exceeded α there was a tendency to under-production and under population; if they were equal there was an optimum population.

Economic density of population is conceived as the ratio between the total rural population and the total economic area of the countryside. Economic area has been obtained by converting all the economic resources into land units. This has been done by calculating, in the first place, an average yield (in terms of rupees) per acre of agricultural land for each district and for the whole region. In the second place, the value of the total agricultural production for each district may be calculated on the basis of price denoted in the season and crop report. Lastly, the total population may be divided by total economic area to obtain an economic density of population.

Instead of the above general index of production, Simon also proposed (a) the general index of consumption, and (b) the general index of all economic activities. Further denominators suggest (c) average income per person, (d) standard of living and (e) available diets.

Measures of Dispersion of Population

Various statistical methods may be employed in the study of population dispersion, which is complementary to the study of central tendency. The standard distance deviation is found by dividing the mean centre (ΣPr^2) by the total population (P) and then by taking the square root of the resulting mean square[40]:

$$S_r = \sqrt{\frac{(\Sigma Pr^2)}{P}}$$

The second method is employed in the derivation of the standard distance deviation. Isolines of aggregate squared-distance deviations are concentric circles about the mean centre, regardless of population distribution.

The mean distance deviation is likewise the counterpart of the mean deviation of linear statistics, and it measures the arithmetic mean distance of individuals from the median centre. It is determined by dividing the medium centre by the total population.

$$Md_r = \frac{(\Sigma P_r)}{P}$$

The first method is employed in deriving the mean distance deviation and it is represented by isolines of aggregate travel distance. The latter are not concentric circles unless the population distribution exerts circular symmetry.

Measuring the Concentration of Population

This study deals with methods of defining the mean distance in between communities and population, as well as with establishing other indicants of the concentration of population. The mean distance of the population (D') is calculated similar to the mean distance of communities (D), but a weighted average is used (according to the relative weight of the population on both extremities of the investigated distance). The distances are measured on the basis of the trigonometric net. The mean distance of the population is calculated from the following formula[41]

$$D^1 = \frac{\Sigma\, d_i,\, o_i}{\Sigma\, o_i}\ , \text{ where } o^i = \frac{OP + OK}{2}\ ,$$

d_i is the individual distances measured; OP is the number of inhabitants of the settlement at the beginning, and OK expresses the number of inhabitants of the settlement at the end of the corresponding distance.

Further, the indicant of the relative concentration of population is being calculated : $R = \frac{D^1}{T^1}$, where T equals the theoretic average distance between the inhabitants (on the assumption of an even distribution of people). This method facilitates to compare the relative concentration of population.

Index of Metropolitan Concentration

An index of metropolitan concentration was applied to see how their distributional pattern fitted into the general population. The index expressed as a percentage ratio of the Chinese in a given LGA to that of the total population in the same LGA[42]:

$$\frac{\text{LGA's\% share of SMA's Chinese}}{\text{LGA's\% share of SMA's Total}}$$

Based on the 1954 census, the 100 householders were then distributed between the concentric zones on a proportional allocation basis with the following formula applied :

$$\frac{\text{Number of Chinese in Zone}}{\text{Total Chinese in SMA}} \times 100$$

where SMA = Sydney Metropolitan Area.

LGA = Local Government Area.

Concentration of Population

To study the distribution of population of any area unit during (1961-71) location coefficient (Deshpande, 1941, pp. 268-86; Sastri 1942, pp. 110-14) can be used with the help of the formula:

$$LC = \frac{P_i}{A_i}$$

where P_i = percentage of the total population growth during 1961-71 in ith area, A_i = percentage of area contained in the ith areal unit. Location coefficient is a measure of concentration of distribution in a region. If the LC for a district is 1, then the share of the growth of population is what it should have in respect of area. The higher the LC value, the higher is the concentration of population growth in the district.

The Index of concentration (Chorley and Haggett, 1970, p. 241) can be calculated with the help of formula [43]:

$$D = \frac{1}{2} \sum_{i=1}^{K} \left\{ X_i - Y_i \right\}$$

where X_i = percentage of the population growth in the ith areal sub-unit, Y_i = percentage of the area contained in that unit. The O index shows that the population growth is uniformly distributed. The higher the index, the higher is the concentration.

Location quotient is a method which shows the relationship between population growth in the decade and size of the population at the beginning of the decade. The formula is as follows:

$$LQ = \frac{X_i}{Y_i}$$

where X_i = percentage of the population growth during 1961-71 in the ith area and

Y_i = percentage of the total population in the ith areal sub-unit. Index 1 shows the equal distribution which is less than 1 indicates an under representation of the percentage of population growth.

Index of Segregation

An index of segregation (Is) shows the extent to which the specified sub-group is residentially separated from the rest of the population.

The formula is : $Is = 1 - \frac{ID}{\frac{\Sigma Xa_i}{\Sigma Xni}}$ where : ΣXai is specified sub-group ID is distance and Xni is rest of the population.

Population Densities of Urban Fields as Statistical Trend Models

Investigation of the general spatial character of population densities within urban regions have led to some well known formulations. Clark (1951) suggested a negative exponential curve of declining density with distance from the city centre with the form[44]

$$Dx = d_0e^{-bx}$$

where Dx = population density of distance x,

d_0e = interpolated down-town density,

x = radial distance from the city centre,

b = the density gradient (Berry, 1963)[45]

Subsequent work by Newling (1967) suggested a modified form of this model which allows for the observed lower densities were the centre of cities, viz.

$$dx = d_0e^{bx-cx^2}$$

In this form the curve of density is convex with the maximum height at some distance from the down-town centre. A more general statement of population density-surfaces is given by Tobler (1967) for a cylindrical model

$$D = P/\pi R^2$$

a conical model $D = (3P/\pi R^2) - (3Px/\pi R^3)$

a parabolic model

$$D = (6P/\pi R^4)\ (X - R)^2$$

and a Cosine model

$$D = \left[MP/R^2 (\pi^2-4) \right] . \left[1-\cos \left(\frac{\pi x}{R}\right)\right]$$

In Tobler's formulation the total population of a city (P) and its radius (R) is linked by $R = .0219 \ P^{0.44}$

while

$$P = 2\pi \int_o^R D dx_1$$

In all the foregoing models, the surfaces were assumed to be isotropic. Inspection of density surfaces within existing urban areas suggests this assumption is unjustified and that a more general model is appropriate. One widely used as isotropic model, the trend surface model, decomposes a variable Z (e.g., population density) into spatial trends and local error terms, viz.

$$Z_{ij} = \eta(u_i, v_j) + \epsilon_{ij}$$

where Z_{ij} = population density at location (i, j),

$\eta(u_i, v_j)$ = Spatial trend in terms of an orthogonal (u, v) coordinate system,

and ϵ_{ij} = local error term.

The spatial trend equation is conveniently described by a set of polynomial equations of increasing order, e.g.,

Linear : $Z_{ij} = a_0 + a_1 u = a_2 v$

Quadratic : $Z_{ij} = a_0 + a_1 u + a_2 v + a_3 u^2 + a_4 uv + a_5 v^2$

Cubic : $Z_{ij} = a_0 + a_1 u + a_2 v + a_3 u^2 + a_4 uv + a_5 v^2 + a_6 u^3 + a_7 uv^2 + a_8 uv^2 + a_9 v^3$

Higher order surfaces carry the ability to model increasingly more complex density surfaces. Thus linear equations can describe only 'shed-roof' type surfaces distinguished by differences in azimuth and regional dip, while quadratic surfaces can accommodate a single extremum (i.e., closed contour). The relationship between the number of extrema (E) and the order of the surface (k) is given by $E_k = (k\text{-}1)^2$ Computer programmes for calculating surfaces up to $k = 15$ exist but the rapid increase in the number of terms in the equation and the problems of evaluating the significance of the trends described suggest diminishing returns with successively higher-order surfaces. Thus where recurrent periodicities in population densities are observed or inferred the polynomial model is less relevant than its harmonic counterpart in which variations in *Zij* are modelled in terms of the sine-cosine functions of the double Fourier model. This may be expressed as

$$Z_{ij} = F(U,V) + \epsilon_{ij}$$

where $F(U,V)$ is the Fourier trend.

The mathematical restrictions implicit in the polynomial and double Fourier models make the portrayal of relatively simple spatial structures difficult. Assume that population density declines from its maximum value near the city centre in a series of waves with increasing wavelength but decreasing amplitude (an assumption not wholly at variance with classic central place settlement theory). Neither polynomial or double Fourier models can accommodate the combination of trends efficiently, but a heuristic combination of the two models may prove useful. James (1967) has suggested a model with the form

$$Z_{ij} = bx + c \sin \frac{2\pi x}{d+ex}$$

where Z_{ij}, b, and x are as defined above and where

c=amplitude of waves,

d=initial wavelength (i.e., near city centre),

and e= rate of change of the wavelength.

For general linear models the least squares function may be minimized through the solution of simultaneous equations but with non-linear functions approximations using steepest descent technique are suggested.

The sensitivity of polynomial and double Fourier series to boundary conditions and the difficulty of using such models extrapolate beyond the spatial margins of the data has also *bd* to the investigation of multiple isotropic models. Casetti and Semple (1968)[46] identify spatial trends by reference to a series of optimal points of origin. An optimal origin (xo) is defined as the location yielding the maximum correlation between the actual Z_{ij} series and distances between Xo and all other (U,V) locations. Since population density surfaces are rarely linear cones and commonly non-linear mounds and hollows, the logarithmic and reciprocal transformations of distance terms are suggested. The logarithmic transformation yields the expression

$Z_{ij} = \exp \alpha_1 \text{-} B_1 x_{1ij} + \alpha_2 \text{-} B x_{2ij} + \ldots\ldots + \alpha_n \text{-} B x_{nij}$ and the reciprocal transformation yields

$$Z_{ij} = \alpha_1 + \frac{B_1}{X_{1ij}{}^{+1}} + \alpha_2 + \frac{B_2}{X_{2ij}{}^{+1}} + \ldots + \alpha_n + \frac{B_n}{X_{nij}{}^{+1}}$$

where αK and BK are regression coefficients, and K_{ij} is the distance of the term with coordinates i and j from the Kth optimal origin, values predicted by the regression equation from the first optimal origin are regarded as a spatial trend, while the residuals from this trend yield a new spatial series within which a new optimal origin may be located, a new spatial trend computed, and a new residual population generated.

A wide range of alternative models to the simple isotropic density decline curves of urban population density are currently being investigated by geographer. Empirical studies using polynomial trends in the analyses of Australian urban populations (Fairbairn and Robinson, 1967),[47] in population trends in southern Ontario (Casetti and Semple, 1968), and in interregional comparisions of density patterns (Haggett, 1968) have already been published. The simple and arguably oversimplified models used there are currently being supplanted both by more rigorous spectral models (Rayner, 1967) and by the adaptation of polynomial and double Fourier models to meet the needs of discontinous distributions (James, 1968).

The decline of population densities towards the outer suburb of the cities follows a simple mathematical equation of exponential decline [48] :

$$Y = A\bar{e}\ bx$$

where x is the distance in miles from the centre of the city; Y is the density of resident population in thousands per square mile; b is a coefficient measuring the degree of overcrowding in the centre of the city.

Economic Factors and Absolute Numbers of Population

Ackerman Formula

After understanding the economy and physical resources of an area within the context of its history and culture, can we explain the population numbers and distribution within the area ? Central place theory has made a promising start in accounting for relative disposition of people and functions (though the theory is most successful where physical, cultural and historical diversities are ignored). Can the economic approach tell us anything about the absolute dimensions of a population. That is to say, can the size of a community be accounted for if enough is known about its economic characteristics and its economic history ?[49]

One of the few serious attempts to reconsider this approach is a formula proposed by Edward Ackerman.[50]

$$P=RQ\,(TASt)+ ES + T_r \pm F-W$$

where P is the number of people.

S is the amount of resources.

R is the amount of resources.

Q is the factor for natural quality of resources.

T is the physical technology.

A is the administrative techniques factor.

St is the resource stability factor.

W is frugality element (wastage or intensity of use).

F is the institutional advantage and friction loss element consequent upon institutional characteristics of society.

Es is scale economic element (size of territory etc.).

Ty is the resources added in tradi.

It would be difficult, though not quite impossible, to quantify some of the factors of this equation, and equally difficult to gather all the basic information for a given area and period. Nevertheless, a serious attempt to apply the Ackerman approach or something resembling it might seem worthwhile. Unfortunately basic flaws in a formula like this make it unworkable. At least three major elements in the equation population size, the resource technology factor and standard of living.

Estimation of the Threshold Populations

'Population threshold' or 'entry level' of various central place functions as the forefront of research in settlement geography, has been defined as the minimum number of people required to support one establishment of functional type. If it is assumed that the population of a central place is large compared with the population of its hinterland, then the threshold population for a particular function can be estimated by using population data for centres.

The Estimation of Thresholds

A number of techniques have been suggested as methods for determing the threshold population values for particular functions. A technique that, for two reasons, would appear to be useful, is simple regression.[51] In the first place, simple regression provides a method for estimating the 'average' or expected values for a set of data that are subject to error. Secondly, it provides a descriptive device that is useful for economic planning.

The simple regression model can be stated symbolically as :

$$PE_i = f(P_i)$$

Where PE_i is the number of establishments of the Fth function type found in the ith central place. Empirical data, obtained from the area of southern Ontario indicate that the relationship of each function takes the following form :

$log\ (FE_i) = log\ a + b\ log\ (P_i)$ which, as has been takes the antilogarithmic form :

$$(PE_i) = a\,(P_i)^b$$

An example of the empirical data is presented for dental services.

The threshold populations for the selected functions are considered to be at the intersection of the least squares line with the abscissa, as is illustrated in the logarithm for dental services.[52] This point represents the expected size of central place for a situation where the number of establishments is equal to unity. These threshold values are given in Table 15.12.

TABLE 15.12 : **Threshold Populations, Parameters and Correlation Coefficients for Selected Functions in Southern Ontario**

Functions	Threshold	A	b	r
General Food Store	65	0.1013	.542	.90
General Clothing	85	0.0936	.534	.89
Banks	727	0.0335	.582	.97
Lawyers	742	0.0105	.689	.90
Dentists	1,734	0.0014	.882	.93
Opticians	2,890	0.0045	.678	.92

Source : After Yeats M.H.

According to the above Table, the data suggest that the minimum number of people required to support one dentist is about 1,734 but two dentists require on the average, 3,800 persons in a central place. This is because the threshold values indicate the minimum number of people it takes to support one establishment and therefore the establishment can handle more than this minimum number.

Measurement of Population Threshold by Reed-Muench Method

The study of 'range or tolerance' of bioassay (there is a lower limit at which none of the animals tested dies and an upper limit at which all animals tested die) and the 'entry zone' of settlement studies suggests that the statistical methods used in the biological science may be applicable to the other. The median lethal dose or LD_{50} as proposed by Bancroft[53] may be used for this concept of a median lethal dose one of a median population threshold or PT_{50}. Hagget and Gunawardena[54] have suggested a simple method for its calculation and applied it on the settlements of Ceylon.

The Reed-Muench Method

The Reed Muench method[55] was put forward in 1938 to calculate the median values for the zone of tolerance. It has the advantage of rapid calculation by either algebraic or graphical means and of having been widely used for preliminary studies in bioassay. In this way, in settlement geography our problem is to determine the population threshold for a function (F_i) among a group of settlements of varied sizes. Table 15.13 illustrates such a simple situation. Fifty settlements are selected for study, ten being drawn at random from each of five population strata. For the lowest stratum (settlements with less than 200 inhabitants) all places are without the function F_i; for highest stratum (settlements with more than 800 inhabitants) all places have the function F_i. Between these two extremes the proportion without declines as that with increases. Cumulative summing of the two columns gives values for one index Ag indicating absence of the function F_i at this and greater levels and for an index P_s indicating presence of the function F_i at this and smaller levels. From these indices a final value may be computed of the proportion of settlements with the function F_i at each of the five population levels :

$$\frac{100.P_s}{P_s+A_g}$$

values for this proportion are given in the final row of the Table below.

TABLE 15.13 : **Determination of Median Population Threshold PT_{50} for Function F_i**

Population Level	Stratum				
	1	2	3	4	5
	00 199	200 399	400 599	600 799	800 999
Number of Settlements	10	10	10	10	10
With Function *Fi*	0	3	4	6	10
Without Function *Fi*	10	7	6	4	0
With Function *Fi* absent at this and greater levels (*Ag*)	27	17	10	4	0
With Function *Fi* present at this and smaller levels (*Ps*)	0	3	7	13	23
Proportion of Settlements with Function *Fi* present (Per cent)	0	15	41	76	100

With populations between 600 and 799 the proportion has risen to 76 per cent. Clearly the median value lies somewhere between these values. The exact position of the 50 per cent point can be estimated as :

$$\frac{50-41}{76-41}=0.26$$

That is the median value lies 26 per cent of the distance between the midpoint of stratum 3 and stratum 4. With an inter stratum distance of 200 inhabitants the median value is :

$$500+0.26\,(200) = 552 = PT_{50}$$

The median population threshold for function F_i is thus a settlement with about 550 inhabitants.

Potential Population Surfaces

The concept of gravity models can be extended to the calculation of potential population surfaces. Assume a set of towns ($1, 2, 3, \ldots\ldots, n$) each of a known population size ($M_1, M_2, M_3, \ldots\ldots, M_n$). Then the population potential for the first town is [56]:

$$P_1=a\frac{M_1}{(\frac{1}{2}D_{1\cdot})}b + a\frac{M_2}{D_{12}{}^6}+\ldots\ldots\ldots+a\frac{M_n}{D_{1n}{}^6}$$

for the second town is

$$P_2=a\frac{M_1}{\frac{b}{D_{21}}} + a\frac{M_2}{(\frac{1}{2}D_{2\cdot})^b}+\ldots\ldots\ldots+a\frac{M_n}{\frac{b}{D_{2n}}}$$

for the third town is

$$P_3=a\frac{M_1}{\frac{b}{D_{31}}} + a\frac{M_2}{\frac{b}{D_{32}}} + a\frac{M_3}{(\frac{1}{2}D_{3\cdot})^b}+\ldots+a\frac{M_n}{\frac{b}{D_{3n}}}$$

and so on; where P_1 is the population potential of town 1; D_{12} is the distance between town 1 and town 2; D_1 is the distance between town 1 and its nearest town; a is an empirical constant and b is a distance component (assumed to have a value of 2 in the original gravity model).

The nearest-neighbour component c is also introduced so we can take into account the effect of the population of each town on itself. The potential values ($P_1, P_2, P_3, \ldots\ldots, P_n$) determine the flow potential for each town. Potential maps are usually presented by drawing contour lines of equal potential expressed in percentages of the town with the highest potential.

Distance-Decay Curves

One of the simplest and most common ways of describing curves that relate flows and distance is with a Pareto function :

$$F=aD^{-b}$$

where F=the flow, D=the distance, and a and b are constants. Geographers are especially interested in the value for the constant b. Low b values indicate a curve with a gentle slope with flows extending over a wide area, whereas high b values indicate a sharp decrease with distance so that flows are confined to a limited area. This formula was used extensively by a group of Swedish geographers in studies of migration between regions on a large variety of geographic scales as far back as the nineteenth century. Their findings showed b values going from as low as -0.4 to as high as -3.3. The mean value for all the studies was just below -2. This figure would suggest that

$$F = a^{D-2}$$

which we can rewrite as

$$F=a\frac{1}{D^2}$$

Apparently, spatial interaction falls off inversely with the square of distance, i.e., the size of a flow at 20 km is likely to be only one quarter of that at a 10 km distance. This inverse square relationship is analogous to that used by physicists in estimating gravitational attraction.

Interaction Model to Determine the Functional Regions

To describe the distance function of various kinds of interaction, the well known gravity model has been introduced :

$$I_{ij} = \frac{m_i\, m_j}{d_{ij}a}$$

In which I_{ij} = The interaction between the places i and j, m_i and m_j = The "masses" of places i and j, measured in some meaningful way,

d_{ij}=The distance between i and j, measured in some meaningful way.

a=an exponent.

From this formula, the location of the hinterland boundary between the centres A and B can be derived, understood as the locus of points with equal amounts of interaction to A and B. The boundary has the form of a circle, whose centre C is placed on the prolongation of the line AB, in the distance from A [57]:

$$AC = \frac{\frac{a}{2}\sqrt{mA.\ 1}}{\frac{a}{2}\sqrt{m\ A} \div \frac{a}{2}\sqrt{mB}}$$

The radius of the circle is :

$$r = \frac{1}{\frac{a}{2}\sqrt{mA} \div \frac{a}{2}\sqrt{}} \qquad \frac{a}{2}\sqrt{mA.mB}$$

in which mA and mB=The "masses" of the centres A and B,

1=The distance between A and B

a=The exponent.

Ratio of Rural Population to Basic Rural Population

In fact the fluctuations of the ratio reflect very closely the variations of the density of landward population and it is a very erratic approximation to the rural.

If we represent the ratio by q, the landward population by P_1, the number engaged in farming industry, and the number dependent on farming, by f and F respectively, then,[58]

$F=r, f,$

and $q=P^1/F=P^1/r, f.$

Hence, mathematically q and r are essentially correlated.

There is a certain amount of concentration of q in the neighbourhood of the value 2. The notable fact is the high degree of variability of that quantity, which ranges from about 1.5 to 6, whereas r rarely fall below 2 or rises above 3. The line AA' is very closely fitted by mean values of r.

The Index of Dissimilarity

The index of dissimilarity is the sum of the absolute differences between the two populations taken area by area.[59] The basic form of the index of dissimilarity may be applied to the wide variety of phenomena depending on the definition given to the two populations. One of its most well-known guises is the index of concentration, a measure of the degree of correspondence between population units and area.

In the hypothetical example (Table 15.14) the index of dissimilarity between the 'A' and 'B' populations is 25 per cent.

TABLE 15.14 : **Hypothetical Data for Computation of Index of Dissimilarity**

Area	'A' Population	'B' Population	Difference
1	15%	10%	5%
2	20	10	10
3	20	20	0
4	30	20	10
5	15	40	25
Total	100	100	50

The formula for dissimilarity Index is like index of concentration.

$$ID = \frac{1}{2} \sum_{i=1}^{k} \left| X_i - Y_i \right|$$

Where X_i is the percentage of the 'X' population in the 'i' th sub-unit, Y_i is the percentage of 'Y' population in the 'i' th sub-unit and the Σ being over all the K sub-units making up the given universe of territory.

Conclusion

In population studies the quantitative methods used are large in number which are mostly related with the distribution of population, migration, spatial interactions of population besides model building and data analysis through statistical techniques.

REFERENCES

1. A.T.A. Learmonth, and L.S. Bhat (Eds.), *An Atlas of Resources*, Asia Publishing House, Calcutta, Vol. 1, 1964, p. 244.
2. F.J. Monkhouse, and H.R. Wilkinson, *Maps and Diagrams*, Methuen and Co. Ltd., London, 1972.
3. J.P. Gibbs, *Urban Research Methods*, Von Nostrand East-West Press, New York, 1961, p. 107.
4. A.S. Boughey, *Ecology of Populations*, Macmillan, New York, 1968, Chapter 2.
5. Ibid.
6. J.I. Clarke, *Population Geography*, Pergamon Press, London, 1966, p. 138.
7. Ibid, p. 138.
8. A.T.A. Learmonth, and L.S. Bhat, (Eds.), *An Atlas of Resources*, Asia Publishing House, Calcutta, Vol. 1, 1964, p. 244.
9. Colin Clark, *Population Growth and Land Use*, Macmillan, New York, 1967, p. 176.
10. L.T. Ruzicka, and T. Kanitkar, "Infant Mortality in Greater Bombay", *Demography India*, No., 1, 1973, p. 45.
11. N.P. Saxena, Diurnal Population Mobility within the Service Area of Meerut City, U.P, *The Geographer*, Vol. XIII, 1966, pp. 75-82 (77).
12. S.N. Agrawala, *A Demographic Study of Six Urbanizing Villages*, Asia Publishing House, Bombay, 1970, p. 120.
13. G.M. Howe, A National Atlas of Disease Mortality in the United Kingdom, *The Geographical Journal*, Vol. 130, Part 1, 1964, p. 20.
14. S.N. Agarwala, *A Demographic Study of Six Urbanizing Villages*, Asia Publishing House, Bombay, 1970, p. 120.
15. Ibid, p. 143.
16. L. Brillouin, *Science and Information Theory*, Academic Press, New York, 1956.
17. C.E. Shannon, and W. Weaver, *The Mathematical Theory of Communication*, University of Illinois, Urbana, 1947.
18. C.R. Pelto, "Mapping of Multi-Component System", *Journal of Geology*, Vol. 62, 1954, pp. 501-11.
19. Har Prasad, "Isoentropy of Demographic Partition: A Spatial Parameter for Area Development", *The National Geographical Journal of India*, Vol. XVII; Part 4, 1971, pp. 190-193.
20. G.U. Yule, *Journal of Royal Statistical Society*, Vol. 88, 1925, pp. 1ff.

21. T.J.D. Fair, and N.M. Shaffer, "Population Pattern and Policies in South Africa, 1951-1960", *Economic Geography*, Vol. 40, 1964, No. B; pp. 229-70.

22. R.N. Prasad, Migration, Lecture No. 8, 9, 10 and 11, mimeograph paper, Summer Institute in Statistics, 1977.

23. P. Haggett, *Locational Analysis in Human Geography*, Arnold, London, 1965, pp. 36-37.

24. M. Dalton and J.M. Seamon, "The Distribution of Commonwealth Immigrants in the London Borough of Ealing", *Transactions of the Council of British Geographers*, No. 58, 1973, p. 33.

25. S.K. Vaitaknass, "A Delimitation of Migratory-Gravitation Zones of Cities in the Lithuanian SSR", *Soviet Geography*, No. 1, Vol. XV, 1974, p. 27.

26. United States Bureau of Censuss, "Internal Migration and Short Distance Mobility", *Methods and Materials of Demography*, Vol. 2, p. 656.

27. M. Tachi and M. Oyama, "Economic Population of Internal Migration of Population and its Determinant", *Annual Report of the Institute of Population Problems*, Tokyo, 1962, pp. 31-35.

28. M.J. Libbee and D.E. Sopher, "Marriage Migration in Rural India", Paper Prepared for International Geographical Union; Symposium on Internal Migration; University of Edmonton, July/August, 1972, pp. 17-23.

29. W. Borejko, "Studies on the Effectiveness of Migrations", 21st International Geographical Congress Abstract of Papers, 1968, p. 223.

30. P. Haggett, op.cit; 1965, p. 45.

31. Kingsley Davis, *Human Society*, New York, The Macmillan Co., 1949, pp. 551-52.

32. J.P. Gibbs, *Urban Research Method*, East-West Press Pvt. Ltd., New Delhi, 1966, pp. 564-568.

33. W.J. Wadycki, "A Note on Opportunity Costs and Migration Analysis", *The Annals of Regional Science*, Vol. VIII, No. 1, 1974, p. 110.

34. Harold McConnell, "Spatial Variability of College Environment as Function of Migration Potential", *The Professional Geographer*, Vol. XVII, No. 6, 1965, pp. 29-37.

35. Clarke John, I., *Population Geography*, Pergamon Press, London, 1966, p. 31.

36. P. Haggett, *Locational Analysis in Human Geography*, Arnold, 1965, p. 230.

37. P.S. Shastri, "Potential Model and the Spatial Distribution of Population in Vidarbha, 1971", *Geographical Review of India*, Vol. 36, No. 2, 1974, pp. 142-143.

38. R.J. Chorley, and P. Haggett, (Edt.), *Frontiers in Geographical Teaching*, Methuen and Co. Ltd., London, 1970.

39. Clarke John, I., *Population Geography*, Pergamon Press, London, 1966, p. 31.

40. Ibid, pp. 37-38.

41. Z. Murdych, *On Methods of Measuring the Concentration of Population*, Sbornik, University of Larlova, Praha, 1970, p. 60.
42. S.E. Teo, "A Preliminary Study of the Cihnese Community in Sydney : A Basis for the Study of Social Change", *Australian Geographer*, Vol. XI, No. 6, 1971, pp. 579-592.
43. Sabita Mitra, and D.K. Singh, "Index of Population Concentration and Temporal Redistribution in the Brahmaputra Valley", *Geographical Review of India*, Vol. XXXIII, No. 1, 1971, pp. 15-22.
44. B.J.L. Berry, et al. *Geographical Review*, Vol. 53, 1963, pp. 389-405.
45. E. Casetti, and R.K. Semple, Michigan Int. University Comm. Math. Geographical Disc. Papers-11, 1968.
46. C. Clark, *Journal of Royal Statistical Society*, 1951, Series A. 114, pp. 490-496.
47. K.J. Fairbairn, and G. Robinson, *Australian Geographical Studies*, Vol. 5, 1967, pp. 125-134.
48. John Stewart, Q., "A Measure of the Influence of Population at Distance", *Sociometry*, Vol. 5, 1942, pp. 63-71.
49. Willer Zelinsky, A Prologue to Population Geography, Prentice Hall, Inc.; New Jersey 1966, pp. 37-38.
50. P.M. Houser, and O.D. Duncan, (Eds.), *The Study of Population : An Inventory and Appraisal*, University of Chicago, 1954, pp. 621-48.
51. P. Haggett, and K.A. Gunawardena, "Determination of Population Threshold for Settlement Functions by Reed Muench Method", *The Professional Geographer*, Vol. XVI, No. 4, 1964, pp. 6-9.
52. M.H. Yeats, *An Introduction to Quantitative Analysis in Economic Geography*, McGraw Hill Book Company, New York, 1968, pp. 104-107.
53. H. Bancroft, *Introduction to Biostatistics*, Cassell, London, 1457, p. 199.
54. P. Haggett, and K.A. Gunawardena, "Determination of Population Threshold for Settlement Functions by Reed Muench Method", *The Professional Geographer*, Vol. XVI, No.4, 1964, pp. 6-9.
55. S. Seigel, *Non-Parametric Statistics*, McGraw Hill Book Company, Inc.; New York, 1956, p. 22.
56. W. Warntz, "Macro-Geography and Income Fronts, Regional Science Research Institute", Philadelphia, 1965.
57. S. Illeris, "Functional Regions in Denmark About 1960", Collected Papers Denmark, 1968, (Ed. Jacobsen & Jensen), pp. 199-200.
58. A. Stevens, "The Distribution of the Rural Populations of Great Britain", Transactions and Papers, 1946, No. 11, p. 24.
59. R.J. Chorley, and P. Haggett, *Frontiers in Geographical Teaching*, Methuen and Co., Ltd., London, 1970, pp. 240-247.

SIXTEEN

R. B. Mandal

RURAL SETTLEMENT GEOGRAPHY

SETTLEMENTS of the countryside are the mainstay of human society where the irresistible hazards of climate and the scorching heat of the sun impede human living. Rural settlement is the only element in human geography where we may really search and try to locate the genesis of settlement geography. In this way, rural settlements are the topographic expression of the grouping and arrangement of two fundamental elements of human geography, houses and highways.[1]

There are two sets of variables which affect settlement types in rural areas. We may call them agglomerating and deglomerating factors. These parameters are both physical and cultural and lead to compact and dispersed settlements according to the relative influence of centrifugal and centripetal forces.

In most countries of the world two methods of rural settlement classification have been generally used, i.e. personal observation and quantitative methods determining the types of rural settlements. The significance of some of these quantitative methods for measuring the indices of agglomeration and disperson have been discussed below.

Index of Dispersion of Rural Settlements

Woeikof[2] of Russia in 1909 devised a formula as $K = \frac{P}{H}$,

where K is the index, P is the number of inhabitants and H is the number of inhabited places. This however, gives merely the average population size of rural settlements, irrespective of settlement dispersion. In the same way, Kristofferson[3] of Sweden in 1924 classified the rural settlements below 50 metres spacing as dispersed and more than 50 metres as compact. In 1926 Lefevre of Belgium[4] used the number of houses per square kilometre grouped into classes of 1-10, 11-25, 26-50, 51-100, 101-250, 251-500, 500-1000, and 1000 + as the basis of classification and dispersion of settlement types. But this method only gives the density classes of houses, and not the types of settlements.

French scientist Bernhard[5] in 1931 devised another formula, $K = \frac{S \times M}{N^2}$. Where K is the index of concentration, S is the total number of houses, M is the area and N is the number of settlements. This formula, however, does not distinguish closely between the settlement variations and K can be identical in value with different patterns.

Another successful formula for measuring dispersion is that of Demangeon[6] published in 1933. He expressed the settlement pattern as $K = \frac{E \times N}{T}$, where K is the coefficient required, E is the population of the commune except for that of the chief nucleated settlement, N is the number of isolated settlements, and T is the total population.

Zierhoffer[7] in 1934 devised another formula on the assumption that the degree of dispersion of settlements will increase area per dwelling. The degree of dispersion is given by :

$$R = \frac{P \times S}{d_i} \times K,$$

where R is the degree of dispersion, P is the average area per dwelling, S is the number of house groups in commune, d_i is the total number of inhabitants and the constant K is 0.0005. This method, however, does not give sufficient contrast to the value of R in intermediate categories between complete dispersion and absolute nucleation.

In 1938 Pawlowski[8] of Poland while studying the concentration of settlements derives the following formula :

$$C = \frac{A}{AS} \text{ and } Mc = \frac{AS}{S},$$

where C is index of concentration, A is 25 km^2, A is area occupied by settlements, Mc is mean index of concentration, S is number of settlements.

A Belgian scientist, Debouverie[9] in 1943 tried to relate regional variations of settlements with the standard concept of concentration. His formula is as follows :

$$K = \frac{X \times L}{H},$$

where K is the index of concentration, X is the minimum number of dwellings per settlement, L is the number of settlement units and H is the total number of dwellings in the commune. This formula is useful in distinguishing types of nucleation.

In 1950 Kant[10] of Estonia presented an Indigen formula to measure the concentration of settlements more precisely. The formula is as follows : .

$$X = \frac{1}{M} \quad \frac{A}{D},$$

where X is the relative interval between two settlements, $\frac{1}{M}$ is map scale, A is area, D is density of habitations.

Based on the formula of Demangeon, Houston[11] of France devised a method in 1953. The formula is as follows :

$$K = \frac{S \times N}{T—E},$$

where K is the index of dispersion, S is the area of commune, N is the number of isolated settlements, T is the total population and E is the population of commune minus its chief place or urban centres.

In 1957 Tanioka[12] of Japan devised a new formula to measure the index of dispersion of rural settlements. The formula is as follows :

$$M = \frac{S \times T}{S \times N},$$

where M is index of dispersion, S is area of map sheet (1/50,000 scale), T is per cent of land cultivated, S is average agriculture area worked per family, N is average number of rural houses per village.

In 1962 K.H. Stone[13] devised a formula to measure the continuous and discontinuous settlements in any region. The measure is as follows ;

$$R \text{ or } Fz = Nh + Nr,$$

where R is a measure of continuous or discontinuous settlement, Fz is fringe zone in discontinuous settlement region, Nh is pattern of permanent residences within 3 miles long 1 – 6 major directions from any one permanent residence, Nr is number (1—4) of inter-regional and local routes of transport within 10-20 miles of each permanent residence.

Inouye[14] of Japan in 1964 devised a measure of dispersion of rural settlements on the basis of 50 metres spacing. In this measure isolated means less than 3 houses within 50 metres of each other, concentrated means more than half of the houses in a village distributed into one group within 50 metres range, condensed means more than half of the houses in a village centralized into one core.

Singh, R.B.[15] in India expressed the following formula in 1969 to measure the compactness and dispersal of rural settlements.

$$Hn = Vn = \text{compact}$$

$$Hn = Vn + 1 \text{ to } Vn \times 2 = \text{Semi compact}$$

$$Hn = Vn \times 2 + 1 \text{ to } \frac{Oun \times 2}{3} = \text{Hamletted}$$

$$Hn = \frac{Oun \times 2 + 1}{3} \text{ to } Oun = \text{Dispersed}$$

where Hn represents the number of hamlets, Vn is the number of villages in an administrative unit and Oun is the number of occupance units in the same area.

In 1972 Mandal[16] in India devised the following measure of dispersal of rural settlements.

$$1 = \frac{\text{Average Population Size of Tola}}{\text{Average Spacing of Tola}}$$

where 1 is index of dispersal. Index of dispersal values grouped according to media and quartile values (Highest value shows the agglomeration while lowest dispersal).

In 1973 Prasad, A.[17] while studying the rural settlements of Chota Nagpur Plateau devised the following formula of descriptive nature after considering the 'deviation from standard index'.

Ten highly compact villages from the adjoining parts of Gaya district have been selected. On the basis of village maps of settlements concerned, the actual built up area in each case was reduced to the percentage of the total area within the homestead boundary. The percentage of ten villages have been averaged. This average is expected as the standard index of agglomeration for Chota Nagpur as a whole. To determine the agglomeration quotient similar percentage value for any other cluster is divided by the standard index and is then multiplied by 100. If it is 100, the cluster has the highest degree of agglomeration and compactness. If it is more than 100, the agglomeration is of exceptionally high order. The quotient under 100 indicates a decrease in agglomeration tendency and a proportionate increase in deglomeration tendency. Fifty is the critical value. As the quotient declines from 50, the settlements are to be classed as dispersed. Dispersal can be easily determined by subtracting the agglomeration quotient from 100. The remainder is dispersal quotient for a cluster.

In 1977 Mandal[18] while studying the "planned development of rural settlements in North Bihar" devised a measure of dispersal index to determine types of rural settlements. The formula is as follows :

$$\text{Dispersal} = \frac{\begin{matrix}\text{Total rural Population} \\ \text{of an anchal}\end{matrix} \times \begin{matrix}\text{Number of tolas} \\ \text{in an anchal}\end{matrix}}{(\text{Total Rural Area})^2}$$

To determine the validity of the above formula, the relationship between the rural area of anchals in thousand square kilometres and the number of tolas multiplied by rural population of an anchal was plotted on the X and Y axes respectively. Both axes were on logarithmic scale and data were plotted for all the 241 anchals of North Bihar. For this set of data, the product moment coefficient of correlation value comes to $r=0.56$ and regression value of $Y=10.30+0.14\ X$.

On the basis of the above idea, with the help of the dispersal index of each and every anchal, the difference between the logarithm of the highest value of 1507 and the lowest value of 76 has been taken. The logarithmic difference has been divided by three so as to get three categories of rural settlement types. This is because in North Bihar the population density is so high that nowhere dispersed settlements have actually been found except in a few localities in the Kosi basin as well as some in the forested and rugged terrain of north-west Champaran. Elsewhere all rural settlements are either of compact or of semi-compact type. The geometric progression of the above logarithmic difference comes to 1.88081, 2.31323, 2.74565 and 3.17808; and their anti-logarithmic values 76, 206, 557 and 1507 (index values) respectively.

The density of habitations (d) is the ratio of two quantities—C and A—the number of habitations (c) belonging to a given area (A):

$$d = C/A$$

In order to prepare a map the density value should be arranged serially either on the basis of Quartile, Octile or Standard deviation. Other methods of grouping can also be used.

Methods for Measuring Spacing of Settlements

Geography is a discipline in distance. Thus the spacing of rural settlement is largely governed by their size. Large rural settlements would be widely spaced, while smaller settlements would be closely spaced.[19] These spatial patterns are constantly changing in time and space which has resulted in an increase in the size of centres and reduction in their spacing. There are methods which may be employed in measuring the spacing of settlements.

Method : *1*

Walenty Winid, a Polish geographer, assumed that settlements are ordinarily placed at the intersection points of a rectangular grid, and the best figure to use for inter-settlement distance is equal to the square route of the area divided by the number of settlements within that area.[20]

$$D=\sqrt{\frac{A}{N}}$$

Here D is average distance between settlements, A is the total area of the region under study, N is the number of settlements.

Method : 2

According to Barnes and Robinson,[21] if the number of settlements and the area of the region are known, we can determine the approximate distance between each under its nearest neighbour and would be under conditions of an even spatial distribution. It can be computed by a simple formula,

$$Hd = 1.11\sqrt{\frac{A}{N}},$$

Here Hd is the hypothetical distance, A is the area of the region, N is the number of settlements, 1.11 is the constant value. In this formula A/N indicates the average area that belongs to a settlement. In representing this area by a hexagon, the distance D of a settlement from its six neighbours, is equal to twice the height of equilateral triangle obtained by linking the corners of the hexagon to its centre. The area of the hexagon is[22]

$$A/N = \tfrac{1}{2}\sqrt{3}.D^2$$

Consequently

$$D = \sqrt{\frac{2A}{n\sqrt{3}}} = 1.11\sqrt{\frac{A}{N}}$$

Method : 3

In 1944, Mather [23] devised a formula to measure the spacing of settlements. The formula is as follows :

$$D = 1.07\sqrt{\frac{A}{N}}.$$

where D is average distance of a settlement to its six nearest neighbours, A is the total area of study and N is the number of settlements.

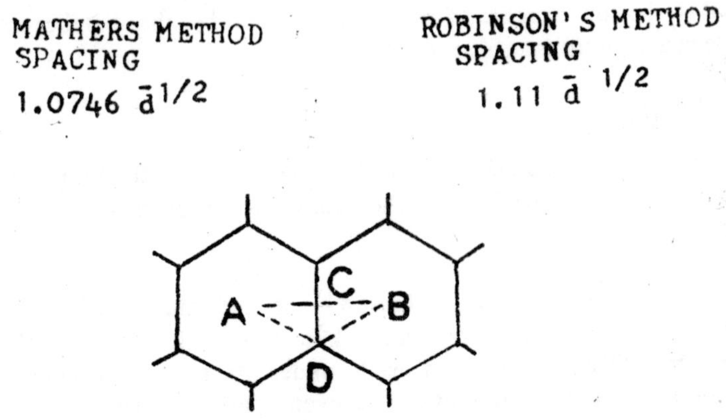

FIG. 16.1 : **Hexagonal Arrangement of Rural Settlements**

The verification about the correctness of Barnes-Robinson and Mather methods were presented by Rana P.B. Singh.[24] He endeavoured to solve this disputable method in this way :

If in Figs. 16.1 and 16.2 A and B are the two points, then AB will be their interpoint distance. Take first the ADC (right angled at C).

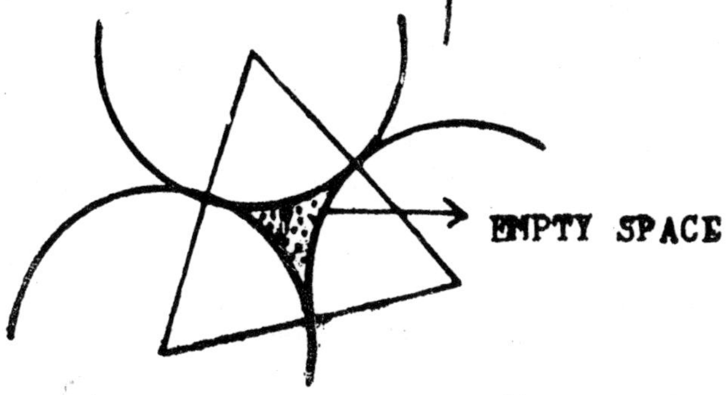

FIG. 16. 2 : **Circular Arrangement of Rural Settlements (Mukerji's Method)**

$$\text{Total area of } ADC = \left(\frac{AC.DC}{2}\right)$$

Let $AD = X$,

and it is known that $AC = \sqrt{\frac{3}{2}}\, X$ and

$DC = \frac{X}{2}$; and AB, the required inter-village distance $(D) = \sqrt{3X}$.

$$\text{Total area of } \Delta\, ADC = \sqrt{\frac{3}{2}} \times \frac{X}{2} \cdot \frac{1}{2}$$

$$= \sqrt{\frac{3}{8}}\, X^2$$

The area of the hexagon should correspond to the average area of a village i.e. : A/N;

$$\frac{3\sqrt{3}}{2} \cdot X^2 = A/N$$

$$\text{or} \quad 3X^2 = A/N. \frac{2}{\sqrt{3}}$$

$$= \frac{2A}{\sqrt{3N}} \text{ or } 1.1547 \left(\frac{A}{N} \right)$$

$$\therefore \sqrt{3X^2} = \sqrt{1.1547} \cdot \sqrt{A/N}$$

$$\therefore \sqrt{3X} = 1.0746 \sqrt{A/N}$$

Thus it is proved that inter-village distances in hexagonal arrangement will be $1.0746 \sqrt{A/N}$ not $1.11 \sqrt{A/N}$ as formulated by Robinson and Barnes, and $1.1284 \sqrt{A/N}$ by Mukerji. As these measures present overlapped area, so they are wrong in their propositions.

Method : *4*

Like Robinson and Barnes, A.B. Mukerji[25] assumes that the average area of village will be a circle $(\pi r)^2$. So the transformation as of average village area into circle, results in the theoretical distance between two villages. According to him, area of the village will be πr^3,

$$\text{Now, if } \pi r^2 = \frac{A}{N}; \; r^2 = \frac{A}{N x \pi}$$

$$\text{or} \quad r = 2 \sqrt{A/N\pi}$$

$$\text{Thus the inter-village distance } (D) = 2 \sqrt{A/N\pi}$$

$$= 1.1284 \sqrt{A/N}$$

In this method empty space is left in between three circles, so this method is also not suitable for measuring inter-village distance. Thus, it is apparent that the theoretical spacing between settlements will be $1.0746 \sqrt{A/N}$, not $1.11 \sqrt{A/N}$; nor $1.1282 \sqrt{A/N}$.

Finally the formula used by A.B. Mukerji for the computation of spacing in miles of rural settlements in an anchal or taluka is as follows :

$$S = 2x \sqrt{\frac{A}{N \times \pi}}$$

here S is spacing in miles, A is the rural area of an anchal and N is the number of rural settlements.

Method : 5

In the given co-ordinates of any two points on a plane, the distance between them can be easily computed by using the Pythagoras Theorem. This question can be generalized to measure distances between points in a dimensional space for given coordinates and is as follows [26]:

$$D_{ij} = \sqrt{\sum_{z=1}^{n} (Xiz\text{-}Xjz)^2}$$

here D_{ij} is the distance between points i and j in a space of n dimensions with X_{iz} and X_{jz} being measures along the zth co-ordinate. This is the simplest distance measure and has been used by Berry in 1961.

Method : 6

If (X_1, Y_1, Z_1) and (X_2, Y_2, Z_2) be the rectangular coordinates of two points P_1, P_2 respectively and O is the origin points then [27]:

$$\vec{OP_1} = x_1i + y_1j + y_1j + z_1K$$

$$\vec{OP_2} = x_2i + y_2j \div z_2k$$

$$\vec{P_1P_2} = \vec{OP_2} - \vec{OP_1} = (x_2 - x_1)i + (y_2 - y_1)j + (z_2 - z_1)K$$

Hence $\vec{P_1P_2}/\vec{P_1P_2}/= \sqrt{(x_2 - x_1)^2 + (y_2 - y_1)^2 + (z_2 - z_1)^2}$

This will give the distance between the points P_1P_2.

Method : 7

As early as in 1901, Bannerman used the following formula for calculating the distance between each village [28] :

$$2\left[\frac{\cos 30^\circ}{3}\right]\frac{1}{2} \times \left[\frac{\text{Area}}{\text{No. of villages and towns}}\right]\frac{1}{2}$$

$$= \frac{\log 2}{3} - \frac{\log 3}{4} - \frac{\log \text{areality}}{2}$$

$$= 0.0312347 + \frac{\log \text{areality}}{2}$$

where 0.0312347 is a constant.

Method : *8*

According to Clark and Evans, the nearest scale can be referred to as an *R* scale, and the individual *R* value can be calculated as follows[29] :

$$R=\frac{\bar{r}A}{\bar{r}E}\ ,\ \text{where}\ \bar{r}A=\frac{\Sigma r}{N}$$ or the mean of the series of distances to nearest neighbours $\bar{r}$ being the distance from each point to its nearest neighbour (actual) and

$$\bar{r}E=\frac{1}{2\sqrt{P}}\ \text{or}\ \bar{r}E=\frac{1}{2\sqrt{\frac{N}{A}}}$$

where the expected mean distance ($\bar{r}E$) to the nearest neighbour in an infinitely large random distribution of density *P* (*E* meaning exepected *d*) and $P=\frac{\text{No. of settlement}}{\text{Total area}}$

The *R* value should range between 0.2.1491 meaning maximum dispersal at the end, with unity (1.00) indicating the presence of a random observed distribution, while the value near 0 indicate clustering of settlements. This index of *Rn* values can be correlated with variance (*V*) for further testing, which can be calculated with the help of following formula[30] :

$$V=(4-\pi/4d\pi)$$

In case the $\bar{r}E$ value is larger than the *V*, the distribution is termed regular; when the *V* is larger than $\bar{r}E$, it is termed clustered and the term random is applied to the case when *V* and $\bar{r}E$ are equal, i.e., the variance mean ratio is one.

The test of the significance of $\bar{r}E$ was made about the hypothesis of randomness on the basis of Standard error ($\sigma\bar{r}E$)[31] which is as follows :

$$\sigma\bar{r}E=\frac{0.26136}{\sqrt{(n_d)}}$$

where *n* is the number of total villages in the areal unit and *d* represents density of villages per square kilometre.

In the nearest neighbour analysis, the differences between the observed and expected frequencies were analysed by Chi-square tests were made because it was assumed that the distances were not normally distributed. Thompson[32] has provided a suitable procedure of Chi-square test and it has been reviewed by King.[33] The values of Chi-square are given by [34]:

$$2\pi d\tfrac{1}{2} \sum_{i=1}^{n} r_{ij}$$

where n is number of points in the pattern, d is the observed density, r_{ij} is the order nearest neighbour distance for i and is distributed as Chi-square with (1-2n) degree of freedom.
The index of standard variate of normal curve (C) can be calculated as :

$$C=(\bar{r}O-\bar{r}E)/O\bar{r}E$$

where $\bar{r}O$ is observed value and $\bar{r}E$ is expected. The upper and lower ranges of random matching at the 95 per cent probability will be computed with the help of $O\bar{r}E$

$$\text{i.e.}=\frac{(2\sigma\bar{r}E \pm \bar{r}E)}{\bar{r}E}$$

The application of these tests has been made by Rana P.B. Singh in his paper related to distribution of rural settlements.

Method : 9
King suggested the formula for calculating the R value in nearest neighbour analysis as follows [35]:

$$R=2\bar{D}\sqrt{\left(\frac{N}{A}\right)}$$

where $\bar{D}$ is average of the observed distances between each point and its nearest neighbour, N is the number of points and A is the area under study.

Method : *10*
In 1967, Birch had applied nearest neighbour measures in Midwest of U.S.A.[36] The formula is as follows :

$$R_1=\frac{\bar{r}A}{(A/N)-\frac{1}{2}}$$

where R_1 is index of randomness of regularity or agglomeration of farmsteads, $\bar{r}A$ is the observed mean distance between each point

TABLE 16.1 : **Results of Nearest Neighbour Measures**

Class of urban places	No. of urban places	Density per sq. mile	Average observed distance in miles $\bar{r}_A$	Average expected distance in miles $\bar{r}_E$	Pattern Statistics R	Chi-square χ^2	Degree of freedom	Significant
Hamlets	59	0.0127	4.881	4.442	1.098	194.484	118.0	Yes
Villages	93	0.0200	3.487	3.538	1.126	314.393	186.0	Yes
Towns	37	0.0079	5.931	5.609	1.057	117.363	74.0	Yes
Cities	14	0.0030	10.407	9.119	1.141	47.934	28.0	Yes

Source : B. Thakur, Indian Geographical Studies, 1973.

and its nearest neighbour, A is the area, and N is the number of rural settlements.

Method : 11

In 1975 a modification in nearest neighbour technique has been presented by Pinder and Witherick.[37]

According to him nearest neighbour analysis is generally used in two-dimensional point patterns. No exploration has been done for distribution of points along a line. The method presented by Evans and Dacey is simple but its result is misleading especially when the number of points in the pattern is low.

In the normal analysis of nearest-neighbour by Clark and Evans method, the formula for predicting the average distance between randomly distributed nearest-neighbours ($\bar{r}E$) is as follows :

$$0.5 \sqrt{(a/n)}$$

where a is the area studied and n is the number of points. This method is designed to fit in two dimensional patterns.

In order to predict the mean nearest-neighbour distance in a unidimensional situation, the average distance between randomly distributed linear nearest neighbours $L\bar{r}E$ may be derived from the following formula :

$$L\bar{r}E = 0.5\left[\frac{L}{n-1}\right]$$

Where L is the line length and n is the number of points. This is the first stage of this linear nearest neighbour technique. This indicates that the predicted random average is half the mean distance that would separate the points if they were distributed with regularity throughout the line.

In this table, for a line of 1000 units comparison of predicted and computer generated experimental values of $L\bar{r}E$ have been given. This provides extremely close correspondence between the two sets of means as given in the second and the third columns, and further emphasis is put forward by the last column, where the discrepancies between the predicted and experimental means are presented.

TABLE 16.2 : **Comparison of Predicted and Experimental Values of $L\bar{r}E$**

No. of points per Pattern	Predicted Mean 0.5 (L/n—1)	Experimental Mean	Predicted Mean Less Experimental Mean
10	55.56	55.89	—0.33
20	26.32	26.62	—0.30
30	17.24	17.57	—0.33
40	12.82	12.74	0.08
50	10.20	10.21	—0.01
60	8.48	8.34	0.14
70	7.25	7.12	0.13
80	6.33	6.23	0.10
90	5.62	5.59	0.13
100	5.05	5.04	0.01

Source : After Pinder and Witherick, Geography, 1975.

Further, linear nearest neighbour statistic (*LRn*) can be defined as :

$$\frac{\bar{D}_{obs}}{L\bar{r}E} \text{ or } \frac{\bar{D}_{obs}}{0.5\,(L/n-1)}$$

Where $\bar{D}_{obs}$ is average observed distance between nearest neighbours, L is length of the line studied, n is the number of settlements. In case *LRn* value is 1.00 this suggests random situation. Less than 1.00 values are indicative of clustering while more than 1.00 shows the presence of some regularity. In a perfect regular situation *LRn* equals 2.00 and this is half the observed mean distance between settlements :

Method : *12*

Order Nearest Neighbour Analysis

Dacey has given the formal derivation of poisson probability law for two dimensional random point process. He assumed that those neighbours whose distance is greater than the parent point-distance from the boundary of the map, the number and distance of such neighbours should be ignored from the study. The formula through which the mean observed distance up to four neighbours can be calculated is as follows[38] :

$$\bar{r}_j = \frac{1}{N} \sum_{i=1}^{n} r_{ji}$$

Where n is the number of the points in the area, $\bar{r}_j$ is the mean distance from i to any other point, i is any point, and r_{ij} is the distance between i and jth nearest point.

Dacey has generalised these concepts of order distances to nearest neighbour to handle the description of pattern in multi-dimensional space. The non-standardized mean value of distance of urban places in Bihar upto four neighbours is given below :

TABLE 16.3 : **Order Distance Statistics for Urban Places in Bihar, 1971**

(Measurement in Miles)

Order j	n	$\sum_{i=1}^{n} r_{ij}$	$\bar{r}_j$
1	163	1361.00	8.35
2	143	1693.75	11.84
3	133	1894.50	14.24
4	127	1997.75	15.83

The above observed mean distances can be used in comparison with the hypothetical estimated distance for pattern interpretation.

Method : *13*

Regional Method of Nearest Neighbour Analysis

Dacey and Tung have put forward alternative approach to the description of point patterns is known as regional method. They divided the map around each point into six K equal sectors. The distances from the central point to six neighbours in each sector are measured and 'these observed distances are used to classify the regions in some systematic fashion and a separate mean is obtained for each kind of region'.[39] The formula used is as follows :

$$\bar{r}_{ik} = \frac{\left(\sum_{i=1}^{k} \bar{r}_{ik}\right)}{n}$$

Where i is the central point, 1 is the set of n points for which the measurements are made, n is the number of points in a region.

He again assumed that those neighbours whose distances are greater than the parent point distance from the boundary of the map than the number and distance of such neighbours can be deleted.

Method : 14

Bachi's Standard Distance Statistics

Bachi's standard distance is a measure of the dispersion of a set of points in respect of two right angled axes. It may be expressed as :

$$d_{ij}=\{(x_i-x_j)^2+(y_i-y_j)\}^{\frac{1}{2}}$$

where x and y are two coordinates, and d_{ij} is the standard distance.

The above formula will give us distance between two points in Euclidian Space (Fig. 11), For the n set of points n^2 of these distances can be calculated, so that the distance between the two points is itself included and every standard distance is measured twice, once for i and secondly for j. For this set of n^2 distance, Bachi suggested the mean quadratic distance, as.

$$D = \left[\left(\frac{\sum_{i=1}^{n} \sum_{i=1}^{n} d_{ij}^2}{n^2}\right)\right]^{\frac{1}{2}}$$

Then, the mean centre of the distribution have the coordination as :

$$x = \frac{\left(\sum_{i=1}^{n} x_i\right)}{n} \quad \text{and } \bar{y} = \frac{\left(\sum_{i=1}^{n} y_i\right)}{n}$$

$$\text{then } D = \frac{\left(2 \sum_{i=1}^{n} d^2_{ic}\right)^{\frac{1}{2}}}{n}$$

where d_{ic} is the distance from point i to the mean centre

$$C = \{ (x_i - \bar{x})^2 + (y_i - \bar{y})^2 \} ^{\frac{1}{2}}$$

then the standard distance is[40] :

$$d = \left[\frac{\sum_{i=1}^{n} d^2_{ic}}{n} \right]$$

Further "Bachi notes that the square of distance (d^2) for the entire area is equal to the sum of a 'within-regions' squared distance (d^2_b)".[41]

$$\text{Then } d^2 = d^2_w + d^2_b$$

$$= \frac{\sum_{j=1}^{k} \sum_{i=1}^{nj} [(x_{ij} - \bar{x}_j)^2 + (y_{ij} - \bar{y}_j)]^2}{n} + \frac{\sum_{j=1}^{k} n_j [(\bar{x}_j - \bar{x})^2 + (\bar{y}_j - \bar{y})]^2}{n}$$

Where (x_{ij}, y_{ij}) are the coordinates of the ith point in the jth region

$(\bar{x}_j, \bar{y}_j)$ is the regional mean centre,

$(\bar{x}, \bar{y})$ is the mean centre for the whole distribution

(d^2b/d^2) is the ratio, provides a crude measure of significance of the regional divisions. Ratio 1 is the maximum differentiation.

where *0* shows the same mean centre as exists in the total pattern.

Computer Function to Measure Arc Distances

Method: 18

The calculation of the arc distance comes from spherical trigonometry. It assumes that the earth is a sphere. The necessary formula for determining the arc distance between points *A* and *B* is given by Robinson. The formula is as follows[42] :

$$\text{Cos } d = (\text{Sin } a \text{ Sin } b) + (\text{Cos } a.\ \text{Cos } b.\ \text{Cos } p)$$

here d = arc distance between *A* and *B*.

a = latitude of *A*,

b = latitude of B,

p = degrees of longitude between A and B.

Using this formula to calculate distances has proved satisfactory. Accuracy has been good (calculated distance between cities on opposite sides of the Pacific have varied only a few miles from distances quoted by air lines) and seems to be limited more by point measurement than by computer rounding errors or the off spherical shape of the earth[43].

If all settlements are equally spaced over a region, a uniform distribution is obtained. This distance (δ) is obtained by the relation,

$$\delta = \sqrt{\frac{A}{N}}$$

where A is the total area, and N is the number of settlements. The degree of concentration (DC) of settlements (nc) is given by

$$DC = \frac{\delta}{S},$$

where S is the average separation between settlements. The quality nCS^2 is designated as the derived area (ac) of the concentration. We get a level of concentration (LC) expressed in the form,

$$LC = \left(\frac{\delta}{S}\right) \quad \frac{nc}{N} = D_c{}^2 \quad \frac{nc}{N}$$

It should be noted that any quantity of LC in fact represents a set of values with varying δ and nc; hence its designation as 'level'. A number of concentrations in a region can be grouped together to form agglomerations. If the distance (A_s) between two contiguous concentrations is less than or equal to, the sum of the average separations of settlement, within the concentrations, thus the concentration can be grouped to form an agglomeration. Two contiguous concentrations are brought together under an agglomeration if

$$A_s \leqslant S_1 + S_2$$

where S_1 and S_2 refer to the average separations of settlements in concentrations. To demarcate the dispersion group, the distance relevant to dispersion (S) is derived as follows :

$$\Delta = \sqrt{\frac{A - \Sigma ac}{N - NC}}$$

where N is total number of settlements, NC is the number of concentrated settlements, and Σac is the sum of the areas of all concentrations. As in the case of concentrations, the degree of dispersion,

$$Dd = \frac{C}{\Delta},$$

where C is the average separation in a dispersion group. The area of dispersion,

$$ad = ndC^2$$

where nd represents the number of settlements involved in the dispersion group. The level of dispersion,

$$L_d = \left(\frac{c}{\Delta}\right)^2 \cdot \frac{nd}{N} = D_d = \frac{nd}{N}$$

The change in any settlement is generally introduced by transport linkage. Inversely in this formula the rate of transportation is determined by the change in a settlement. The formula is as follows[44] :

$$Rn = \frac{Dn}{Tn}$$

where Rn = rate of transportation

Dn = degree of transportation

Tn = time unit of transition in a settlement

$$\therefore Tn = \frac{Dn}{Rn} \quad \begin{matrix} \rightarrow \text{Length of phase of development} \\ \rightarrow \text{Rate of transformation in } n\text{th phase.} \end{matrix}$$

$$\therefore T = \sum \frac{D}{R}$$

$$\therefore T = \sum_{n-1}^{n} \frac{Dn}{Rn}$$

where T is the time required for attaining total development through transformation.

To express the change in number of lower order central places mathematically, the average rate of change can be calculated with the help of formula[45] :

$$b = \frac{P_n - P_o}{n}$$

and the arthmetic approximation to an average rate is given by

$$r^1 = \frac{b}{\frac{1}{2}(P_o + P_n)} \times 100$$

Where P_o is the initial number of centres, P_n is the number of centres at the end of the period (n year), n is the time in years, and b is the annual amount of change.

In a geometric series, the frequency distribution increases or decreases at the same rate over each unit of time, e.g. each year. If this constant rate of change is represented by r and the initial number is represented by P_o, then after n years the final frequency of central places is given by :

$$P_n = P_o (1+r)n$$

with the observed data that is available, we get

$$1 + r = \sqrt[n]{\frac{P_n}{P_o}}$$

$$\text{or } \log(1 + r) = \frac{\log P_n - \log P_o}{n}$$

$$\text{or } \frac{\log\left(\frac{P_n}{P_o}\right)}{n}$$

$$= \tfrac{1}{2}[(\log 342 - \log 239)] = \tfrac{1}{2}[(2.53403 - 2.37840)]$$

$$= 0.03112 \text{ taking antilog } 1+r = 0.10748$$

$r = .10748$ i.e., the rate of change is 10.75 per cent per annum.

Regional Peculiarities in the Dynamics of Rural Settlements

The visual evaluation of the degree of change in setting patterns may be supplemented by an objective measure of the magnitude of change in the form of the sum of deviations of the 1970 distribution curve from the 1959 curve. For this purpose use was made of the simple index K_r (the coefficient of redistribution of population among size classes during the study period). If the distribution of population by size classes did not change, then $K_r = 0$; the index will increase with the scale of structural changes in the network of places. The coefficient can be calculated as follows[46] :

$$K_r = \sqrt{\sum_{P=1}^{n} (P_{70} - P_{59})^2}$$

where n is the number of size classes, P_{70} is the percentage of the 1970 rural population living in places of particular size class; P_{59} the same for the 1959 rural population.

The formula applied for size of the rural settlements is as follows :

$$S = \frac{P}{N}$$

where S is the average size of population per village, P is the rural population of the anchal, and N is the number of villages in the anchal.

Another possibility is as follows :

$$S = \frac{\text{Number of Houses in an Anchal}}{\text{Number of Settlements in an Anchal}}$$

Information may be collected from Census returns.

Measurement of Shape of Rural Settlements

Method: 1

In order to determine the shape, a circle with its circumference touching the two extreme points of settlement was drawn. Now the pattern seems to be obvious in its geometrical perspective in comparison with the circle. The formula for the measurement of this is[47] :

$$MC = \frac{100 \times A_a}{\pi r^2}$$

where MC is the measure of circularity in shape; A_a is the actual area contained in the settlements and πr^2 is the area of circle of settlement $= A$.

Method: 2

The shape of settlements can also be measured with Simmon's measure. He devised a method based upon area of a settlement and its outer boundary. The measure is defined as the ratio of the sum of distances of all settlements arranged in a regular network within the boundary of the residential area of a settlement to its centre. The formula for measuring shape is as follows[48] :

$$S = \frac{\sum_{i=1}^{n} R_i}{\sum_{i=1}^{n} r_i}$$

Where R_i is the distance for the settlement area, and r_i is the area for the circle. Here for a given settlement with given centre and boundary; S is a function of a and of type of network of settlements.

Method: 3

According to Boyce and Clark, the formula for measuring the outer boundary of a settlement is as follows[49] :

$$S = \sum_{i=1}^{n} \left| \frac{R_i}{\sum_{i=1}^{n} R_i} . 100 \frac{100}{n} \right|$$

Where S is the measure and the R_i (i=1, 2.........n) are the lengths of radial lines measured from the settlement centre to the boundary along equally spaced radii.

Method: 4

The shape of rural settlement is theoretically circular at its peripheries for the distance between the central point and the edge of the territory is always shorter and quicker in space and time in comparison with any other geometrical shape of the settlement. But circles either overlap or leave gaps between them. The hexagonal pattern of rural settlements, therefore, is the ideal shape of a circle. Settlements in the area are neither equally spaced nor have equal size. The distance that should actually exist among settlements in a particular region with homogeneous character throughout, has been called as 'mean actual distance' (D_a). It has derived from the following equation.[50]

$$D_a = \frac{\Sigma Ps_1 + Ps_2 + Ps_3 + \ldots\ldots Psn}{\Sigma N}$$

Where Ps stand for the distance of possible straight line joining the different villages and N stands for total number of lines joining n villages. In case the same number of settlements are supposed to be arranged on a hexagonal pattern in the same region occupying the same area, their inter-distance, as it is called Hypothetical Distance (D_h), is derived from the following equation[51] :

$$D_h = \sqrt{\frac{\Sigma A_1 + A_2 + A_3 + \ldots\ldots + An}{\Sigma N}}$$

Where A denotes the area of settlements up to nth number in Km^2 and N stands for the number of settlements.

In case D_a equal D_h, there is a likelihood of equi-spaced normal settlement pattern. If D_a is more than D_h, there will be a tendency towards dispersed pattern and if D_a is less than D_h, the settlements will adopt a concentrated pattern. Hence, the above mentioned technique will help in determining the spacing, regularity and type of settlements in the study area. Now D_h being the ideal distance, is supposed to be 100 per cent and the regularity index (I_r) is derived from the following equation:

$$I_r = 100 - \frac{D_a \times 100}{D_h}$$

The indices derived out of this formulla will both be either positive or negative. The negative indices will suggest concentration while the positive, dispersion.

The smaller urban settlements or the larger village have been described variously as rurban settlements. A rurban centre is essentially a local focus of activities and is marked by commercial, industrial, administrative and cultural institutions. Centres having at least three out of nine central functions (viz., education, medical, communication, finance, transport, administrative, trade, recreation and percentage of workers engaged in tertiary occupations) have been considered as rurban centres. Classification of urban centres on this basis from the list of villages is easy, but a problem arises in the selection of urban centres from the list of towns. Both towns and cities have functions of a higher order as compared to the rural, which creates confusion. In the selection of rurban centres from the list of towns, the following formula may be used.[52]

$$RI = PC \times \frac{t_w \times 100}{Tw}$$

where RI = Rurban index

PC = Total population of the centre

t_w = Workers engaged in tertiary occupations in that centre, and

TW = Total number of workers in that centre.

After calculating the value of RI by giving proper weighting to above mentioned 9 functions for all towns, the mean was determined. On accepting the mean as the line of demarcation, all towns having RI indices less than mean were classified as rurban centres.

Mean Centres of Rural Settlements

Using the idea of 'mean centres of population' of Hart,[53] D.K. Singh and others have written a paper on rural settlements of Orissa which deals with the mean centres of rural settlements. Their idea about this method is as follows:

Rural settlement types and patterns give the simplest description of arrangement of settlements according to the physiography, drainage, transport and communication system. But this would not help in assessing the grouping of settlements around the centrally located important rural settlements which might help in representing the mean centre of settlements for the State as a whole in order to act as the point of diffusion for rural development innovations.

In all, 7724 rural settlements with functional centrality have been considered to find out their mean centres at various levels dividing the whole state into 20×20 mile grid. The co-ordinates to the mean centre $(\bar{x}_c, \bar{y}_c)$ have been calculated using the expression[54] :

$$\bar{X}c = \frac{\Sigma (x_i f_i)}{n}$$

and $\overline{Y}c = \frac{\Sigma (y_i f_i)}{n}$

Where xi and yi are the coordinates of ith point in the area and n is the total number of settlements in the State, while f_i represents the frequency of settlements in the individual grid along the x axis and y axis separately. The mean centre for functionally important rural settlements of Orissa is found out to be represented by—

$$\overline{X}c = 236.3$$

$$\overline{Y}c = 181.6$$

It is observed that though the coastal area and the north-western area have highest concentrations of functionally important rural settlements, the frequency for x_i as well as y_i clearly indicate a thin representation in the centre and the arrangement of frequency depicts a bimodel concentration.[55]

The hypothesis, that the environmental factors are also responsible for the pattern variation of settlements, is tested by Chi-square analysis[56] (Table 17.4). The calculated value (13.07) of Chi-square is greater than the tabulated value (11.07) for 5 degree of freedom but significant only at 5 per cent level and poses a weak challenge to our hypothesis. It may be inferred that the physical factors are not so strong, though important, as to influence the pattern of settlements to a great extent in the study area, but the nearness to urban centre has exerted a stronger influence to the pattern of settlement distribution. The formula for Chi-square test is as follows:

$$\text{Chi-square} = \frac{(D-E)^2}{E}$$

Where E stands for expected number of settlements, D stands for number of settlements.

The degree of freedom : $n-1 = 6 - 1 = 5$. The Table 4.4. as given will clarify the calculation of Chi-square test.

TABLE 16.4 : **Chi-Square Test**

Sl. No.	Factors affecting rural settlements	Observed number of settlements (D)	Expected number of settlements (E) 278/6= 46.33	$(D-E)2$	$\frac{(D-E)^2}{E}$
1.	Plains	48	46	4	0.80
2.	Hills	56	46	100	2.10
3.	Ravines	38	46	64	1.40
4.	Shrubby Forests	58	46	144	3.10
5.	Reserved Forests	49	46	9	0.19
6.	Around Urban Centres	29	46	289	6.20
	Total Observations	278	276	610	13.07

Thus, it can be said that most of the statistics used in rural settlement geography are concerned with the formulation of types of rural settlements, spacing of settlements along with the density, size and functional hierarchy. Only some of the methods which are simpler in relation to others are presented here for the students of geography, sociology, anthropology, economics and statistics.

REFERENCES

1. Brunhes, *Human Geography*, New York, Rand McNally, 1952, p. 152.
2. A. Woiekof, "Le groupement de la population rurale en Russie", *Annales de Geographic*, Vol. 18, 1909, pp. 13-23.
3. A. Kristofferson, Landskapsleildens foran dringar i norra och ostra delen av Fars Harad, Lund, 1924, p. 141.
4. M.A. Lefevre, *L'Habitat Rural en Belgique*, Liege, 1926, p. 114.
5. J. Bernhard, "Une formule Pour a Cartographie de l' Habitat rural avec

application en department de youne", *Comptes Rendu*, 1931, Vol. III, pp. 108-117.

6. A. Demongeon, "Une Carte de l' Habitat", *Annales de Geographie*, Vol. 42. 1933, pp. 225-232.
7. A. Zierhoffer, "Sus Une Formule Servant a Exprimer la Dispersion et la Concentration Absolue de l'Habitat Rural", *Comptes Rendu*, 1934, Vol. III, pp. 410-415.
8. S. Pawlowski, "Encore Une Methode de Representation Cartographique Generale de l'Habitat Rural", *Comptes Rendu*, 1938, Vol. 2, pp. 129-30.
9. A.H. Debouverie, "Pour un Cartographie de l'Habitat Belgique", Bulletin, *Societe Belge de Etudes Geographiques*, Vol. 13, 1943, pp. 146-196.
10. E. Kant, Quelque Problems Concernant la Representation de la Densite des Habitations Rurales, *Lund Studies in Geography*, Series B. No. 2, 1950, pp. 1-9.
11. J.M. Houston, *A Social Geography of Europe*, London, 1961, p. 83.
12. T. Tanioka, "Differenciation Regional des Types de l'Habitat au Japan", in *Proceedings of the International Geographical Union Conference*, Tokyo, 1957, pp. 503-512.
13. K.H. Stone, "Swedish Fringes of Settlement", *Annals of the Association of American Geographers*, Vol. 52, 1962, p. 374.
14. S. Inouye, "Settlement Measurement in Japan", in Abstracts of Papers, *XX International Geographical Congress*, London, 1964, p. 308.
15. R.B. Singh, "Rural Settlement Types and their Distribution", *The National Geographical Journal of India*, Vol. XV, 1969, p. 100.
16. R.B. Mandal, *Geography of Rural Settlements in Bihar Plain*, (unpublished Ph.D. Thesis) Patna, University, Department of Geography, 1972, p. 148.
17. A. Prasad, *Chota Nagpur: Geography of Rural Settlements*, Ranchi University, 1973, pp. 314-315.
18. R.B. Mandal, *Planned Development of Rural Settlements*, New Delhi, Concept, 1979, pp. 117-18.
19. P. Haggett, *Locational Analysis in Human Geography*, London, Edward Arnold, 1966, pp. 107-13.
20. G.T. Trewartha, "The Unincorporated Hamlet: One Element of the American Settlement Fabric", *Annals of the Association of American Geographers*, Vol. 33, 1943, pp. 32-81.
21. J.A. Barnes and A.H. Robinson, "A New Method for the Representation of Dispersed Rural Population", *Geographical Review*, Vol. 30, 1940, pp. 134-137.
22. Edgar Kant, "Some Notes on the Representation of the Density of Rural Habitations", *The Indian Geographical Journal*, Vol. 30, 1955, pp. 1-8.
23. E.C. Mather, "A Linear-Distance Map of Farm Population in the U.S.", *Annals of the Association of American Geographers*, Vol. 34, 1944, pp. 173-80.
24. R.B. Singh, "Pattern Analysis of Rural Settlement Distribution and their Types in Saran Plain: A Quantitative Approach", *The National Geographical Journal of India*, Vol. 20, 1974, p. 126.

25. A.B. Mukerji, "Spacing of Rural Settlements in Andhra Pradesh: A Spatial Interpretation", *Geographical Outlook*, Vol. VI, 1969, p. 3.
26. C. Board et al. (Ed.), *Progress in Geography*, London, Edward Arnold, Vol. 2, 1970, p. 14.
27. R.S. Lall and A.N. Mehra, *A Text Book of Vector Analysis*, Lucknow, Prakashan Kendra, 1965, p. 14.
28. A.D. Bannerman, Census of India, 1901, Part 1, Lucknow Government Press, 1902, Quoted by R.C. Sharma: *Settlement Geography of Indian Desert*, New Delhi, Kumar Brothers, 1972, p. 107.
29. P.J. Clark and F.C. Evans, "Distance to Nearest Neighbour as a Measure of Spatial Relationships on Population", *Ecology*, Vol. 35, 1954, pp. 445-453.
30. M.F. Dacey, "Order Distance in an Inhomogeneous Random Point Pattern", *Canadian Geographer*, Vol. 9, 1965, pp. 141-153.
31. P.J. Clark and F.C. Evans, op. cit.
32. H.R. Tompson, "Distribution of Distance to nth Neighbour in a Population of Randomly Distributed Individuals", *Ecology*, Vol. 37, 1956, pp. 391-94.
33. L.J. King, *Statistical Analysis of Geography*, New Jersey, Prentice Hall, 1969.
34. B. Thakur, "Nearest Neighbour Analysis as a Measure of Urban Place Patterns", *Indian Geographical Studies*, No. 1, 1973, p. 66.
35. Peter Ambrose, *Analytical Human Geography*, London, Longmans, 1969, p. 119.
36. B.P. Birch, "The Measurement of Dispersed Patterns of Settlements", *Tijdschrift voor Economische en Sociale Geographie*, Vol. 58, 1967, pp. 68-75.
37. D.A. Pinder and M.E. Witherick, "A Modification of Nearest Neighbour Analysis for use in Linear Situations", *Geography*, Vol. 60, No. 266, 1975, pp. 16-23.
38. L.J. King, op. cit., p. 89.
39. M.F. Dacey and T. Tung, "The Identification of Randomness in Point Patterns", *Journal of Regional Science Association*, Vol. 4, 1962, p. 85.
40. R. Bachi, "Standard Distance Measures and Related Methods for Spatial Analysis", Papers, *Regional Science Association*, Vol. 10, pp. 83-102.
41. L.J. King, op. cit., pp. 92-93.
42. A.H. Robinson, *Elements of Cartography*, (Ed. 2), New York, John Wiley, 1964, pp. 16-23.
43. W.J. Craig, "A Useful Computer Function to Measure Distances", *Professional Geographer*, Vol. XXIII, 1931, pp. 157-58.
44. K.K. Khatu, "Significance of the Study of Rural Settlements in their Transitional Stages", *Indian Geographical Journal*, Vol. XLIV, No. 3-4, 1969, pp. 52-63(61).
45. R.B. Mandal, *Planned Development of Rural Settlements*, New Delhi, Concept, 1981.
46. S.A. Kovalay, "Regional Peculiarities in the Dynamics of Rural Settlement in the USSR (1959-70)", *Soviet Geography*, Vol. XV, 1974, p. 5.
47. Jack P. Gibbs, *Urban Research Method*, New Delhi, 1966, p. 101.

48. J.W. Simmons, Relationships Between the Population Density Pattern and Site of Cities, (Unpublished M.A. Thesis), University of Chicago, 1962, Quoted by B.J.L. Berry, etc,: "Urban Population Densities: Structure and Change", *Geographical Review*, Vol. 53, 1963, pp. 389-405.
49. R.R. Boyce and W.A.V. Clark, The Concept of Shape in Geography", *Geographical Review*, Vol. 54, 1954, pp. 561-62.
50. I. Singh and V.K. Asthana, "Spatial Distribution of Rural Settlements in the Foot Hill Zone of Haryana", Paper Presented at Varanasi Symposium, December, 1975, p. 3-4.
51. L.J. King, op. cit., p. 103.
52. R.B. Mandal, "Rurban Centres in Bihar Plain", *Indian Geographical Studies*, No. 1, 1973, p. 16.
53. J.F. Hart, "Central Tendency in Areal Distribution", *Economic Geography*, Vol. 30, 1954, pp. 48-49.
54. W.H. Theakstone and C. Harrison, *The Analysis of Geographical Data*, London, Heinemann, 1971, p. 59.
55. D.K. Singh, B. Routray and S. Jena, "Spatial Analysis of Rural Settlement Pattern in Orissa", Paper Presented at Varanasi Symposium, December, 1975, pp. 5-6.
56. C.R. Pathak and R.K. Dhabai, "Rural Settlement Structure of Mewar Rajasthan", I.I.T. Kharagpur, Silver Jubilee Symposium on Rural Community Development, April, 1976.

SEVENTEEN

R. B. Mandal

URBAN GEOGRAPHY

IN URBAN geography the advancement of statistical techniques has established a respectable position especially in the field of metropolitan hinterland organisation, retail gravitation and potential models. Other applied techniques are Lösch's market area equilibrium, functional classification of urban centres, measures of centrality index, mathematical models of urban hierarchy, shape index and several other fields have been statistically analysed in recent years. Hence this chapter on urban geography deals with various methods of applied statistics which have been used by different authors from time to time in measuring the characteristics of urbanization and urban centres.

Methodology

There are two customary indices to study the level and trends of urbanization in a country/state, etc. They are:

Degree of Urbanization

It is the percentage of total population living in urban areas. The formula for computing it is:

$$\alpha = \frac{U}{T} \times 100$$

where α= Degree of urbanization,

U= Urban population,

T= Total population.

In accordance with the international comparison, this index is further defined in two ways:

(a) Instead of total urban population, population living in cities (places with 100,000 and more inhabitants) is considered as urban population.

(b) In place of total urban population, the population living in urban places of at least 20,000 inhabitants is considered as urban.

Rate and Speed of Urbanization

It is a simple arithmetic growth of degree of urbanization between two or more censuses. It is calculated in the following way:

$$\text{Rate of urbanization} = \frac{\alpha_2 - \alpha_1}{\alpha_1} \times 100$$

where α_1 and α_2 are the degrees of urbanization at two points of time.

"There are two ways of measuring urban growth. One is the instantaneous method—which ascertains the population in all urban categories at each and every census, tracing the changes in each class regardless of the cities that make it up. The other is the continuous method which begins with particular cities and traces the subsequent expansion. The instantaneous given class that were not in that class". These two methods have been used to study urban population growth by different categories of towns considering 101 towns of 1951, none of which was declassified in the next censuses.

The concept of National Growth Rate Method has been adopted to study the level of migration in cities and towns. The method describes that if $P^0{}_T$ and $P'{}_T$ be the national population at and the beginning and end of the intercensal period, respectively $P^0{}_1\ P^0{}_2\ P^0{}_3 \ldots\ldots\ldots P^0{}_i$ be the population of the geographic subdivisions at the beginning of the period and $P'_1\ P'_2 \ldots\ldots$ their populations at the end of the period, then the estimated migration rate m_i for area i is given by

$$m_i = \frac{P_i' - P^0{}_i}{P^0{}_i} = \frac{P'{}_T - P^0{}_T}{P^0{}_T} \times K,$$

where K is taken as 100 or 1000.

In this concept, the growth rate of a town has been assumed as equal to the growth rate of the district in which the town is situated. The difference between the actual growth rate and the assumed growth rate of the town has been considered as the net migration rate of the town. The net migration so obtained was multiplied by the base year population in order to get the volume of migration for each town. These were summed up and divided by the base year total population of the towns of the size class and multiplied by 100 in order to get the net migration rate for each category of towns and cities.

Laws of Retail Gravitation

If the number of workers for each town is known, we can calculate the market area around that centre with reference to other central places in the neighbourhood. This can be done by applying the laws of retail gravitation developed by Reilly (1931).[1]

According to the break point equation of retail gravitation, the trade area boundary between two centres A and B is, in kilometres from B, equal to:

$$\frac{\textit{Distance between } A \textit{ and } B}{1 + \sqrt{\dfrac{\text{Size of } A}{\text{Size of } B}}}$$

Thus for Patna and Muzaffarpur which are cities selling goods of a higher order the break point is determined as follows :

Distance between A and B=38 miles.

Size of A (Patna)=27.097 workers in trade.

Size of B (Muzaffarpur)=9606 workers in trade.

The distance from Muzaffarpur (B) to the break point of market areas is :

$$= \frac{38}{1 + \sqrt{\dfrac{27097}{9606}}} = 14.45 \text{ miles}$$

On the basis of above formula, it has been found that cities and district headquarters from large retail market zones, which broadly attest to their regional importance as we know. On the other hand, for goods of lower order, their influence gets somewhat restricted because of the ring of small towns, which surrounds them. This is natural because if alternatives are the same the consumers tend to select the nearest centre even though it is much lower in the retail hierarchy.

Law of Retail Gravitation within Metropolitan Regions

There is, however, no such thing as an absolute breaking point within metropolitan regions. The breaking point formula simply gives the point at which the proportion of consumers located around the breaking point splits equally between two competing alternatives of differing attractiveness. Reilly's original law of retail gravitation[2] becomes relevant. "Two centres attract trade from intermediate places approximately in direct proportion to the size of the centres and in inverse proportion to the square of the distances from these two centres to the intermediate place", or

$$\frac{TA}{TB}=\frac{PA}{PB}\left(\frac{DB}{DA}\right)^2$$

where TA, TB = Proportions of trade from the intermediate place attracting by centres

A and B

P_A, P_B = Size of A and B

D_A, D_B = distances of A and B from the intermediate place.

This model is deterministic; it says that T_A / T_B is exactly the proportion of trade that will move in one direction or another. Yet it is known that within a city individual behaviour is probabilistic.

The Validity of Reilly's Law in Australia—Some Preliminary Considerations

The retail gravitation is strongest toward large cities; weakens with decreasing size, and finally peters out at the periphery of a city's tributary area. Where the tributary areas of two competing cities overlapped it was found that the external retail trade was shared in accordance with Reilly's Law. Stated as a general formula:[3]

$$\frac{B_a}{B_b}=\left[\frac{B_a}{P_b}\right]\times{}^{N}\left[\frac{D_b}{D_a}\right]^n$$

Where $B_a : B_b$ is the proportion of the retail trade attracted by city A and city B respectively, $P_a : P_b$ the ratio of populations of the two cities, and D_a and D_b are the distances of any intermediate town under consideration from city A and B respectively.

In this equation the two unknowns are N the exponent of populations and n the exponent of the inverse distance. Reilly found from his data that the exponent N was the first power. That is, if a city of 50,000 has 5,000 credit-account customers, a city of 100,000 would typically attract 10,000 credit account customers. This left n, the exponent of inverse distances, as the only unknown, and in solving the equation for each of the 225 cases studied, values of n from 0 – 12.5 were calculated. It was observed from frequency tabulations, however, that the model value for n was between 1.51 to 2.5 which led to the conclusion that exponent of inverse distances is nearer the second power than to any other even power. Subsequent studies have indicated broad agreement with these findings.

Thus Reilly's Law becomes :

$$\frac{B_a}{B_b} = \left[\frac{P_a}{P_b}\right] \times \left[\frac{D_b}{D_a}\right]^2 \qquad \text{... (1)}$$

A second formula is derived from this equation, and is used to determine the "breaking point" or boundary from which retail trade form an intermediate centre attracted towards two cities equal; half going to city A and half to city B. This second formula is :

$$\frac{\textit{Distance from city A to B}}{1 + \sqrt{\dfrac{\text{Population of city } A}{\text{Population of city } B}}} \qquad \text{... (2)}$$

According to first formula : The population of Sydney (A) according to the 1954 census is 1610,5800, and that of Brisbane (B) is 469,000. Grafton, the intermediate centre selected, is 447 miles from Sydney and 224 miles from Brisbane. Then substituting in the first formula, we have

$$\frac{B_a}{B_a} = \frac{1,610,580}{469,000} \times \left[\frac{224}{447}\right]^2 = 0.8619$$

Thus Sydney, according to Reilly's Law, should attract 0.88 times as much retail trade from Grafton as Brisbane does. Then also, the comparative percentages of trade attracted by Sydney and Brisbane respectively are :

$$\text{Sydney: } \frac{0.86 \times 100}{1.86}, \text{ or } 46.3 \text{ per cent}$$

$$\text{Brisbane : } \frac{1.00 \times 100}{1.86}, \text{ or } 53.7 \text{ per cent}$$

It is apparent that, in theory, the retail gravitation of Grafton towards Brisbane, is somewhat stronger than toward Sydney.

Law of Retail Gravitaion and Profitability Model

In densely built-up areas consumers have considerable business centres of differing attractiveness available within the maximum distances where they are willing to travel. They will visit none exclusively, but each at some time and with some probability. One probability model, developed by Huff in the Reilly tradition, appears to work well. Where PA_1 is the probability that a consumer located at A will visit centre 1, and there are r differing shopping opportunities of sizes $S_1 \ldots S_r$ located at travel times $T_{a1} \ldots TA_r$ from A, the model is

$$PA_1 = \frac{\dfrac{S_1}{TA_1}}{\sum_{i=1}^{r} \dfrac{S\alpha i}{\dfrac{S\alpha}{A_i}}}$$

and of course $\sum_i PA_i = 1.0$, and α is a parameter which varies for different levels of the hierarchy.

Applying this model to a variety of residential areas will result in a set of probabilities of visiting each centre for each area. Probability contours can be drawn for any centre linking areas which have the same probability of being visited. Berry shows three places where only two way choices are made, so the points of indifference are the tangents of the 0.5 probability contours. A third indifference point is the meeting of the 0.33 contours from each of the three centres; there one finds equality of the three-way interaction probabilities.

The Market Area for a Good

Since travel is equal in all directions, it is possible to draw a demand cone around the store, according to which quantity consumed drops off with distance because of price increases to the consumer due to increased transport cost and quantity consumed is zero. This is the maximum possible economic reach of the store. The "ideal" trade area of the store is a perfect circle with radius r.

Total quantity of good X consumed by customers within this maximum reach r may be obtained by calculating the area, D, beneath the demand cone. Since quantity demanded, q, varies in response to level of retail price plus transport cost, $P+m_t$, D is found by integrating the function $q = f(P + m_t)$ out to the maximum radius r, and multiplying by population density S.

$$D_i = S \int_o^{2\pi} \left[\int_o^{m=r} f \ (P_i + m_t) \ mdm \right] d\phi$$

If this calculation is repeated for a variety of different store selling prices, P_i, cones of varying heights and maximum radii will result, and different levels of total demand D_i may be calculated. If these values of P_i and D_i are plotted in a graph and a line is fitted to the results, an aggregate demand curve D for the market area can be drawn (Berry, B. J. L.).[4]

Generalization of Losch's Market Area Equilibrium

In the above formula, the market area for a good, the shape of that function was not specified and the shape of the firm's cost curve was not defined.

Let both the cost and the individual demand function be linear :

$$<= F + Cu$$

where c is fixed costs, u is number of units sold, and c is variable cost per unit, and

$$qi = g - h \ (P_i + m_t)$$

Where qi=is quantity consumed by the family, P_i is at store price, t is transport costs per mile, and m is number of miles of residence from the store.

Therefore, for a circular market area,

$$D_i = S \int_{o}^{2\pi} \left\{ \int_{o}^{m=r} [g-h\ (P_i+m_t)]\ mdm \right\} d\,\phi$$

where S is population densities and r is maximum radius consumers will travel. Similarly, for market areas of other figures displaying bilateral symmetry,

$$D_i = 2\,n\,S \int_{o}^{\pi/n} \left\{ \int_{o}^{m/\cos\phi} \left[g-h\ (P_i + m_t) \right] mdm \right\} d\,\phi$$

where n is number of sides of the figure and m is distance from center of figure to midpoint of its sides.

Since a firm's total profit R is

$$R = P_i\ D_i - (F + CD_i)$$
$$= D_i\ (P_i - c) - F$$

The results of evaluating the integrals for a (Δ), square (□), hexagonand circle (O) are

$$R\Delta = 6s\ r^2 \left(\frac{\sqrt{3}}{2}\ g\ \frac{\sqrt{3}}{2}\ hp_i - .7069\ htr \right) (P_i - c) - F$$

$$R\Box = 8S\ r^2 \left(\frac{g}{2} - \frac{hp_i}{2} - \ .3848\ htr \right) (P_i - c) - F$$

$$\underset{\text{(Hexagon)}}{R} = 12\ S\ r^2 \left(\frac{g}{2\sqrt{3}} - \frac{hp_i}{2\sqrt{3}} - .2027\ htr \right) (P_i - c) - F$$

$$RO = 2\pi\ S\ r^2 \left(\frac{g}{2} - \frac{hp_i}{2} - \frac{htr}{3} \right) (P_i - c) - F$$

Given population densities S, and the demand curve for individuals $g\text{-}h\ (P_i + mt)$, plus cost function of the business $F + Cu$, the above equations indicate that the firm's profits will depend upon price (P_i) and the maximum distance consumers will travel (r). In fact, solving the relevant different equations with respect to P_i and m yields the information contained in (Table 17.1). For the profit maximizing distance m, the price charged will be the same whatever the market area form:

$$P_i = \frac{g}{2h} + \frac{c}{2} - 0.25 \left(\frac{g\text{-}hc}{ht} \right)$$

TABLE 17.1: Solutions to Market Area Equations

	Shape of Area			
	Triangle	Square	Hexagon	Circle
Profit Maximizing m	$.5434\left(\frac{g-hc}{ht}\right)$	$.6534\left(\frac{g-hc}{ht}\right)$	$.7121\left(\frac{g-hc}{ht}\right)$	$.7501\left(\frac{g-hc}{ht}\right)$
Maximum Profit	$.0959\left[\frac{(g-hc)^4}{g^3\ t2}\right]S\text{-}F/$	$.1067\left[\frac{(g-hc)^4}{g^3\ t2}\right]S\text{-}F/$	$.1098\left[\frac{(g-hc)^4}{g^3\ t2}\right]S\text{-}F/$	$.1104\left[\frac{(g-hc)^4}{g^3\ t2}\right]S\text{-}F$
Profit Minimizing m	$1.0863\left(\frac{g-hc}{ht}\right)$	$1.3070\left(\frac{g-hc}{ht}\right)$	$1.4242\left(\frac{g-hc}{ht}\right)$	$.15000\left(\frac{g-hc}{ht}\right)$
Minimum Profit	—F	—F	—F	—F

Source : Mills, E.S, and Lav Michael, R. "A Model of Market Areas with Free Entry", *Journal of Political Economy*, Vol. 72, 1964, pp. 278-290.

However, maximum profit will not be the same: They are least for the triangle and most for the circle, further proof of the contention that firms prefer circular market areas.

If the second row of the Table 17.1 is examined, it will be seen that each of the equation has the form

$$\left\{Z\left[\frac{(cg\text{-}hc)^4}{g^3\ t^2}\right]S\text{-}F\right\}$$

and Z is .0459 (triangle), .1067 (square), .1098 (hexagon), and .1104 (circle). For zero profits, the left side of the equation must equal the right side, fixed costs F. It is possible to find a level of fixed costs F such that no firms with hexagonal market areas can break even. Lösch's contention that the set of market areas must completely fill the plain, and that there be no unserved consumers, is thus invalid. If F is less than

$$\left\{.1098\left[\frac{(cg\text{-}hc)}{g^3\,t^2}\right]S\right\},$$

Lösch's contention is correct. If it is not then same unserved area will remain.

An Alternative Demand Curve

Baumol and Ide[5] set the solution for a single good within context of a set of goods. Baumol and Ide define the following:

$P(T)$ = The probability that a consumer will find the goods he needs to make his trip a success in the centre that offers T goods.

mt = The cost of travelling m miles to the centre at t per mile.

$C_n\sqrt{T}$ = The difficulty of shopping, emerging out of the size and congestion, defined as costs proportional to the square root of the size of centre T.

C_i = The opportunity costs of alternative activities foregone, including other shopping opportunities.

The consumer, it is argued, will not shop at a store unless $f(t, m)=e\,[P(T)]-l\,(mt + (n\sqrt{T}+C_i)$ is positive. Constants e and l are, respectively, the subjective weighting functions assigned by the consumer to his expectation of success (e will also be affected by price levels, since the probability of success is the probability of achieving wants at a price), and his feelings concerning costs involved (l). Solving for the maximum distance consumers are willing to shop yields:

$$r = \frac{e}{lt} P(T) - \frac{l}{t}(C_n\sqrt{T} + C_i)$$

Increased variety is important to consumers up to a point. The minimum number of items needed to induce a consumer to a centre must increase with distance from the centre, m, but the optimum variety is independent of distance.

Suppose that the store's sales depend directly upon the number of individuals who can be induced to shop there, and that at any distance m, the function $f(T, m)$ gives the proportion of consumers who will shop there. Then, with a uniform plain of population density S, the aggregate demand will be—

$$D_i = S\int_o^{2\pi}\left\{\int_o^r \left[(ep(T) - l(C_n\sqrt{T} + C_i) - ltm\right] mdm\right\} d\phi$$

$$= 2\pi\, Sr^2\left[\frac{ep(T) - l(C_n\sqrt{T} + C_i)}{2} - lt\frac{r}{3}\right]$$

But

$$r = \frac{ep(T) - l(C_n\sqrt{T} + C_i)}{lt}$$

Therefore, substituting:

$$D_i = \frac{1}{3}\, lt\,\pi\, Sr^3$$

If it is assumed that population densities are not uniform, but vary from a point of maximum concentration S at the store inversely with distance S/m, then the results become—

$$D_i = S\int_o^{2\pi}\left\{\int_o^r\left[ep(T) - l(C_n\sqrt{T} + C_i) - ltm\right] dm\right\} d\phi$$

$$= 2\pi\, S\left\{ep(T) - l(C_n\sqrt{T} + C_i)\, m - \frac{ltm^2}{2}\right\}\int_o^r$$

$$= lt\,\pi\, Sr^2$$

In both cases aggregate demand is a function of maximum distance consumers are willing to travel, r, but on the uniform plain it varies as the cube of distance, whereas if population densities drop inversely with distance from the centre, it varies with the square of the distance.[6]

Prices will generally lower in larger centre and lesser transport cost due to multi-purpose trip of the nature.

Methods for Functional Classification of Urban Centres

Method: 1

Mattila and Thompson[7] have advanced a new method for the classification of urban centres where they have considered the index of surplus workers, both in absolute and relative form to assess the local business of an industry. Rather than relying on an index which indirectly measures the surplus workers, it was calculated by taking the difference of local employment and country's share of per head employment as denoted by the formula:

$$e_i = \frac{e_t E_i}{E_t}$$

where e_i = local function employment,

e_t = local total employment,

E_i = national function employment,

E_t = national total employment.

where $e_i = e_t \frac{E_i}{E_t}$

This denotes that the industry is not specialized and there is a balance between local and regional production and consumption. When e_i is more than $e_t \frac{E_i}{E_t}$, that means the industry has extra workers. But this index of surplus workers, in its absolute form, though highly suited for intra city industry comparisons is not at all suited for intra-city industry comparisons. Therefore, it was converted into a relative form by taking the ratio of surplus workers to total employment or total number of surplus workers, Then the formula comes to either:

$$\frac{e_i - \frac{e_t . E_i}{E_t}}{e_t} \quad \text{or} \quad \frac{e_i - \frac{e_t . E_i}{E_t}}{E\left[e_i - \frac{e_t . E_i}{E_t} \right]}$$

They also find out the index of service workers by substracting total surplus workers from total employment. This method is suitable both for inter-city and intra-city comparisons, we can say, whether *A* function is more basic to the city *X* than *B* function in city *Y*. But the method does not denote the intensity of specialisation for the towns as a whole.

Method: 2

According to Mahamaya Mukerji if we take into consideration the *i* function in a town, then the proportion of all [the workers engaged in the ith function in the total working population is $\frac{E_i}{E_t}$ (where E_i = state function employment and E_t = state employment). If the same proportion were holding good for a particular town, the number of workers engaged in the ith function would be $\frac{et.\ E_i}{E_t}$.

But the actual number of workers, engaged in the ith function in the locality is e_i (e_t = local total and e_i = local function employment). Hence, the excess number of workers engaged in that function in the locality is:

$$\frac{e_i - \frac{e_t.\ E_t}{E_t} = P_i \text{ (say)}}{e_t = \frac{E_i}{E_t}}$$

Now these P_1, P_2 .... P_i etc. values do not have equal significance for all the functions.

Therefore, it will be appropriate to attach importance proportional to the percentage of workers engaged in the ith function in the locality. Thus, the formula for finding out functional index becomes,

$$\left(\frac{e_i - \frac{e_t.\ E_i}{E_t}}{e_t.\ \frac{E_i}{E_t}}\right)\left(\frac{e_i}{e_t} \times 100\right)$$

For finding out specialisation index, all the functional index values are added. Naturally the specialization of a town having the mean value in a function would be *O*. The categories are then grouped in arithmetic progression at certain intervals.[8]

Method: 3

In functional classification of towns L.N. Ram and V. N. P. Sinha (1972)[9] have solved the problem of predominance of major towns over smaller towns. Some of the urban centres, due to certain economic and cultural growth, over-shadow the neighbouring towns. To eliminate this difficulty, the local significance of trade function has been weighted by the consideration of town's regional significance because the hierarchy of an urban centre not only depends on its local specialisation but also on its regional importance. In order to integrate the two, the percentage of trade workers in individual urban centres has been multiplied by its regional percentage. This gives a specialisation value, the range of which is too high to admit mathematical analysis. In order to avoid this exaggeration, the square-root of the resultant figures have been calculated. Thus this formula enables us to find out the inter-town and intra-town significance of trade functions.

The formula on which the significance of trade centres has been calculated is given below.[10]

$$F.\,I. = \sqrt{\frac{NW \times 100}{TW} \times \frac{NW \times 100}{TN}}$$

Where $F.\,I.$ = Functional Index

NW = Number of workers in trade and commerce of a town.

TW = Total workers of that centre.

TN = Total number of workers in trade and commerce for all towns of the region.

Calculation of Centrality Value

Method: 1

Christaller takes the telephone and derives the degree of centrality of a place, that is, amount of service performed apart from those relating to the need of its own inhabitants by means of the following formula.[11]

Centrality of a place	Number of telephones in place	Number of inhabitants in place	×	Number of telephones in area served by place

or

$$C.\,I. = T_2 - E_2 \frac{T_g}{E_g}$$

Method: 2

Godlund has put forward the formula for finding the centrality as.[12]

$$C_i = \frac{S_t.\ 100}{P_t.}$$

Where $S_t.$ stands for persons employed in retail trade; P_t population of the region and C_i the desired centrality.

Centrality Index

Method : 3

Centrality index as devised by W. Christaller in connection with settlement studies, based on telephone services:[13]

$$C = \left(t - \frac{PT}{P}\right)$$

Where C = index of centrality, t = number of telephones in a town, P = population of town, T = number of telephones in whole region, P = population of whole region.

Method: 4

For the calculation of centrality value Sven Godlund, in his study of Swedish towns suggested a formula based on the relationship between the total population of a town (F) and the number of people actively occupied in retail trade and services (D).

$$\text{Index} = \frac{D}{F} \times 100$$

In a regional centre this index will exceed 6.5.

Method: 5

To calculate an index of centrality using the expression:

$$\text{Index of centrality} = \text{inward calls} \frac{\text{inward calls}}{\text{outward calls}}$$

The index is thus based on two things: the total volume of incoming telephone traffic, and the extent to which intensity of incoming traffic exceeds that of outbound traffic. On the map the size of the circle is proportionate to the index value, and the shading of each circle is proportionate to the ratio of inward to outward traffic.[14]

Method: 6

The measure the Centrality Index the central functions selected are trade, retail services and a few other service functions. These functions were selected because they are primarily consumer oriented and were available for all sizes of urban centres. The calculation is based on Davies (1967) centrality index.[15]

$$C = \frac{t}{T} \times 100$$

where C = The location coefficient of function t,
t = One functional unit of functions,
T = Total number of functional units of function t in the whole area.

The functional index was applied and all urban places were ranked from highest to lowest according to functional index values. In order to define order of urban places, in hierarchical level are large city, town, village and hamlet.

Method: 7

Kashi Nath Singh has put forward a more feasible proposition in the shape of a formula as:[16]

$$C = \frac{N.\ 100}{P}$$

where N stands for the number dependent upon commerce; P for the total regional population in commerce and C the desired centrality.

Method: 8

According to M.S. Vishwanath[17] the formula of K. N. Singh can be applied in calculating the centrality value with a slight modification. The formula stands as:

$$C = \frac{N.\ 100}{P.}$$

Where N stands for the employment force in trade and commerce, transport and communication. Similarly P stands for the total regional population engaged in trade, commerce, transport and communication. The state employment force in these as per cent of total population.

$$\left(\frac{P \times 100}{\text{Total State Population}}\right)$$

stands at 2.108. In his study, the bigger centres like cities are not taken into consideration, the centrality figure stands at a very low mark. The centrality thus derived is plotted against the population size of the centre and the equation of the line of regression was calculated as $Y = 0.000012 + 0.0109$.

Method: 9

Ashok Dutt[18] has developed the following formula to determine the hierarchic level of a market.

Composite Hierarchic Level (CHL) =

$$\frac{1000.X_i}{A.} + \frac{100.Y_i}{B}$$

When $A = \Sigma$ Consumers of markets,

$B = \Sigma$ Consumers catchment areas of markets,

X_i = Number of consumers shopping in market i,

Y_i = Catchment area of market i.

Composite Hierarchic Level (CHL) values were assigned to all markets for which data were available.

Method: 10

A measure similar to that of Davies relative certainty index was adopted by O.P. Singh[19] making use of commercial population of centres. He used the following formula to calculate relative and absolute centrality index :

$$R_{p_1} = \frac{P_{c1} - \left(Pt_1 \frac{R_c}{R_1} \right)}{\sum P_c - \left(\Sigma Pt_1 \frac{R_c}{Rt} \right)} 100$$

where P_{c1} = Commercial Population of the Place or Centre,

P_{t1} = Total Population of the Place,

R_c = Regional Commercial Population,

R_t = Total Regional Population,

ΣP_c = Summation of P_{c1}, P_{c2} . . . P_{cn} and

ΣP_t = Summation of P_{t1}, P_{t_2} . . . P_{tn}

Absolute centrality was calculated thus

$$A_{c1} = \frac{P_b}{100000}$$

where P_b = Basic population of the place. In order to make a comparison between the total size and the centrality of a centre as the size index was also calculated this way :

$$S.I. = \frac{P_{c1} \cdot 100}{\Sigma Pt}$$

Method: 11

Onkar Singh [20] used the commercial worker data for his hierarchical study of U.P. towns and the final centrality index computed as follows :

$$\frac{E \times 100}{T}$$

where, E = excess commercial worker population of the town and

T = total excess commercial worker population of all the urban centres of the state.

Method: 12

The relevant centrality measure proposed by N.P. Saxena and R.C. Tyagi is as follows :[21]

$$C = A \frac{P \times A_r}{P_r}$$

where, C = Centrality measure;

A = Total number of workers engaged in central functions at a centre,

P = Population of the centre,

A_r = Aggregate of workers engaged in central occupations in the region; and

P_r = Population of a region.

Relative Centrality Index

Method: 13

The Relative Centrality Index (R.C.I.) of a centre can, therefore. be calculated by the following formula :[22]

$$R.C.I. = \frac{P_{c1} - \left(P_{t1} \frac{R_c}{R_t} \right)}{\Sigma P_c = \left(\Sigma P_t \, . \, \frac{R_c}{R_t} \right)} \times 100$$

where P_{c_1} = Commercial Population of the place of centre,

P_{t1} = Total population of the place,

R_c = Regional commercial population,

R_t = Total regional commercial,

ΣP_c = Summation of P_{c1}, P_{c2}, . . . P_{cn}, and

ΣP_t = Summation of P_{t1}, P_{t2}, . . .P_{tn}.

Here the numerator of the fraction stands for the region-serving commercial population of the place; while the denominator, for service provisioning commercial population of the region. $\frac{R_c}{R_t}$ denotes the number or fraction required on an average to fulfil commercial requirements.

$P_{t\,1} \cdot \frac{R_c}{R_t}$ gives to us a figure which can serve only the local internal population of the centre and is, therefore, to be subtracted from the total. The corresponding fraction below i.e. $\Sigma P_t \frac{R_c}{R_t}$ indicates the same for the region (i.e., for all the service centres of a region considered together).

The Absolute Centrality Index (ACI) may be calculated in the following way :

$$ACI = \frac{P_b}{100{,}000}$$

When P_b = Basic population of the place. In order to make a comparison between the total size and the centrality of a centre the Size Index (SI) may also be calculated in this way :

$$SI = \frac{Pt_1 \, . \, 100}{\Sigma \, Pt}$$

ACI of a place can be calculated by dividing the total basic population (non-agricultural) by 100,000 in order, to lessen the size of the figure by around one so that an easier grasp and calculation can be had and secondly one lakh population city is considered as norm among higher and lower ranks of centres.

Method : 14

A further step toward such a general theory was recently taken with the development of a model that describes many of the features of central place systems in the aggregate incorporates systematic links between the aggregate relations and the more traditional picture of a steplike central place hierarchy that may be found in more local studies :

Parts of the aggregate model are as follows[23] :

Definitions :

P_c = population of the central place.

P_{cx} = population of the tributary area.

D_m = maximum distance consumers travel to a central place of a given size.

A = area of trade area.

C = number of central functions performed by a centre.

Q = gross population density.

E = number of establishments located in centre.

Identities :

$P_t = P_c + P_{ex}$ ...(1.1)

$A = f\,(D_m)$ ...(1.2)

$P_{ex} = AQ_{ex} = f(D_m)\,Q_{ex}$ ...(1.3)

$P_t = AQ_t = f(D_m)\,Q_t$ ...(1.4)

Structural equation[17] :

$\log P_c = a_1 + b_{1c}$ ...(2.1)

$\log D_m = a_4 + b_{4c}$ ...(2.2)

Implications from 1.1 — 2.2

$\log P_{ex} = \log Q_{ex} + \log \{f[\log^{-1}(a^4 + b^4 c)]\}$ ...(3.1)

$\log A = \log \{f[\log^{-1}\left(a4 - \frac{b4a1}{b1}\right) P_c b4/b1]\}$ ...(3.2)

$\log P_t = \log Q_t + \log\{f[\log^{-1}\left(a4 - \frac{b4a1}{b1}\right) P_c b4/b1]\}$ ...(3.3)

The implications 3.1 and 3.3 have been verified empirically. Fits to both structural equations and implications have coefficients of determination exceeding 0.84 in seven different study areas.

Although continuous functional relationships are described by the model the comparative studies also show that, if small areas are studied, then the observable patterns are those of the classical steplike hierarchy of central places nested trade areas, and predictable consumer travel behaviour.

A set of inequalities applied in conjunction with 3.2 and 3.3 provides the link between these elemental patterns and the aggregative findings, which at first glance might seem to be at odds :

Villages

$\log A < 10.4 - 2.67 \log P$ ...(4.1)

Towns

$\log A < 9.3 - 2.067 \log P$ ...(4.2)

Cities

$\log A < 22.25 - 4.75 \log P$ ...(4.3)

From these, the maximum trade area of centres of any rank may be obtained, after equations 3.2 and 3.3 have provided overall relationships between areas and populations served at given gross densities.

Mathematical Models of the Hierarchy

First of all a mathematical model of the hierarchy was given by Martin Beckmann,[24] starting in the manner of Lösch with assumptions of a uniform plain and a network of market areas for a set of lowest-level centres providing the lowest order good. He then hypothesized that the relationship of population of a centre (P_c) and the total population (P_t) served by that centre, comprising the trade area population (P_r) plus the centre's population (P_c), was :

$$P_c = B\,(P_c + P_r)$$

Which can be written

$$P_c = \left[\frac{B}{1\text{-}B}\right] P_r$$

and $B/(1-B)$ is the urban multiplier.

His next step, departing from Lösch, assumed a Christaller-type urban hierarchy of constant bifurcation ratio K, although, in contrast to Christaller, built from the lowest level upwards. For this hierarchy he wrote

$$P_{tw} = P_{cw} + KP_t\,(w-1)$$

This equation states that the total population served by a centre of level w, where w increases from a lowest level of $w=1$, equals its own population plus the total population served by the K centres of the next lower level that it dominates. In Christaller's $K=3$ case, this would mean that the total population served by a centre of the second level in the hierarchy equals its own population and the population it serves at the first level, plus the total population served by the two centres of lowest level that nest into its trade area. Stated more generally, where P_{r1} is the total population in one of the basic lowest-level trade areas.

$$P_t^{\,w} = \frac{K_w - 1P_{r1}}{(1-B)^w}$$

and

$$P_{cw} = \frac{Bk^w - 1P_{r1}}{(1-B)^w}$$

Both city size and population served increase exponentially as level in the hierarchy increases, a fact matching exactly the empirical observation that the number of business types in a town progresses with level of centre in the form $T_{cw}=dk^{w-1}$, d being a factor varying with population densities. All characteristics of central places thus for studies also vary exponentially with level of centre in the hierarchy; total population served, population of centre, number of establishments, traffic generated, number of different types of business provided, and so on. Many graphs indicating this have already been presented.

This kind of variation implies that the percentage change in any one of these variables is constant with movement from one step of the hierarchy to another. That is, if F indicates establishments in a centre and C is a constant,

$$\frac{dE}{Edw}=C$$

and

$$W=C-1 \log E$$

A similar expression could be presented for each of the other characteristics. Where functions of this form are found, hierarchical organization must be present and the hierarchy must have a regular bifurcation ratio.

Gravity Potential Models

The earlier gravity potential models were based on the primitive notions of "populations" as well as "distance". The former was measured in terms of absolute number while the later in terms of geodetic distance. An outline of these models is given below :

The next primitive model can be written in this form :

$$A_i=\frac{P_i}{d_{ij}}$$

where A_i=Full of attractive power of town,

P_i=Population of town,

D_{ij}=Distance of town $_i$ from $_j$.

This simple notion was inherent in demographic force equation suggested by Stewart, when on the analogy of Newton's gravitational law, he suggested.

$$F = G \frac{P_i P_j}{d^2_{ij}}$$

where F=Demographic force,

G=Constant,

P_i=Population of town i,

P_j=Population of town j,

d^2_{ij}=Square of distance between towns i and j.

His equations for demographic energy and demographic potential were essentially modified forms of this equation. To qualify absolute population and distance, weights were suggested—

$$I_{ij} = G \frac{(W_i P_i)(W_j P_j)}{d_{ij}b}$$

for both. The equation so derived is more realistic but it fails to solve the basic issues involved in the very concept of both population and distance. The P_i and P_j as well as b values have not been defined properly and it is left completely to the individual researcher to give values as they want.

The formula used in this model to depict the influence of any town will be[25] :

$$I_i = \frac{P_i \, a\beta\lambda}{d_{ij} \, X.Y.}$$

where I_i=Propensity index,

P_i=Population of town i,

$a\beta\lambda$=Weights on population

d_{ij}=Distance between towns i and j,

$X.Y$=Weights on distance.

$$\text{Mandya } (I_i) \text{ value} = \frac{2.5 \times 1 \times 11 \times 3}{45 \times 1 \times 1} = 1.8.$$

(Population of Mysore 25 lakhs, weight for population 1, weight for city 11, weight for socio-economic characteristics 3, distance between Mandya-Mysore 45 kms, cost Index i and time Index 1).

$$\text{Propensity to none towards Bangalore} \Rightarrow \frac{12 \times 3 \times 9 \times 4}{94+2+2} = 3.4$$

Distance between Bangalore and Mysore=140 kms.

$$\text{Bangalore's influence} = \frac{140 \times 3.4}{5.2} = 91.5$$

$$\text{Mysore's influence} = \frac{140 \times 1.8}{5.2} = 98.5 \text{ kms.}$$

From this we conclude that Mandya lies within the sphere of Mysore's influence as it is only 45 kms from Mysore.

The Accessibility Potential Variable

If the *CBD* is the only work-place, if the journey to work is the only significant trip the household member takes, and if travel to and from the *CBD* is equally easy in any direction, then the distance from any site R_i to *CBD* provides an index of the site's accessibility. If these conditions are not met, then this distance variable is an imperfect proxy for accessibility. It is quite obvious that the necessary conditions are violated in all metropolitan areas and especially in Los Angeles. The .following potential model variable makes allowances for many workplaces and is suggested as a possible improvement to dlstance alone.

$$P_i = \sum_{j=i}^{N} \frac{E_j}{\overset{c}{a+b\ D_{ij}}}$$

Where P is the accessibility potential of the i th site, E_j is the total employment in the jth workplace, D_{ij} is the distance between site i and j; a, b and c are parameters; and the summation is overall of the N work places in the community. Under this formula the employment accessibility potential of site i varies directly with the employment opportunities. The workplace and residence zones may, of course, be aggregates of sites rather than individual lots; in fact census tracts were used in this study.

Retail Market Potential Model

$$S_{ij} = \frac{\left[C_i = \frac{Fj}{(d_{ij})^a}\right] \left[C_i = \frac{F_j}{(d_{ij})^a}\right]}{\left[\sum_{K=1}^{r} \frac{F_j}{(d_{jk})^a}\right] \left[\frac{F_j}{(d_{ij})^a} + \ldots + \frac{F_n}{(d_{in})^a}\right]}$$

where S_{ij}=Consumer retail expenditures of population in zone i spent at zone j,

C_i=Total consumer retail expenditure of the population in zone i,

F_j=Size of the retail activity in zone j,

a=An exponent applied to the distance variable.

The above model states the retail centre in zone j (F_j). Attracts consumer dollars (S_{ij})

(a) In direct proportion to the consumer expenditure, c_i

(b) Indirect proportion to its size F_j

(c) In inverse proportion to the distance to the consumers $(d_{ij})^a$ and

(d) In inverse proportion to competition $\sum_{k=1}^{n} \frac{FK}{(d_{jk})^a}$

Above equation can be modified to state the consumer expenditures available in all zones of the region that would possibly be spent in zone j (Retail Centres) F_j.

$$S_j = \sum_{i=1}^{n} \frac{\dfrac{F_j}{(d_{ij})^a}}{\displaystyle\sum_{k=1}^{n} \dfrac{F_k}{(d_{jk})^a}}$$

Where S_j=Total Sales in the retail centre F_j.

A Dualistic Model of Urban Growth

The hypothesis that after accounting for the change in local employment related to the size of the export base and its ability to support additional local workers over time, the residual change in local employment can be explained by a change in export employment as hypothesized by traditional base theory.[26]

The above discussion give's rise to the following model of urban growth. Population (P) can be considered as a simple function of employment (E). Thus,

$$P_t = \lambda\,(E_t) \qquad ...(1)$$

As posited by traditional economic base theory, total income (Y) is equal to export income (Y_x) plus local income (Y_1) :

$$Y = Y_x + Y_1 \qquad \ldots(2)$$

and total employment (E) is equal to export employment (X) plus local employment (L).

$$E = X + L \qquad \ldots(2a)$$

Assuming a fixed proportion production function, neutral technological change, and constant prices, export employment at time t (X_t) can be expressed in terms of export income (Y_{xt}), the initial level of productivity (α_o), and the rate of increase of productivity (e^{-mt}).

$$X_t = Y_{xt}.\ \alpha_o\ e^{-mt} \qquad \ldots(3)$$

Under similar assumptions of local employment (L_t) can be expressed in terms of local income (Y_{1t}), the initial level of productivity in the local sector (Y_o), and the rate of increase of productivity in the local sector (e^{-mt}). Thus

$$L_t = Y_{1t}\ Y_o e^{-mt} \qquad \ldots(4)$$

According to traditional base theory, local income is related to export income through a multiplier. However, over time at least two factors tend to increase income in the local sector relative to income in the export sector. One is the relatively high income elasticity of demand for the out-put of the local sector combined with the increasing per capita income which could be expected to accompany rising productivity. Another factor of some significance is the secular change in consumption patterns brought by such factors as increasing urbanization, which tends to favour the local sector. For simplicity it is assumed that the combined effects are related through a positive Q to time.

Thus

$$Y_{1t} = Y_{xt}.\ \beta_o\ e^{Qi} \qquad \ldots(5)$$

where β_o is the initial income multiplier. Local employment can be expressed in terms of export employment by substituting equation (5) in equation (4)

$$L_t = Y_{xt}.\ \beta_o\ Y_o e^{(Q-n)t} \quad (4a)$$

Restating equation (3) in terms of employment

$$Y_{xt}=\frac{X_t \cdot e^{mt}}{\alpha_o} \qquad \text{...(3a)}$$

and substituting (3*a*) in (4*a*) yields

$$L_t=X_t \frac{\beta_o.Y_o}{\alpha_o} \cdot Q\ (m+Q-n)^t \qquad \text{...(6)}$$

or

$$L_t=X_t \cdot K_o \cdot e^{at} \qquad \text{...(6a)}$$

where $K_o=\frac{\beta_o Y_o}{\alpha_o}$ and $a=(m+Q-n)$

The dualistic approach of urban growth is contained in equation (6*a*), which shows that employment in the local sector can be generated by two forces. One is the traditional product of the change in export employment and the base multiplier (K). The second force given by e^{at} is the combined impact of differential rates of productivity growth in the export and local sectors and the relatively high income elasticity of demand for local goods and services.

If equation (6*a*) accurately reflects the process of urban growth it should be possible to isolate the change attributable to "*a*".

In this study "*a*" was estimated by solving equation (6*a*) since the other values, L_t, K_o, X_t and t were known for each city. Stating equation (6*a*) in *log* form yields :

$$\log L_t=\log K_o+\log X_t+\bar{a}t.$$

Then $$\bar{a}=\frac{\log L_t-\log K_o-\log X_t}{t} \qquad (6b)$$

For the 39 cities in this study equation (6*b*) was solved using

L_t=local employment in 1961,

X_t=export employment in 1961,

K_o=employment multiplier in 1951,

$t=10$ *This gave,*

$$\bar{a}=.03396;\ \alpha=.009889;\ \frac{\alpha}{a}=.001579;$$

where "*a*" is significantly different from zero.

The foregoing indicates that for the 39 cities included in this study the ability of export workers to support local workers grew at a compound annual rate of approximately 3.4 per cent per year over the decade 1951-1961. In other words, over the decade each 100 export workers gave size to a growth of approximately 40 workers in the local sector.

Shape Index

Method : 1

Haggett[27] has developed and applied two indices to measure the shape of countries in Brazil ; they are the contact number (C) and the shape index (S). The contact number indicates the number of contacts between areas. When C is 4 the area is a quadrilateral, when C is 5 the area is pentagon and when C is 6 the area is a hexagon. The shape index is derived from the formula[28] :

$$S = \frac{1.27\ A}{1^2}$$

Where A is the area and l is the long axis. The theoretical range of S is from zero to one. When S is 1.0 the area is circular in form, as its deviates from a circle S approaches zero. The value of S for a four sided regular polygon is 0.64 and for a rectangular hexagon it is 0.83.

Method : 2

Boyce and Clark have jointly postulated the 'Radial Distance Method' through which the shape index of CBD is worked out by measuring the radial distances from a selected point to the circumference of geometrical form. Having applied this method the shape index of the CBD of the Kaval towns has been calculated with the help of the following formula[29] :

$$\text{Shape Index} = \sum_{1}^{n} \left(\frac{r}{\Sigma n} - \frac{100—100}{n} \right)^1$$

where r stands for radials extending outward from a central node and n for number of radials.

TABLE 17.2 : **Shape Index of the CBD of Kaval Towns**

Name of City	Shape Index
Kanpur	29.86
Agra	45.31
Varanasi	24.27
Allahabad	74.53
Lucknow	19.47

Measures of Shape Distortion of Urban Places

Geography has always contained a mathematical element though the emphasis placed upon it has varied from period to period. As Haggett has pointed out, the geometrical tradition was basic to the original Greek conception of the subject, and many of the more successful attempts at geographical models have stemmed from this type of analysis. As examples of such work, he cites the contributions of Christaller, Lewis Wooldridge, Hagerstrand and Breisemeister. Branches of mathematics that are being used in geography today include topology, graph theory, linear programming and various statistical and stochastic subjects as markov chains.[30]

To quote Haggett again, 'Models are made necessary by the complexity of reality'.[31] Cities are very complex structures and so the urban geography is making increasing use of mathematical models of cities. The basic model of the 'ideal city' is taken by various authors to be circularly symmetric about its centre. Unfortunately, real cities usually deviate from this ideal due to a variety of economic, physical or sociological factors.

The problem thus arises of measuring this deviation to determine whether or not it is significantly high. The first measure we discuss is that suggested by Boyce and Clark.[32] This is based upon the shape of the outer boundary of the city and is defined by the equation :

$$I = \sum_{1}^{n} \left| \frac{R_i}{\Sigma R_i} 100 - \frac{100}{n} \right|$$

where I is the measure and the R_i $(i=1, 2 \ldots\ldots\ldots\ldots n)$ are the lengths of radial lines measured from the city centre to the boundary along equally spaced radii.

For a given city with given centre and boundary, the Boyce-Clark measure is a function both of n and of the direction of the R_i. Obviously n should be as large as possible so that this measure is to be regarded as an approximation to a limiting measure. If $R(Q)$ denotes the distance from the city centre to the boundary in the direction Q, where $R(Q)$ is of course assumed single valued and if we put

$$\bar{R} = \frac{1}{2\pi} \int_{0}^{2\pi} R(Q)dQ$$

$\bar{R}$ = The arithmetic mean of $R(Q)$ then for Boyce and Clark's measure, using the elementary theory of the Riemann integral,

$$\lim_{n \to \infty} \left| = \frac{100}{2\pi} \; \frac{1}{\bar{R}} \int_{0}^{2\pi} \right| R(Q) - \bar{R} \Big| \, dQ$$

In statistical terms, if we consider Q to be uniformly distributed over $(0, 2\pi)$, then the limiting value of Boyce's measure is 100 times the ratio of the mean variation of the random variable $R(Q)$ to its mean.

Simmons[33] devised a measure based more upon the area of a city than upon its boundary. His measure is defined as the ratio of the sum of distances of points, arranged in a regular network within the boundary of the city's urbanized area from the city centre to the sum of distances of points in the same regular network from the centre of a circle of the same area as the city. In symbols, and using J for Simmons measure,

$$J = \frac{\sum_{1}^{n} R_i}{\sum_{1}^{n} r_i}$$

where R_i are the distances for the true city, and the r_i for the circle. Here for a given city with given centre and boundary, J is a function of n and of type of network, so again we look for the limiting form to remove this dependence.

Now as we have a regular network, we may imagine small areas attached to the points of the network, one area for each point and all areas of the same measure, say δx, δy.

$$\lim_{n\to\infty} \int = \lim_{n\to\infty} \frac{\sum_{1}^{n} R_i}{\sum_{1}^{n} r_i} = \frac{\int \sqrt{X^2 + Y^2}\ \alpha x \,.\, \alpha y}{\frac{2}{3} \frac{1}{\sqrt{\pi}} A^{\frac{3}{2}}}$$

where the top integral is taken over the city's urbanised area, and A is the magnitude of this area in suitable units. The origin of the rectangular co-ordinates x, y must of course be taken at the given centre.

Both I and J deal with the geometric shape of the city in relation to its centre and depend on the definitions of centre and outer boundary.

However, useful as the Boyce, Clark and Simmons measure are, there still remains the difficulty that a city's urbanized area may be fairly circular about its centre although the distribution of population within this area is far from circularly symmetric.

Now a city is circularly symmetric about its centre; and only if :

$$d(r_i\ Q) = \frac{1}{2\pi} \int_{o}^{2\pi} d(r_i\ Q)\ dQ$$

where r, Q are the polar co-ordinates with the city centre as pole and $d(r_1\ Q)$ is the population density at point $(r_1\ Q)$. If we adopt a least squares criterion, then the deviation from symmetry D, say, may be defined by the equation :

$$D = \int_{o}^{2\pi} \int_{o}^{\infty} \left(d(r_1\ Q) - \frac{1}{2\pi}\ d(r_1\ Q)\ dQ \right)^2 rdr\ Q$$

$$= \int_{o}^{2\pi} \int_{o}^{\infty} d^2 (r_1\ Q)\ r\ dr\ dQ - \frac{1}{2\pi}$$

$$\int_{o}^{\infty} \left(\int_{o}^{2\pi} d(r_1\ Q)\ dQ \right)^2 rdr$$

As an example suppose Clark's law applies. Then[7]:

$$\frac{1}{2\pi} \int_{0}^{2\pi} d\,(r_1\,Q)\,dQ = Ae^{-br}$$

where A and b are constants so:

$$\tfrac{1}{2}\pi \int_{0}^{\infty} \left(\int_{0}^{2\pi} d\,(r_1\,Q)\;dQ \right)^2 rdr = \frac{P^2\;b^2}{8\pi}$$

Where P is the total population of the city, and r_1Q are the polar co-ordinates with the city, centre as pole and d is the population density at point (r_1Q). By using the common simplifying assumption that the city extends to infinity in all directions.

In practice :

$$\int_{0}^{2\pi} \int_{0}^{\infty} d^2\,(r_1\,Q)\;rd\;\;rd\,Q = \sum_i P_i^2/a_i$$

where P_i and a_i are respectively the population and area of the ith census district.

Hence :

$$D = \frac{\sum_i P_i^2}{a_i = \dfrac{P_2\,b^2}{8\pi}}$$

On the assumption that Clark's law applies that without serious numerical error, the city may be considered to stretch to infinity in all directions.

Another more statistical approach may be based upon chi-square statistic,[34] which may be used to compare the observed distribution of population with that expected in a circularly symmetric city. As an example of how this may be done imagine m concentric circles drawn with common centre at the city centre. Let R_i $(i=1, \ldots\ldots\, m)$ be the radius of the ith circle :

$$(O < R_1 < R_2 < \ldots\ldots \quad < R_m - 1 \;< R_m).$$

Now let n half rays :

$$r > O,\ Q = Q_j,\ J = 1, 2, \ldots\ldots\ n,\ O = Q_1$$

$$< Q_2 < \ldots . < Q_n - 1 \;< Q_n < 2\pi \text{ be drawn}$$

Also let $Ro = O$ and $Q_{nt} = 2\pi$.

Let the *i, jth* area A_{ij} be the area determined by

$$R_i - 1 < r < R_i,\ i = 1, 2, \ldots\ldots\ m \text{ and}$$

$$Q_j < Q < Q_j + 1,\ J = 1, 2, \ldots\ldots\ n$$

Let R_{ij} be the population of the *i jth* area.

According to the hypothesis of circular symmetry, we should have :

$$P_{ij} = \frac{Q_j + 1 - Q_j}{2\pi} \sum_{j=1}^{n} P_{ij}$$

$$= T_{ij} \text{ say}$$

The T_{ij} are the theoretically expected populations, the P_{ij} are actually observed populations, so the chi-squared statistic is :

$$\chi^2 = \sum_{i=j} (P_{ij} - T_{ij})^2 / T_{ij}$$

This will be approximately distribution in a chi-squared distribution with $m.\ n-m$ degree of freedom and the usual chi-squared test may be applied.

Shop Rent Index

William Olso (1940) had applied the shop rent index for the *CBD* delimitation of Stockholm. He had considered the ratio of

$$= \frac{\text{Total Shops Rent}}{\text{Total length of frontage of shops.}}$$

Delimitation of CBD

The most recent and scientific system for delimiting the *CBD* is the Murphy-Vance Method. In this system, a block is considered part of the CBD if it fulfills three requirements[35] :

Central Business Height Index

The Central Business Index (*CBHI*) of block is a ratio in which the numerator is the square frontage of floor space devoted for typical *CBD* activities retail stores and offices, excluded are wholesale outlets, storage, manufacturing plants other than those of newspapers, churches and residences and the denominator is the ground area in square feet.

$$CBHI = T_s \quad \frac{\text{Block Inventory (Floor space used for typical) } CB \text{ USES}}{\text{Total space (ground floor)}}$$

Central Business Intensity Index

(*CBII*) of 50%. This is also a ratio in which the numerator is the same as in *CBHI* but the denominator is the square frontage of all floor space in the block.

$$CBH = F \quad \frac{\text{Block Inventory (Floor space used for) typical } CB \text{ uses}}{\text{(Total space for all the uses)}}$$

Total Height Index

$$THI = t \quad \frac{\text{(Total Floor Space)}}{\text{(Ground Floor Area)}}$$

Determination and Verification of Umland by Statistical Data

Method: 1

The Umland of a town can be verified with the empirical statistical formula of Tnominen[36] which states that the range of influence between two cities varies in accordance with the number of retail stores. The formula is as follows :

$$\frac{\text{Number of retail stores in city } D \times 100}{\text{Number of retail stores in city } D \times \text{No. of retail stores in city } A}$$

= Proportional range of influence of city *D* as compared to city *A*. For our purposes, we changed the number of stores to number of non-agricultural labour force employed in a city. As the number of retail stores of Dacca[37] or of the ring town are not available we have changed this to the number of non-agricultural labour force as they represent purely urban population.

$$\frac{\text{No. of non-agricultural labour force in Dacca} \times 100}{\text{No. of non-agricultural labour force in Dacca} \times \text{No. of non-agricultural labour force in city } A}$$

= Proportional range of influence of Dacca as compared to city *A*.

On applying this formula it is found that the sphere of influence of Dacca so determined, compares favourably with the umland boundary arrived at by field investigations. This formula may be gainfully employed for the approximates delimitation of umland of other urban centres.

Method : *2*

Mahadeva and Jayasankar, of Mysore school of geographers[38] have used a mathematical model for the delimitation of the potential umland of Mysore city. In their model, they have modified the simple and well known gravitational model of Newton. Underlying the model, the basic assumption is the amount of interaction between two cities is directly proportional to the number of people living in those cities and inversely proportional to the intervening distance.[39]

$$1_i = \frac{P_i \cdot a\beta_l}{d_{ij} \cdot X_v}$$

Where 1_i = Propensity Index of two i

P_i = Population of town i

$a\beta^l$ = Weights on population,

d_{ij} = Distance between town i and j,

X_v = Weights on distance.

They have also demonstrated the operation of the model by delimiting the umland of Mysore and have suggested that further refinement can be made by transforming the population and distance data using appropriate values for a, β and l, or by using other variables such as income, goods produced, etc. in place of population. Similarly the distance can be modified by using cost, time or other variables.

Typology of Towns Based on Commuting Patterns

In attempting to arrive at a typology of towns based on commuting patterns, the technique used by the Polish urban planner St. Lewinski.[40] Using the total number of employed city dwellers (M), the number of sub-urban commuters employed in the city (P) and the number of city commuters employed in the suburbs (W), Lewinski arrived at a coefficient of employment.

$$Z = \frac{M+P+W}{M} \cdot 100$$

and the coefficient of commutation mobility

$$R = \frac{P+W}{M} \cdot 100$$

Four types were distinguished.[41] Working towns are distinguished by relatively high levels of commuting into the city; residential towns, on the other hand, show a predominance of out commuting. Exchange towns are distinguished by highly intensive two-way commuting, both out of the city and into the city; isolated towns display the lowest levels of commuting relative to employment.

Spatial Distribution of Cities and their Association with the Urban Centres

Before we discuss the distribution of cities and urban centres, let us examine the association between spatial distribution of cities and urban centres which materially influence the distribution of cities in a region. As it is "urban ratio and on urban population density (that is on total urban population divided by the total place in which towns and cities are scattered),[42] the spacing of either cities or major towns depends. The spatial association of phenomena is essential, because it reflects in turn the geographical phenomena either conductive or repulsive to the growth and development of the same.

The table calculated on the method described by Gibbs[43] is as follows :

$$L_a = 100 - [\,\Sigma\ 1XY^1 / 2\,]$$

where L_a. is the locational association and X and Y are the (1) per cent of total number of cities in the sphere of influence of each city and (2) per cent of total number of towns and cities in the sphere of influence of each city respectively.

Table 17.3 indicates the locational association of cities with the urban centres on the formula noted above :

The total L_a value for the State comes to 79.4, calculated by dividing the total of column six by two and then the quotient is substracted from 100.0.

The total number of towns included within the sphere of

TABLE 17.3 : **Locational Association of Cities**

Name of towns	No. of cities in each city's sphere of influence	No. of cities as per cent of total of column 2.	No. of urban centres in each city's sphere of influence	No. of urban centres as per cent of column 4.	Difference between X and Y	$X-Y/2$	L_a
1	2	3	4	5	6	7	8
Patna	4	27.3	41	36.6	10.3	5.15	94.8
Jamshedpur	2	13.3	19	16.9	3.6	1.8	98.2
Gaya	2	13.3	16	13.9	0.6	0.3	99.7
Bhagalpur	1	6.4	12	10.7	4.3	2.15	97.7
Ranchi	1	6.4	9	8.9	0.6	0.3	99.7
Muzaffarpur	3	20.0	7	6.0	13.8	6.8	93.2
Darbhanga	2	13.3	8	7.0	6.2	3.1	96.9

influence of different cities comes to 112. This means that many urban centres are overlapped by one or many cities simultaneously, depending upon the size, spacing and functions. The percentage of total number of urban centres falling within the umlands of cities is as follows :

Land Use Trends in Forty-Eight American Cities

We shall see later how gross densities (gross density is defined as the city's population or employment divided by the entire land area of the city) compare with the net densities. But first, the relationship between gross and net densities ought to be derived. Let L stand for land in a given use category, E the number of people employed on the land and T the total land area of the city. Then, gross employment density can be written as follows.[44]

$$E/T = (E/L)\, L/T \qquad \ldots(1)$$

Taking the total differential

$$d\,(E/T) = (E/L)\,(TdL - LdT)\,/T^2 + (L/T)\,d\,(E/L). \qquad \ldots(2)$$

The percentage change in gross density can be obtained for small changes by dividing each term of the equation by E/T :

$$\frac{d\,(E/T)}{E/T} = \frac{dL}{L} - \frac{dT}{T} + \frac{d\,(E/L)}{E/L} \qquad \ldots(3)$$

In words, the percentage change in gross density equals the percentage change in land in use minus the percentage change in total land area plus the percentage change in net density. Percentage changes in mean gross densities can now be shown.

Land Use	22 Cities	12 Cities
Residential	— 7.8	— 4.4
Industrial	— 5.1	— 5.9
Commercial	— 4.0	+ 4.3

The Determinants of Residential Land Values

The study is an empirical investigation of residential land values, and its structure and methodology are straightforward. First, a very simple theoretical model of land values is developed and used as a frame of reference within which to study the empirical data. The underlying model, descriebd in more detail in the following sections, assumes that the value (V) of a particular urban area site is functionally related to its accessibility to economic activities (P), to its amenities (A), to its topography (T), to its present and future use U—i.e. industrial, commercial or residential, and to certain historical factors that affect its utilization (H). Provided these variables can be quantified, the basic land value model of the ith site may be expressed as[45]

$$V_i = f(P_i, A_i, T_i, U_i, H_i) \quad \text{...(1)}$$

Since it is likely that accessibility, amenity and topography will react differently on the value of land in different uses, the basic equation is restated as

$$VR_i = f_1\,(P_i, A_i, T_i, H_i) \quad \text{...}(1_a)$$

$$VC_i = f_2\,(P_i, A_i, T_i, H_i) \quad \text{...}(1_b)$$

$$V1_i = f_3\,(P_i, A_i, T_i, H_i) \quad \text{...}(1_c)$$

where the subscripts R, C, and 1 designate residential, commercial and industrial land respectively.

Locational Quotients are computed by

(1)

$$\frac{X_i / \Sigma X_i}{N_i / \Sigma N_i} \cdot 100$$

where X_i equals the number in occupation X residing in zone i and N represents the total numbers of all occupations residing in zone i.

(2) The index of dissimilarity is computed by :

$$\sum_{i=1}^{n} \frac{(X_i / \Sigma X_i) - (Y_i / \Sigma Y_i)}{2} \cdot 100$$

where X_i represents one occupation and Y_i represents another occupation residing in zone i.

(3) The coefficient of geographic association is computed by :

$$1 - \sum_{i=1}^{n} \frac{(X_i / \Sigma X_i) - (N_i / \Sigma N_i)}{2}$$

where X_i and N_i are defined in (1) above reference. The coefficient has a range from zero to one.

Index of Surplus Workers in Urban Areas

The national proportion employed in any occupation is obtained and the number expected to be so employed in a town of any size is calculated. If the actual employment when compared with this expected total shows a surplus, then this represents basic employment. This is the principle used in the clearest statement of this concept by J. M. Mattiala and W. R. Thompson who propose an 'Index of Surplus Workers'.[46] (J. M. Mattiala and W. R. Thompson (1955). The measurement of the economic base of the Metropolitan Area; Land Economics; 31; 215.)

$$S=e_i-\frac{e_t}{E_t}E_i$$

where e_i=The regional employment in the industry concerned,

e_t=The total regional employment in all industries and economic activity,

E_i=The national employment in the industry concerned,

E_t=The total national employment in all industries and economic activity.

This index can be used in a classification of towns by the following procedure. The index is calculated for all occupational groups in the town and all positive values are to talled giving an aggregate of surplus workers for the town. The percentages of the aggregates formed by each occupational surplus are calculated and ranked. These rankings are then used as a basis for functional classification.

(2) To some extent Ullman and Dacey minimize the difficulties since the variation of the minimum with city size is taken into account. This non-basic ratio is subsequently used to classify cities not by comparing one employment sector with another but by measuring the extent to which a city departed from the various minimum, thus providing an Index of Diversity (D) where

$$D=\frac{\Sigma_i\left[\frac{(P_i-M_i)^2}{M_i}\right]}{\left[\frac{\Sigma_i P_i-\Sigma_i M_i}{\Sigma_i M_i}\right]^2}.$$

where i=each of employment groups,

P_i=percentage employed in each of i groups,

M_i=minimum requirement for each group,

Σ_i=the sum of all groups.

The higher this index is the more specialised a town, the lower, the more diversified. It is apparent that this index demonstrates how statistical measures are increasingly employed and how the study of the ways in which towns differ from each other is the real basis of classification. This leads on to the last group of classifications.

Methods of Areal Definition

1. Calculation of ratios :

For each of the blocks a series of ratios or indices can be calculated.[47] These are

(1) Total height index (*HI*)

$$HI = \frac{\text{Total floor space}}{\text{Ground floor space}}$$

This is the height of each block in floors if all the space, whatever its use were to be spread evenly over the whole block. This is general measure for direct mapping of building heights.

(2) Central Business Height Index (*CBHI*)

$$CBHI = \frac{\text{Total } CB \text{ uses floor space}}{\text{Ground floor space}}$$

This is the height of each block in floors if all the central business uses were evenly spread over the whole block.

(3) Central Business Intensity Index (*CBII*)

$$CBII = \frac{\text{Central business floor space}}{\text{Total floor space}} \times \frac{100}{1}$$

This measures the proportion (%) of all available floor space in *CB* uses and this is useful of the ratios.

(4) Central Business Index (*CBI*)

CBI=*CBHI* of 1 Plus *CBHI* of 50 per cent

To define the *CBD* Murphy and Vance took a composite measure as indicated above and called it the *CBI*. **All blocks meeting the requirement were regarded as part of the *CBD*.**

Urban Potentiality

The concept of 'potential' developed by Warntz[48] (1960) and 'gravity model' by Reilly (1961) appear in many forms in social science literature but here we are concerned with the distance formulation and some relevent aspects of urban potentiality.

The potential at any point is defined as[49] :

$$\int \frac{1}{r} DdA, \quad \text{or} \quad \left(\sum_{i=1}^{A} P_i \right) / r$$

where dA is the small element of the area, D is the density for that area; and r is the distance from each area to the point concerned. Similarly the point at which $\int \frac{1}{r} DdA$ is minimized defines the medium centre. In second formula A is the total number of small sub-areas involved, P_i is the population in each sub-areas and r is the distance from the particular point to a representative point in each sub-area.

Retail Stores Spatial Affinities

An initial problem was to devise a method by which significant associations among stores could be identified. It needs a test which would isolate these spatially associated store types which occur together more often than chance would dictate. A Brief summary of the test is given here which is as follows[50] :

First the expected mean number of associations of business type "B" with those types of "A" are obtained. The expectation is based on a random distribution containing the same number of total elements (N). This is given by the formula :

$$\frac{2AB}{N}$$

The variance of the distribution is

$$\frac{2AB\,(2AB+N)\,(C-1)}{N^2\,(N-1)}$$

where C is equal to the total number of stores less the number of A's and B's ($C=N-A-B$). Then the expected number of associations must be compared with the observed number. Finally, in most instances it is necessary to employ normal distribution curves for test of significance. When the expected number of links is large, the normal curve is an appropriate sampling distributions to use. Of course, in that case the following formula would be applied.

$$Z = \frac{\text{Obs} - \text{Exp.}}{\sqrt{\sigma 2}}$$

where Z is the standard variate of normal curve, observed and expected are the number of links, and $\sqrt{\sigma 2}$ is the square root of variance. When the expected number of links is small, the poisson distribution used for tests.

Absolute Population of the Area Served by a Market

Urban and rural markets have various grades determined on the basis of absolute population served by each of the markets in the region. The absolute population of the area served either by urban or rural market may be determined with the help of the following simple equation .[51]

$$PSA = \pi r^2 \left(\frac{P_n}{A}\right)$$

where PSA is the population of the service area of either rural or urban market, r is the radius of the circle representing the field of the market influence, P_n is the rural or urban total population of the region and A is the total area of the region.

Market Influence

$$\text{Radius of the Market influence} = \sqrt{\frac{\text{Total Area of the region} \times \text{Total Population of the market}}{\text{Total Population of the Area}}}$$

Thus the extent of influence is proportional to the total area of the region and the total population of the market on the one hand and the total population of the area on the other. This relationship may be expressed by the following simple equation.[51]

Market Grades According to their Field of Influence

Grades	Field of Influence in km.	Total No.	Average distances within the grades km.
I	Above 20	1	Only one Market
II	15—20	2	45
III	10—15	8	22
IV	5—10	40	16
V	Below-5	65	10

Source : H. M Saxena, Ibid, p. 127.

Two City and Four City Index

Two City Index [52] $1 \frac{P_1}{P_2}$

where P_1 = Population of the first city, P_2 = Population of the second city.

Four City Index $\frac{1 P_i}{P_2+P_3+P_4}$

where P_2, P_3, P_4 is population of next cities.

According to G.K. Zipf's in rank size rule the expected population for all urban centres of the region have been calculated by a formula :

$$P = \frac{L_c}{R_i}$$

where P_i is the population of a given city; R_i, is the rank of the size of the city, L_c is the population of the largest city.

Index of Primacy

Type of Index	Region	Ranchi	Singhbhum	Dhanbad	Palamau	Hazari-bagh	Santal Parganas
Two City Index	1.07	15.00	13.15	4.01	2.12	1.28	1.20
Four City Index	.58	5.32	5.15	2.51	1.24	.47	.61

Source : A. Kumar & N. Sharma, "Spatial Arrangement and Degree of Specialization of Urban Centres", *Geographical Review of India*, Vol. 36, No. 2, 1974, pp. 122-124.

Peculiarities of the System of Urban Places

Such analysis is designed to establish some peculiarities of the system of urban places of Crimea Oblast with the aid of the 'rent-size rule'.

The analysis was based on change over time in the parameters of Yu. V. Medvedkov's formula on the "structure of a system of places". The formula takes the form[53]:

$$P_j = K^{-1} P_1 J{-}a, \ldots\ldots\ldots\ldots$$

where P_j is the population of city of rank J; P_1 is the population of the city of rank 1; a is the system (the tangent of the slope of the regression live joining cities of various sizes arrayed in rank order on double logarithmic paper); j is the rank of the city; $K = \frac{P_1}{P_1*}$ is the "coefficient of primacy" of the main city in system (P_1 is the population of city of rank 1 as determined by the regression line).

TABLE 17.4 : **Parameters of the Medvedkov's Formula for all Urban Places of Crimea**

Year	No. of Places	Slope of regression line	System contrast measure	*R* Value of regression line	Mean error of regression line	Primacy coefficient of Main City
1861	11	—46.7°	—1.06	—0.84	29	0.79
1897	16	—59.5°	—1.70	—0.94	11.5	0.32
1939	22	—58.6°	—1.64	—0.97	6	0.36
1959	38	—58.5°	—1.63	—0.96	8	0.30
1970	50	—57.1°	—1.55	—0.97	6	0.32

Source : After V.R. Popov, Soviet Geography, 1974.

The Measurement of Centralization

Centralized population is one in which the units are clustered around a particular point, measurement must be based on the distance which separates the individual units from the point and not the correspondence between the area and population contained in territorial divisions.

If the number of units in each territorial division (P) is multiplied by the distance (D) separating the approximate geographical centre of the division from a central point and the sum of the products is divided by the total number of units (N), the resulting quotient is the approximate average distance between the individual units and the point. This number indicates the degree to which the population is centralized (Ce), and it is derived from the formula :

$$Ce = \Sigma D.\ P/N$$

The figures in the second and third columns of Table 17.5 the 188,586 residents (N) of George Town would have to travel a total of 181,095 miles ($\Sigma D.P$) for each person to move from his place of residence to the point taken to be the centre of the municipality. When these numbers are inserted in the formula, Ce for George Town is 0.96.

If the residents of George Town had been distributed evenly throughout the municipality in 1947, Ce would have been 1.70. The ratio of this hypothetical value (designated as Ch) to Ce give us a second measure (Cr) which indicates how much the average distance has been reduced as a consequence of the clustering of the population about the centre. This measure for George Town is :

$$Cr = Ce/Ch = .96/1.70 = .44$$

TABLE 17.5 : **Population and Distance from the Centre of the Municipality in George Town, Malaya, 1947**

Division	Distance from the Centre in Miles	Population (P)	D.P.	Population under even distribution	$D.P_e$
1	0.56	22,631	12,673	7,826	4,383
2	0.24	18,804	4,513	3,206	769
3	0.29	22,553	6,540	5,620	1,630
4	00.66	19,828	13,086	10,636	7,020
5	0.75	20,024	15,018	7,827	5,870
6	0.75	21,368	16,026	9,429	7,072
7	1.50	21,808	32,712	17,652	26,478
8	2.06	11,705	24,112	30,098	62,002
9	1.50	7,935	11,903	14,482	22,263
10	1.88	16,067	30,206	27,288	51,301
11	2.44	5,863	14,306	54,162	132,155
All Divisions		188,586	181,095	188,586	320,946

Source : M.V. Tufo, del, Malaya : A Report on the 1947 Census of Population, London : Crown Agents for the Colonies.

If C_{e1} is a measure of the degree to which a population was centralized at one time and C_{e2} is the measure at a later time, then the degree of decentralization (D_z) is :

$D_z = C_{e2} - C_{e1}$, with negative numbers indicating that the distribution has become more centralized during the period. An alternative formula[54]

($D_z = C_{e2}/C_{e1}$) expresses the amount of decentralization relative to the original C_e value, with values less than 1.00 indicating centralization.

The difference between C_{e1} and C_{e2} in any given case may be a function of an expanding boundary and not an actual change in the pattern of distribution. For this reason C_r values should be used express shifts in population distribution independently of an increase in the areal size of the urban area. This measure (D_r) may be computed in two ways :

$$D_r = C_{r2} - C_{i1} \text{ or}$$

$$D_r = C_{r2}/C_{r1}.$$

The first formula expresses the change in distribution in absolute terms (with negative numbers indicating centralization), while the second formula expresses change relative to the original level (values less than 1.00 indicate centralization).

Notions of Consumer Behaviour

It is interesting to consider correspondence between theory of tertiary activity and recent work regarding shopping behaviour by consumers. Baumol and Ide (1956) have produced a simple model displaying the choice variables in the selection of a shopping centre by a consumer. A customer will shop at a centre when his demand function is such that[55]

$$f(N, D) = WP(N) - V(C_d D + C_n \sqrt{N + C_i})$$

is positive. $f(N,D)$ is a measure of the expected net benefit of the consumer from entering a store. It varies with D, his distance from the store, and N, the number items offered for sale at the store. Assumed costs are C_d, a cost of transport assumed proportional to distance $C_n \sqrt{N}$, the assumed costs of actual shopping, and C_i, the opportunity costs of other shopping opportunity foregone. $P(N)$ is the probable satisfaction function, and W and V are the subjective weights assigned by the consumer when he evaluates the size of each element in the equation.

The economic implications from this statement are many. For example, minimum number of items necessary to induce a customer to shop at a given store will increase with D. Maximum shopping distance is given by the equation of the indifference curve which offers the consumer zero net benefit from shopping at a store, and is obtained by setting $(N.D.) = O$ and solving for D to yield

$$D_n = \frac{W}{V C_d} P(N) - \frac{1}{C_d} (C_n \sqrt{N + C_i}).$$

Given the hierarchical spatial system of the theory of tertiary activity : $P(N)$=will be a step function related to levels of goods available at each rank of the hierarchy.

$C_n \sqrt{N}$=will be a like step function.

C_i=will be dependent upon the spatial characteristics of the hierarchical system.

Hence any solution D_n will be a step function related to levels of the hierarchy. The Baumol and Ide system is entirely compatible with theory of tertiary activity. The statement is also true for the development of a total retail sales model since such a model is based upon the previous one of maximum consumer distance, and of notions of maximization of profits (which are of course based upon N). Not only do consumers discriminate among centres hierarchically and spatially but varieties and sales levels and the extent of potential profits are likewise determined.

Beckmann's Summary Model

Beckmann (1958) developed the simplest model of Central Place theory in the following manner. There is a basic level of rural population which is served by a first layer of settlements performing the most elementary production and distribution functions. Market areas of these first settlements are limited by the largest distance tolerable to the rural population served. A fundamental hypothesis is that the size of any city is proportional to the population it serves (including that of the city itself). Therefore, if r is the rural population in the market area of cities of the first layer, K the proportionality factor, and C the city size, then :

$$(1)\ C = K\,(r+c),\ \text{or}$$

$$(2)\ C = \frac{K_r}{1-K}$$

and $K/1-K$ is therefore the effective "Urban Multiplier".

It is assumed that cities of each order (but the lowest) have a fixed number of satellites, S, of the next lower order; it is now possible to derive the size P_m of cities of each order $_m$, and the population P_m served by these.

From the population level hypothesis $P_m = KP_m$ and from the assumption concerning satellites, $P_m = P_m + SP_m - 1$. By substitution it may be shown that :

$$(3)\ P_m = \frac{S_m - 1_r}{(1-K)m}$$

$$(4)\ P_m = \frac{K_s\ m - 1_r}{(1-K)m}$$

where r is size of the basic rural community, K is the ratio of city size to population served, and S is the number of satellites per city.

When $_m$ assumes its largest value, say N, then (3) describes a relationship between population P, the size of the smallest community r, and the number of ranks in the city hierarchy N :

$$(5)\ P = PN = \frac{SN - 1_r}{(1-K)\ N}$$

This may be written in other ways :

$$r = \frac{(1-K)\ N}{S\ N-1}\ P$$

$$N = \frac{\log\,(P_s/r)}{\log\,[S \div (1-K)]}$$

A second relationship exists between the total number of cities T, the number of ranks in the hierarchy N, and the number of S of satellites per city :

$$(6)\ T = 1 + S + S^2 + \ldots\ldots + SN = \frac{SN+1-1}{S-1}$$

To summarize, there are six variables : *K*, *S*, *r*, *P*, *N*, *T*. If certain of them are given, the others are determined by the model.

Burros' Control Ratio Concept

R.H. Burros (1957) presented a theory of static hierarchy which facilitates the generalisation of certain aspects of Beckmann's statement. In the Beckmann model there is a single ratio (s) which describes the number of satellites per city. Burros recognizes that this control ratio (*R* in this system) may have several values between lower and upper bounds, and his model makes explicit recognition of this fact. In the Burros model the variables are:

E=The total number of echelons in the regular hierarchy 1,......*E*.

C_i=The number of centres at echelon *i*

C_e=(or *S*) The number of centres at the lowest echelon *E*.

L=The total number of centres at all echelons.

R=The control ratio, the number of centres at echelon $i+1$. Controlled by centres at echelon *i*.

$Ri \leqslant C_i+1/C_i \leqslant R_2$ and R_1 and R_2 are the lower and upper bounds of the control ratio, respectively

Generally, $C_i=1$ and $CE=S>2$

$1 < R_1 \leqslant C_i+1/C_i \leqslant R_2$

$$L = \sum_{i=1}^{E} C_i$$

It may be proved that

(1) $E_j = 1 + \log S/\log R\ 3-j,\ J=1, 2$ i.e., $E_1 = 1 + \log S/\log R_2$

$E_2 = 1 + \log S/\log R_1$

(2) $L_j = (SR_3\text{-}j-1)/(R_3\text{-}j-1),\ J=1, 2$

i.e., $L_1 = (S\ R_2-1)/(R_2-1)$

$L_2 = (S\ R_1-1)/(R_1-1)$

By applying (1) and (2) it is possible to obtain the range of possible values of E and L. From this one can calculate and draw possible regular hierarchies. Note that R_1 and R_2 relate to the "span of control" in hierarchies, and provide a more flexible definition of the S in the Beckmann model.

The Theory of Tertiary Activity

There are two definitions necessary for the formulation of the theory of tertiary activity.

(1) The range of a good. The range of a good offered from a business centre has an upper limit beyond which the business centre is unable to attract purchasers for the good and a lower limit which incorporates the threshold purchasing power needed for the good to be offered. It thus defines the market area of the business centre for the good.

Threshold

This is some minimum size of a market below which a business centre will have no economic justification to supply a good. This is the point at which sales are large enough only for the firm to earn normal profits. It constitutes a minimum scale (condition of entry) defined by the lower limit to the range of a business centre for the good. Such a minimum scale of purchasing power necessary to support the supply of a central good from a business centre is here termed the threshold sales level for the provision of the good from a centre.

Brink and de Cani's Model

Brink and de Cani (1957) have provided an analogue solution to the problem of locating retail points most efficiently (i.e., in a least cost manner) relative to market areas The solution is also applicable to wholesale-retail relationships. With a little practice it is possible, using this technique to derive optimal solutions for small scale problems by hand with the assistance of little more than a desk calculator. Presentation here is motivated by the obviously valuable relationships to the theory of tertiary activity presented above.

The variables in the Brink and de Cani's model are as follows :

Y_j, a vector representing distribution point j

$Y_j=(Y_{j1}, X_{j2},\ldots\ldots, Y_{ir})$ and $i, j=1, 2, \ldots\ldots\ldots, n$.

$X_i(j)$, a set of vectors representing consumers i within the market area of Y_j

$X_i=(X_{i1}, X_{i2},\ldots\ldots, X_{ir})$

$T_i(j)$, The total demand of $X_i(j)$

$W_{ij}=T_i(j)/\{X_i(j)-Y_j\}$

M_j, The set $[X_j(j)]$ of n_j consumer locations for each

Y_j, a uniquely determined market.

$\{X_i(j)-Y_j\}$ are the cost of transport from j to i in the market area of Y_j

From the variables it is shown that

$$(1)\quad Y_i=\sum_{i=1}^{nj} W_{ij}\times i\,(\mathrm{i})/\sum_{i=1}^{nj} W_{ij}$$

$$(2)\quad \left[X_i\overset{(j)}{-}Y_j\right] < \left[X_j\overset{(j)}{-}Y_k\right]$$

Condition (2) is the nearest neighbour relation for markets M_j and M_k, uniquely distributed. By alternatively meeting conditions (1) and (2), beginning by arbitrarily locating Y_j in cost space, an optimal solution may be iterated. To map back into geographic space from the cost space in which the optimal solution was achieved, the steps are : (*i*) locate Y_i at the $[X_i(j)]$ for which the distance is least; (*ii*) define markets on the later sets; (*iii*) sum over the market sets to obtain total demands. This also gives total inventory at the Y_j.[56]

Rank-Size Relationship of Urban Centres

The frequency distribution of urban centres according to their size and hierarchy, plotted on a double logarithmic graph paper gives a curvilinear trend which is known as the rank-size curve, the least squares equation can be calculated by standard regression analysis.[57] The equation is of the form :

$$Log\ P_i=Log\ P_1{}^{-b}\ Log\ R_i$$

where P_i and R_i are respectively the size (population), and rank of the ith city; P_1 is the size of the largest city; and b is a constant representing the slope of the regression line. This equation may be rewritten as :

$$P_i = P_1 \cdot R_i^{-b}$$

where the symbols have the same meaning as given above. The rank-size regularity as expressed by *Zipf* is :

$$P_r = P_1 / r_1$$

where P_r is an exponent which generally approximates to unity. This suggests that if the population of the largest city is divided by the rank of any city in the same region, the result will approximately be the population of the city whose rank number is used as a divisor.

The rank-size rule is a regular relationship which often exists between absolute size of an urban centre (measured by its population, not area) and its rank (the centre with the largest population in the region ranking 1, the next largest 2, and so on to the smallest town). This relationship may be expressed by the simple formula[58]:

$$P_n = \frac{P_1}{n}$$

where P_n is the population of the n_{th} town in the list of urban centres and P_1 is the population of the region's largest urban centre, i.e., the centre which ranks 1 in the list.

The variety of available choice may be called entropy and is the complements of the degree of ordering. We may consider N people divided among Z towns, each of which has an equal chance of attracting a given population.[59] The number of settlements having a population of i persons is Z_i. The total number of distinguishable ways in which the people can be distributed among the settlements, neglecting the spatial pattern and considering only frequency distribution is :

$$P = Z \; 1 / \overset{N}{\underset{i=o}{1}} Z_i \,! \; (o \leqslant i \leqslant N)$$

where the system is large, its entropy is ;

$$H = log\ P = Z\ log\ Z - \Sigma Z_i\ log\ Z_i$$

where entropy is defined as *log P*.

H is maximized when

$$Z_i=(Z/_n)^{-}e^n,$$

an exponential distribution, n being the mean population per settlement N/Z. In terms of the cumulative distribution function :

$$Z_i \leqslant S=S\,(1-e^{-i}/n),$$

where S is the size of the largest city. In this case H max$=Z$ *log* (*en*). The most probable state of system is that which gives maximum entropy, i.e. when the sum of logarithms is a maximum. This corresponds to a situation in which, given the size of the largest city, the probability of the $q+1$st city having a population which is a given ratio of the qth city is constant.

Expected Population of the Premier City

The expected population of the premier city has been calculated by dividing the total population of all towns by the sum of the reciprocals of their ranks. The formula to calculate the expected population of the premier city (P_i) is as follows :

$$P_i=\frac{\Sigma P}{\Sigma 1/R}$$

where ΣP is the sum of the population of all the urban centres and $\Sigma 1/R$ is the reciprocal of the rank.

Law of Primate City

Primacy is the superlative lead of the largest or primate city over the smaller cities and towns. This may be expressed as a ratio :

Primacy Index (1) $=\frac{P_1}{P_2}$ where P_1 and P_2 are the populations of the first and second largest settlements respectively or :

$$12=\frac{P_1}{P_3}$$

where P_1 and P_3 are the populations of the first and third settlements respectively. It is necessary under the rank-size rule that the primacy index for P_1 and P_2 settlements be 2; for P_1 and P_3 it must be 3 and so on. The investigation of primacy is important in finding out the optimum size of the settlements and whether the size of the smaller settlements is in order with reference to the primate city. Optimum size of the settlements

$$=\frac{\Sigma P}{\sum \frac{1}{R}}$$

where ΣP is the total population of all settlements in the region, and $\sum \frac{1}{R}$ is the rank of a settlement.

The resultant value indicates the expected size of all settlements in the region with reference to the total urban population (ΣP) and the number of urban settlements in the region.

Rank-Size Rule and the Comparison of Christaller and Zipf

Christaller provided a network of cities in homogeneous space while formulating his central place hierarchy. This is the well-known $K=3$ network, with one "primate city" of population K, and $r=1$; three cities of population $K/3$, and $r=2$; nine cities of population $K/9$ and $r=3$; twenty-seven of population $K/27$; etc.[60] This arrangement of city sizes may be compared directly with the rank-size rule provided by Zipf.

In order to make the comparison the steps of the hierarchy of centres in the $K=3$ scheme are taken as ranks, $r=1, 2, 3$, etc., and $P_i=K, K/3, K/9$, etc. A rank-size distribution in the manner of Zipf is formed if the exponent :

$$q=\frac{Log\ (K/r)}{log\ (K/3r^{-1})}$$

Thus, where $r=2$,

$$q=\frac{Log\ (K/2)}{log\ (K/3^{2-1})}$$

and if

$$q.\ log\ P=log\ (K/r),$$

Then

$$log\ P=log\ (K/2)\ .\ \frac{log\ (K/3)}{log\ (K/2)}$$

and $P=K/3$

as required by Christaller's theory.[61]

Implications

Having no difference between Christaller and Zipf, on the level of the basic notions the two schemes are seem to be the same. The class hierarchical scheme of Christaller may take on the rank-size character of Zipf's observations but this consistency requires a rigorous choice of relationships.

REFERENCES

1. W.J. Reilly, *The Law of Retail Gravitation*, New York, 1913.
2. B.J.L. Berry, *Geography of Market Centres and Retail Distribution*, Prentice Hall, Inc., Englewood Cliffs, N.J. 1967, p. 41.
3. R.S. Mathieson, "The Validity of Reilly's Law in Australia Some Preliminary Considerations", *The Australian Geographer*, Vol. VII, No. 1, 1957, pp. 27-32.
4. B.J.L. Berry, *The Geography of Market Centres and Retail Distribution*, Prentice Hall, Inc., New Jersey, 1967, p. 61.
5. W.J. Baumol, and E.A. Ide, "Variety in Retailing", *Management Science*, Vol. 3, 1956, pp. 93-101.
6. R. Cox, and W. Alderson, *Theory of Marketing*, Irwin, 1950, pp. 19-40.
7. J.M. Mattila, and W.R. Thompson, "The Measurement of the Economic Base of the Metropolitan Area", *Land Economics*, Vol. XXXI, No. 3, 1955, pp. 215-228.
8. M. Mukerji, *Functions and Functional Classification of Towns in Bihar*, Unpublished, Ph.D. Thesis, Patna University, 1966, p. 68.
9. L.N. Ram and V.N.P. Sinha, *Functional Classification of Towns of Bihar*, Unpublished Paper Received.
10. V.N.P. Sinha and R.B. Mandal, "Hierarchy of Trade Towns in Bihar", *Indian Geographical Studies*, No. 3, 1974, p. 80.
11. G.N. Singh, *The Central Place Theory*, Manuscript Received.
12. J.E. Brush and H.E. Bracy, "Rural Service Centres in South West Wisconsin and Southern England", (Eds.) Mayer and Kohn, Allahabad, 1967, p, 213; S. Godlund, "The Functions and Growth of Bus Traffic within the Sphere of Urban Influence", *Lund Studies in Geography*, Series B, No. 18, 1956, pp. 13-14.
13. N.G. Jain, "Urban Hierarchy and Telephone Service in Vidarbha", *The National Geographical Journal of India*, Vol. XVII, Part 2-3, 1971, p. 135.
14. B. Thakur, "Nearest Neighbour Analysis : As a Measure of Urban Place Patterns", *Indian Geographical Studies*, No. 1, 1973, pp. 64-65 and W.K.D. Davies, "Centrality and the Central Place Hierarchy", *Urban Studies*, Vol. 4, pp. 61-79.
15. S. Godlund, "The Functions and Growth of Bus Traffic within the Sphere of Urban Influence", *Lund Studies in Geography*, Series B, No. 18, 1956, pp. 13-14.

16. K.N. Singh, *Rural Markets and Urban Centres in U.P.* Unpublished, Ph.D. Thesis, B.H.U., 1962, p. 187.

17. M.S. Vishwanath, "Growth Pattern and Hierarchy of Urban Centres in Mysore", *The Indian Geographical Journal,* Vol. XLVII, 1972, No. 1 and 2, pp. 1-11.

18. Ashok Dutt, "Intracity Hierarchy of Central Places, Calcutta : A Case Study", *The Professional Geographer,* Vol XXI, No. 1, 1969, pp. 18-22.

19. O.P. Singh, "Towards Determining Hierarchy of Service Centres : A Methodology for Central Place Studies", *N.G.J.I.*, Vol. XVII, Part, 4, 1971, pp. 165-177.

20. Onkar Singh, "Hierarchy and Spacing of Towns in U.P.", in R.L. Singh (ed) *Urban Geography in Developing Countries,* Varanasi, 1973, pp. 318-326.

21. N.P. Saxena and R.C. Tyagi, Criteria for Determining Centrality in a Macro-region", *The Geographical Observer,* Vol. 11, March, 1975; pp. 53-64.

22. O.P. Singh, "Towards Determining Hierarchy of Service Centres : A Methodology for Central Place Studies", *The National Geographical Journal of India,* Vol. XVII, Part 4, 1971, pp. 171-172.

23. P.M. Hauser, and L.F. Schnore, *The Study of Urbanization,* John Wiley and Sons, London, 1967, pp. 405-408.

24. M. Beckmann, "City Hierarchies and the Distribution of City Size", *Economic Development and Cultural Change,* Vol. 6, 1958, pp. 243-48.

25. P.D. Mahadeva and D.C. Jaysankar, Concept of a City Region : An Approach with a Case Study, *The Indian Geographical Journal,* Vol. XLIV, 1969, No. 3 and 4, pp. 15-26.

26. A.S. Harvey, "A Dualistic Model of Urban Grow", *The Aunals of Regional Science,* Vol. VIII, 1974, No. 1, pp. 66-68.

27. P. Haggett, *Locational Analysis in Human Geography,* Methuen, London, 1965, pp. 48-55.

28. B.H. Massani, "A Note on Shape", *Professional Geographer,* No. 4, 1970, Vol. XXII, p. 197.

29. H.H. Singh and K.K. Dube, "Central Business Districts of Kaval Towns". *Urban Geography in Developing Countries* (Ed.) R.L. Singh, Varanasi, 1972, p. 247.

30. C.A. Wilkins and J. Shaw, "Measures of Shape Distortion in Urban Geography", *The Australian Geographer,* Vol. XI, 6, 1971, pp. 593-95.

31. P. Hagget, *Locational Analysis in Human Geography,* Arnold, London, 1965, p. 15.

32. P.R. Boyce and W.A.V. Clark, The Concept of Shape in Geography, *Geographical Review,* Vol. 54, 1964, pp. 561-72.

33. J.W. Simmons, Relationships between the Population Density Pattern and Site of Cities, Unpublished M.A. Thesis, University of Chicago, 1972, Quoted by B.J.L. Berry et al. in "Urban Population Densities, Structure and Change", *Geographical Review,* Vol. 53, 1963, pp. 389-405.

34. C. Clark, Urban Population Densities, *Journal of the Royal Statistical Society* Vol. 114, 1951, pp. 490-6.
35. K.M. Singh, *Delimitation of CBD*, Kharagpur, Manuscript received.
36. J.E. Brush, "Urban Hierarchy in Europe", *Geographical Review*, Vol. XLIII, 1953, p. 415.
37. F.J. Khan and M.H. Khan, "Delimitation of Umland of Greater Dacca", *The Oriental Geographer*, Vol. V, 1961, No. 2, p. 118.
38. P.D. Mahadeva, and D.C. Jaysankar "Concept of the City Region : An Approach with a Case Study", 57th Indian Science Congress, Kharagpur, Geography and Geography Section.
39. H.N. Mishra, "Nomenclature and Delimitations of the Umland : An Appraisal", *The Indian Geographical Journal*, Vol. XLVI, No. 122, 1971, pp. 16-18.
40. S. Lewiski, "Commuting as an Element in an Urban Typology", *Pr Zaglad Geograficzuy*, 1966, No. 4.
41. B.S. Khorev, T.K. Smolina and N.G. Sominskaya, "Commuting Patterns in the Sub-Urban Zones of Small and Middle Size Towns", *Soviet Geography*, 1973, No. 1, p. 25.
42. S.P. Sinha, "Distribution of Cities in Bihar", *Geographical Outlook*, Vol. VII, 1970-71, pp. 182-198.
43. J.C.T. Stewart, "The Size and Spacing of Cities", *Geographical Review* Vol. XLVIII, April, 1958 and Gibbs, J.P., Urban Research Method (1961), p. 247.
44. J.H. Hiedercorn and F.R. Edward, Hearle, "Recent Land Use Trends in Forty-Eight Large American Cities", *Internal Structure of the City*, p. 125.
45. F. Brigham Engene, The Determinants of Residential Land Values in Internal Structure of the City, p. 150.
46. L. Ullman, Edward and F. Dacey, Michael, The Minimum Requirements Approach to the Urban Economic Base, in K. Norborg, Editor (1962), *Proceedings of Land Symposium Urban Geography*, 1960, p. 121.
47. Harold Carter, The Study of Urban Geography, Edward Arnold, 1972, pp. 200-201.
48. W. Warntz and D. Neft, "Contributions to a Statistical Methodology for Areal Distributions", *Journal of Regional Science*, Vol. 2, 1960, pp. 45-66.
49. F.J. Monkhouse and H.R. Wilkison, Maps and Diagrams, Methwes and Co. Ltd. London, 1971, pp. 359-361.
50. A. Getis and J.M. Getis, "Retail Stores Spatial Affinities", *Urban Studies* Vol. 5, No. 3, 1968, pp. 317-322.
51. H.M. Saxena, *Geography of Transport and Market Centres : A Case Study of Hadaoti Plateau*, S. Chand and Co. Ltd, New Delhi, 1975, p. 129.
52. J.P. Gibbs, (Ed.), Urban Research Methods, Van Nosterand Co. Bombay, 1966, p. 440.
53. V.R. Popov, "Investigation of the System of Urban Places of Crimea Oblast", *Soviet Geography*, Vol. XV, No. 1, 1974, pp. 19-22.

54. J.P. Gibbs, "Urban Research Methods, Von Nostrand, East-West Press, New Delhi, 1966, p. 246.
55. W.L. Garrison, B.J.L. Berry, D.F, Marble, J.D. Nystuen, and R.L. Morrill, *Studies of High Way Development and Geogrophic Change*, University of Washington Press, Seattle, 1959, pp. 59-60.
56. W.L. Garrison, B.J.L. Berry, D.F. Marble, J.D. Nystuen, and R.L. Morrill, "Studies of Highway Development and Geographic Change", University of Washington Press, Seattle, 1959.
57. R.B. Mandal, "Rank-Size Relationship of Urban Centres in Bihar", *The Indian Geographical Studies*, No. 3, 1974, pp. 41-49.
58. J.V. Marshall, The Location of Service Town: An Approach to the Analysis of Central Systems, University of Toronto Press, Toronto, 1971, p. 49.
59. L. Curry, "The Random Spatial Economy : An Exploration in Settlement Theory", *Annals of the Association of American Geographers*, Vol. 54, 1964, p. 144.
60. A. Lösch, "The Economics of Location", translated by Woglom and Stolper, New Haven, Yale University Press, 1954, p. 131.
61. H.M. Mayer and C.F. Kohn, (Eds.), *Readings in Urban Geography*, Central Book Depot, Allahabad, 1967, p. 234.

EIGHTEEN

R. B. Mandal

INDUSTRIAL GEOGRAPHY

RECENTLY the importance of applied quantitative methods in geography has considerably increased with the industrial development of a nation, when most of the countries are on the threshold of planning. The measures of industrial indices may be meant for large scale or small scale industries, industrial labour force, location of industries, availability of raw materials, trends and rate of production, zone of influence and commuters to the industries. In most of the industrialised countries of the world indices to measure industrial structure are available regarding capital investment, labour welfare, cost of production, details of output raised and power consumed.

Index of Industrial Production

This index is generally used by the Ministry of Commerce and Industry, Government of India. It covers twenty important industries and is compiled and issued monthly with base year 1946. Production figures are taken from Monthly Statistics of Production of Selected Industries of India. The formula used to measure the index of industrial production is as follows[1] :

$$1 = \frac{\Sigma P_1 W_1}{\Sigma W_1}$$

where P_1 is the production relative for the month and W_1 is the weight allotted to the industry.

Among all the industries covered by this index number, cotton is the most important with a weight of 43.49 out of total weight of 100. The industries covered by the index number with assigned weight are as follows :

TABLE 18.1 : **The Industries Covered with Assigned Weight**

Sl. No.	Industries	Weight Assigned
1.	Coal	11.95
2.	Sugar	3.54
3.	Paints and Varnishes	0.61
4.	Cement	0.67
5.	Glass	0.55
6.	Refractories	0.48
7.	Ply-wood	0.15
8.	Paper and Paper Board	1.46
9.	Matches	1.21
10.	Cotton Textiles	43.49
11.	Woollen Manufactures	1.38
12.	Jute	16.53
13.	Chemicals	3.10
14.	Non-ferrous Metals	2.14
15.	Steel	7.16
16.	Bicycles	0.11
17.	Sewing Machines	0.02
18.	Electric Lamps	0.04
19.	Electric Fans	0.35
20.	General Engineering and Electrical Engineering	5.06
	Total	100.00

Source : Asthana, B.N. and Srivastava, S. S, 1962.

The Location Quotient of Industry

The location quotient refers to the concentration of industrial location or production or labour force in a particular region. It means of ratios. A location quotient 1.0 means that a region has neither more nor less of that of the State or national industry than its overall volume of manufacturing would suggest. A quotient over 1.0 indicates that a particular region is industrially developed in relation to others. A quotient less than 1.0 suggests that an industry is less developed in that region.

Location quotient is a measure which can be applied to any region, cities and towns. This method is very useful in comparing different industries with one another. Formula for location quotient is as follows[2] :

Towns Employment in Industry

$$LQ = \frac{\text{Towns employment in all activities}}{\text{Karnatak State employment in industries}}$$

Karnatak State's employment in all activities. For example—the ratio of employment in industries to population for the Karnatak State and Bhadrawati town are taken. The location quotient of industrial employment for a town is determined by dividing its percentage by the State percentage.

$$\text{Towns Ratio} = \frac{\text{Bhadrawati town employment in industry}}{\text{Total Population of Bhadrawati}}$$

$$= \frac{10{,}000}{65{,}800} = 16.5\%$$

$$\text{States Ratio} = \frac{\text{Karnatak State employment in industry}}{\text{Total Population of Karnatak State}}$$

$$= \frac{11{,}30{,}600}{23{,}586{,}800} = 4.8\%$$

The location quotient of Bhadrawati in terms of its population and industrial employment is as follows :

$$\frac{16.5\%}{4.8\%} = 3.4\%$$

In this particular case the location quotient exceeds an index value of 1.0. So Bhadrawati has more than its share of the State's employment strength in industry.

Measurement of Industrial Concentration and Specialization

The measure is more or less the same as location quotient. The measure of concentration and specialization considers the difference between the industrial structure of the local area and that of the region as a whole. Various ways of developing this into a comparative index were tried and finally a 'coefficient of specialization' ($c\,S\,x$) for a given Employment Exchange Area (EEA) was adopted. The same is as follows[3] :

$$c\,S\,x = \tfrac{1}{2} \sum_{i=l}^{n} \left(\frac{100./r}{T_r} - \frac{100./r}{T_x} \right)$$

where 1_x=number employed in industry i in EEA x

T_x=total employed in all industries in EEA x

1_r=number employed in industries i in the region

T_r=total employed in all industries in the region

N=total number of all industries.

All deviations are summed irrespective of sign. Industries in the case are order totals. The use of Minimum List Heading (MLH) total was rejected on the ground that the advantage of a finer breakdown would be offset by the disadvantages inherent in using areas as small as EEA's for such an analysis. Moreover, Standard Industrial Classification (SIC) orders may be better indicators of occupational opportunity than MLH though this aspect has not been pursued in all the cases.

Role of Small Industries and Labour Intensive Technique in Economic Development

So far as the small scale industry is concerned it needs very careful planning especially for a developing country where a large scheme of capital goods production entails a high social cash and sets about a spiral of inflationary prices. In order to make such a scheme successful there should be simultaneous plan for a severe restriction of consumption in the initial phases. This can be explained by the following equation[4] :

$$C_t = Y_o\,(1-i)\,(1-r)\,t$$

$$C'_t = (1+r)\,t\,(1+r^1)\,t$$

where C_t=consumption at time t;

C'_t= change in consumption when rate of growth changes from $(1+r)$ to $(1+r^1)$;

i=share of productive accumulation in national income;

i=share of productive accumulation in the next period;

Y_o=national income;

$(1+r)$=rate of growth of national income;

$(1+r^1)$=rate of growth of national income when productive accumulation is raised from i to r^1.

The above equation can be represented as follows.

$$\log c_t = \log y_o + \log (1-i) + t \log (1+r)$$

$$\log c'_t = \log y_o + \log (1-i^1) + t \log (1+r^1)$$

It is obvious that with a higher rate of growth the level of consumption is less favourable in the initial period OP, but thereafter the level of consumption goes up after the point of intersection Q.

Measure of Accessibility in Industry

The measure of accessibility attempts at quantifying the relationship between any one area and all other areas within the region and this is concerned with external factors. It is based on the gravity model principle, incorporating two variables of size of centres and distance between centres. The argument is simple if an area is located around it with a number of accessibles located there having alternative employment, then it is less vulnerable to its own internal weakness than an area which is relatively isolated.

The index of the surface for any area is obtained thus[5] :

$$A_i = \sum_{j=1}^{n} \frac{E_j}{d_{ij}^2}$$

where A_i = The index of accessibility for EEA i

n = total number of EEAs.

E_j = total employment in EEA. j

d_{ij} = straight line distance between the geographical centroids of i and j.

The higher the index for any area, the more accessible are other centres of employment relative to the area.

Determination of Wages in Industry

Concentration is in the manufacturing sector because the data available allow further breakdown by two-digit industries according to the Standard Industrial Classification. We explain wages (W) in the ith region in the jth industry by the equation[6] :

$$W_{ij} = a + X_i + X_j + X S,$$

where a is the intercept,

X_i is the region effect,

X_j is the industry effect,

and X_s is a variable which measures the percentage of women to the total working force in each observation. The wage rate is regressed upon 24 independent variables of which 23 are dummy variables, the 24th variable is the sex variable.

Change in the Distribution of Manufacturing[7]

Column 1 $= y_d$; column 2 $= y_d - x_d \left[\frac{y}{x} \right]$

$$\text{Column 3} = \Sigma_j \left[y_{di} - x_{di} \ \frac{y_i}{x_i} \right]$$

$$\text{Column 4} = 2\text{—column 3} = \Sigma_i \left(\frac{x_{di}\, y_i}{x_i} \right) - \frac{x_d \cdot y}{x}$$

where x_{di} = employment in 1951 of industry i in division d.

x_d = employment in 1951 of all industries in division d.

x_i = employment in 1951 of industry i in NSW (New South Wales).

X = employment in 1951 of all industries in NSW X_{di}, y_d, y_i and y represent the same values in 1961.

This method shows the employment intensity in a particular time period in industries of any sector.

Index of Growth Industries in Manufacturing

One index was calculated for manufacturing industry, and the simple proportions of employment in extractive industries used as a second index. The proportion of total employment in manufacturing accounted for by these M.L.H.S. was calculated for each area and these proportions were then related to the regional average. The final index was derived thus[8] :

$$GMe = \frac{100\ Pe}{P_r}$$

where Pe = the number of employed in the defined growth industries as a percentage of total manufacturing in EEAe. P_r = The equivalent percentage of the region. GMe = index of growth potential in manufacturing industries in EEAe.

Thus areas with an index greater than 100 have a higher proportion of their total manufacturing employment in national growth industries than the region, which for those with an index of less than 100 the reverse is true.

Employment Change in Industry

The employment change in any industry in the metropolis can be expressed as the sum of three additive components (i) national change effect (N), (ii) composition effect (C) and (iii) differential growth effect (G). The national change effect will refer to the change which would have occurred if the metropolitan employment structure would have been the same as the national. Composition effect will refer to the employment change accounted for by the differences between the metropolitan employment structure and that of the nation.. The differential growth effect will refer to the change resulting from the differences in the rate of growth of employment for any industry in the metropolitan area and the same industry in the nation. The data used earlier in this paper to calculate the urban growth and urban service employment will be used for the analysis of the employment change in Delhi Metropolitan Area (Table 19.2). The following equations have been used to obtain the values of (N), (C) and (G).

$$N = K\,(X_{io}^{*} - X_{io}) \qquad \ldots(1)$$

$$C = K\,(X_{io} - X_{io}^{*}) + X_{ij}\left(\frac{X_{io}^{*}}{X_{io}} - 1\right) \qquad \ldots(2)$$

$$G = X_{ij}^{*} - \left(X_{ij} - \frac{X_{io}^{*}}{X_{io}}\right) \qquad \ldots(3)$$

where N=National change effect,

C=Composition effect,

G=Differential growth effect,

X_{io}=National employment in industry i in 1951,

X_{io}^{*}=National employment in industry i in 1961,

X_{ij}=Metropolitan employment in industry i in 1951,

X_{ij}^{*}=Metropolitan employment in industry i in 1961,

$$\frac{\text{Total employment in D.M.A. in 1951}}{\text{Total employment in India in 1951}} = 0.0143923,$$

TABLE 18.2 : **Delhi Metropolitan Area—Analysis of the Employment Growth**

Industries	Change in Employment	National Change Effect	Composition effect	Differential Growth Effect
Plantation Crops and Field Produce	329	10,053	—10,053	329
Forestry and Logging	546	1,775	— 1,656	427
Fishing	50	1,656	1,646	40
Livestock	15,90	4,3542	—39,681	—2,271
Mining	15,77	5,057	— 4,914	1,434

Source : After R.B. Singh, 1968.

Measurement of Economic Base of Metropolitan Areas by an Index of Local Specialization of Industry

To identify the urban economic base the indices of local specialization unadjusted and adjusted forms should be carved out. The former index is a simple ratio of an industries share of local employment relative to the industries share of the national employment or alternatively the localities share of the national employment and is measured by the following formula [10]:

$$\frac{\frac{e_i}{e_t}}{\frac{E_i}{E_t}} \quad \text{or} \quad \frac{\frac{e_i}{E_i}}{\frac{e_t}{E_t}}$$

where e_i = Local industrial employment,

e_t = Local total employment,

E_i = National industrial employment,

E_t = National total employment.

In essence this index indicates the importance of industry to the locality relative to the importance of the industry to the nation. When the value of index is one, local production per capita is equal to the national production per capita and, therefore, the local production is first sufficient to satisfy local consumption (demand)

and the locality neither exports nor imports. When the value of index is greater than one, it is presumed that the locality has surplus goods or services and exports this surplus. To inter-local specialization one requires a special combination of interrelated assumptions a coefficient of greater than one indicates that the locality's labour force is more heavily specialised to a particular industry than is true of the nation on the average.

The latter, adjusted form of specialization is considered as it prevents the downward bias and eliminates the overlaps between the local and national economy's specialised basic services. The formula considered is :

$$\frac{\dfrac{e_i}{e_t}}{\dfrac{E_i - e_i}{E_t - e_t}}$$

In the unadjusted form the index is of greater than one time specialised and has more than one time labour force in a particular industry than is actually needed. On the other hand, an index of higher than one in the adjusted form overstates the multiple by which the local industry employment exceeds its pro-rate share and understates the relative importance of the industry in the total economy. It is a measurement of the local uniqueness. The adjusted index will always be higher than the unadjusted form when the latter has an index of more than one, but will always be lower than the index one. The unadjusted index of local specialization has the added attraction of not only ranking industries according to degree of local specialization but, in addition, the unadjusted index serves as an indirect measure of "surplus" workers relative to total local industry employment. To demonstrate this characteristic, let us consider the simple hypothetical example in which assumption is that e_i equals 12, e_t equals 120, E_i equals 30 and E_t equals 1,200 then the values of the unadjusted and adjusted index are as follows :

Unadjusted Index of Local Specialization

$$\frac{\dfrac{e_i}{e_t}}{\dfrac{E_i}{E_t}} = \frac{\dfrac{12}{120}}{\dfrac{30}{1,200}} = 4$$

Adjusted Index of Local Specialization

$$\frac{\dfrac{e_i}{e_t}}{\dfrac{E_i - e_i}{E_t - e_t}} = \frac{\dfrac{12}{120}}{\dfrac{30 - 12}{1,200 - 120}}$$

The indices of 4 for the unadjusted index indicates that the locality is 4 times as specialised in industry as the nation as a whole and has 4 times as many workers as are needed for the locality's own consumption of the goods or service.

An Index of Surplus Workers

The index of surplus workers takes into account the role or importance and size of a local industry and indicates the importance of an industry to its locality relative to the importance of other industries to the same locality. This index could be measured by the following formula[11] :

$$S = \frac{e_i - e_t}{E_t} E_i$$

where S represents the absolute number of surplus workers in the industry i . e_i is local industry employment.

E_t is national total employment.

E_i is national industry employment.

The number of surplus workers in any local industry could never exceed or even equal industry's total local employment. In that case e_i itself is limited by the value of E_{ij}. The lower limit of this index is :

$$\frac{e_t}{E_t} E_i,$$

i, e_i; the deficit of local industry workers could never exceed the locality's share of the national industry employment and even this value would be attained only if there were absolutely no local employment in the industry in question. These limits may be summarized as follows :

$$\frac{e_t}{E_t} E_i \lessgtr \left[e_i - \frac{e_t}{E_t} E_i \right] < e_i \leqslant E_i$$

The distinction between the index of local specialization and the index of surplus workers as measures of the "basicness" of an industry in a locality's industrial structure is best demonstrated by recourse to the simply hypothetical example below :

City "A"	"Nation"
e_i 20	E_i 50
e_i 40	E_i 200
e_t 100	E_t 1000
Industry "i"	Industry "j"

$$\text{Index of Local Specialization} \begin{cases} \dfrac{e_i / e_t}{E_i / E_t} = 4.0 \quad \dfrac{e_i / e_t}{E_i / E_t} = 2.0 \end{cases}$$

$$\text{Index of Surplus Workers} \begin{cases} e_i - \dfrac{e_t}{E_t} E_i = 15 \\ e_j - \dfrac{e_t}{E_t} E_j = 20 \end{cases}$$

In this example though city "*A*" appears to be more specialized in industry "*i*" industry "*j*" contributes a greater number of surplus workers to its economic base. This is due to the larger size (locally) of industry "*j*" which more than offsets industry "*i*" in higher proportion of export activity to total activity.

Elasticity of Demand and Supply

The idea of supply and demand has a number of implications on price of industrial goods and the level of industrial development. The rate of response of demand and supply is measured by elasticity.

$$\text{Elasticity of Demand} = \frac{\text{Percentage Change in Quantity Bought}}{\text{Percentage Change in Price}}$$

$$\text{Elasticity of Supply} = \frac{\text{Percentage Change in Quantity Offered}}{\text{Percentage Change in Price}}$$

Elasticity varies from commodity and the living standard of a particular country.

Market Linkages of Manufacturing Centres

The number of linkages in a given market centre is influenced by competition with manufacturers at other locations. Thus four distinct elements are incorporated in the modes-distance decay, size of manufacturing centre, market size, competition and intervening opportunities for customer linkages. The basic model is structured to embody these elements in two stages, the first being defined by the equation[12]:

$$P_{ij} = \frac{\dfrac{A_j}{D_{ij}}}{\displaystyle\sum_{j=1}^{n} \frac{A_j}{D_{ij}^{a}}} \qquad \ldots(1)$$

where P_{ij} is the probability of a customer outlet at market centre "e" establishing a purchasing linkage with a manufacturer at producer location "j" ;

A_j = is the number of manufacturers operating at producer location "j" ;

D_{ij}= is a measure of the economic distance between "e" and "j" ;

a= is an exponent measuring the effect of distance upon market linkages.

The element of relative demand or differential market size is introduced by multiplication of the probability of interaction with each potential market centre by the number of outlets found there, as follows :

$$O_{ij} = P_{ij}, M_i \qquad \ldots(2)$$

where O_{ij} is the number of potential outlets found in market centre "e" which might be expected to interact with manufacturers located at producer location "j" ;

P_{ij}=is defined as in formula (1);

M_i=is the absolute number of outlets in centre 'i'.

This model investigates the presence or absence of market linkages between the manufacturers surveyed and possible market outlets, rather than the flow of goods between places, the measure of market size, "M_i" incorporates all possible consumers of manufactured products.

Orlocis Weighted Similarity Coefficient of Industry

L. Orloci's weighted similarity coefficient has been employed as a suitable index for comparing two establishments of the same industry. This coefficient is written as[13]

$$S_{jk} = \sum_{i=1}^{n} (X_{ij} - \bar{X}_i)\, X_{ik} - \bar{X}_i)$$

where $_j$ and $_k$ are two establishments of the same industry; S_{jk} is the similarity between $_j$ and $_k$, according to their arrays of linkage connection with various suppliers and customers in the conurbation; X_{ij}, X_{ik} are the scores (either 1 or 0) of linkage variable i for each establishment, j and k; and X_i is the average score of the linkage variable i for all the establishments in the sample.

Measures of Level of Industrialization

R.B. Mandal, while studying the manufacturing regions of North Bihar has taken, (a) the number of registered factories and (b) persons employed in manufacturing. His formula to measure the intensity of manufacturing on an anchal level in North Bihar is as follows[14]:

$$X = \frac{\text{No. of factories in a block} \times 100}{\text{No. of factories in North Bihar}}$$

$$Y = \frac{\text{No. of persons employed in manufacturing in a block}}{\text{No. of persons employed in manufacturing in North Bihar}}$$

$$\frac{X+Y}{2} = \text{Manufacturing intensity.}$$

The industrial workers of an industry can be compared with the principal economic activities of workers of the region. Such comparison measures the value of an industry in relation to regional and national levels. Several measures are generally used to understand such comparisons in which 'location quotient' is one of them.

Apportion Model

Based on the investment figures of city vis-a-vis the region, city's growth rate of outputs will be at least that of the region.[15]

Industrial employment at any time T_2

$$= \sum_{c=i}^{w} (1+r_v)\, Ec_i,\, T_1 \frac{VT_2\, ST_1}{ST_2\, VT_1}$$

where r_v=Growth rate of each industry.

$Ec_i\ T_1$=Employment in C city in i industry at time T_1,

VT_1=Investment of the region in time T_1,

VT_2=Invertment of the region in time T_2,

ST_1=Investment of the state in time T_1,

ST_2=Investment of the state in time T_2.

This apportion model shows the change in industrial growth of a particular industry or industries of a city or the region as a whole in relation to previous estimates and with the help of the data of two time periods.

Position of Engineering Industry of Central Bohemian Region : A Statistical Approach

The author[16] characterises the position of the engineering industry in the districts of central Bohemia using the basic indicants: number of employees, rough value of production and value of basic funds. First, he follows, in accordance with those indicants : the share of engineering in the districts' industry and the share of district engineering in the engineering industry of Czech lands. To define the position of engineering among the other branches of industry within an individual district he resorts to aggregate statistical characteristics. He suggests that such a statistical characteristic (defining the *size of the industry*) be the geometric mean calculated according to the following formula :

$$Q_{ij} = 100 \sqrt[3]{\frac{P_{ij}}{\frac{1}{mn}\sum\limits_{i,j} P_{ij}} \cdot \frac{V_{ij}}{\frac{1}{mn}\sum\limits_{i,j} V_{ij}} \cdot \frac{F_{ij}}{\frac{1}{mn}\sum\limits_{i,j} F_{ij}}}$$

for $i=1, 2, \ldots\ldots\ldots\ldots, m$.

$j=1, 2, \ldots\ldots\ldots\ldots, n$, where, Q_{ij} equals the geometric mean of the respective shares, it is a relative coefficient expressing the per cent deviation from the normative value (=100) of the j branch in the i district;

P_{ij}=the annual average of the number of employees;

V_{ij}=annual rough value of production at existing wholesale prices;

F_{ij}=value of basic funds at original purchase prices;

m=number of districts;

n=number of industrial branches;

The result demonstrates that the districts Prague-East and Prague-West have the biggest amounts of engineering industry. The author deals further with the size of engineering plants in individual districts.

Economic Evaluation of Mineral Deposits : In Relation to Industry

Let us assume that we have to make an economic evaluation of iron-ore deposit S. In this case, the change in national labour productivity will be expressed as the change in labour input in the production of a ton of pig iron.

The first phase of our analysis involves calculation of labour input per ton of pig iron if the ore sources of the iron and steel industry were to exclude deposit S. Initially we need to establish national market zones of particular deposits, using a formula proposed by A.V. Turkin [17]:

$$X+Y=Z$$

$$\frac{AK_1+fkA.K_1+ak_1Z+frp.Ha.K_1.Z-BK_2-fkB\,.k_2}{bkz+frp.HB.k_2+ak_1+frp.Ha.k_1}=Y\,(Km)$$

$$\frac{Bk_2+fkB\,.k_2+Bzk^2+frp.Hs.\,K_2.z-Ak_1-fkA\,.\,k_1}{ak_1+frp.Ha.K_1+Bk_2+frp.HB.k_2}=X(Km)$$

where Z is the distance between two contoured deposits :

A=the cost of a ton of ore at the originating rail-road station of deposit No. 1.

B=the cost of a ton of ore at the originating rail-road station of deposit No. 2.

a=the cost of one ton-kilometre from deposit No. 1.

b=the cost of one ton-kilometre from deposit No. 2.

k_1, k_2 ore coefficients reflecting the metal content of the ore.

KA, KB are capital inputs per ton of ore at deposits 1 and 2.
f is the coefficient of capital efficiency in the ore industry;
$f_r p$ is the coefficient of capital efficiency in rail transport;
X and Y are the distance between the deposits and the point of equal costs.

If we take A and B to mean not simply the cost of extraction and concentration of the ore, but the total cost, including that of the recultivation of the land, the conservation of water resources, etc., then the formula may be used of iron ore extraction, especially in case of open-pit mines. The next step would be to calculate the mean weighted labour input per ton of pig iron for the entire industry.

The second phase of analysis requires calculation of labour input if deposit S is included among the ore sources of the industry. As soon as deposit S is put into operation, its market zone will encompass certain users and the nationwide input of labour will change.

In other words, if in the first phase of the analysis the output of pig iron of iron and steel plants were M_1, N_1............Q_1, and unit cost were m_1, n_1,.........q_1, and the nationwide labour input would be :

$$C_1 = \frac{M_1.m_1 + N_1.n_1 + \ldots + Q_1.q_1}{M_1 + N_1 + \ldots + Q_1}$$

Then in the second phase, the input of labour per ton of pig iron would be :

$$C_2 = \frac{M_2.m_2 + N_2.n_2 + \ldots + Q_2.q_2}{M_2 + N_2 + \ldots\ldots + Q_2}$$

The ratio $\frac{C_2}{C_1}$ would thus yield the change in national labour input, and therefore, in the national productivity of labour, that would be the quantitative expression of the economic evaluation of deposit S, since the change from C_1 to C_2 was produced by the commercial development of deposit S.

The formulas collected on industrial geography are not exhaustive. Most of the statistics are related with the industrial production and its change in time, wages to the industrial labourer, and concentration, and dispersal of industries. The measures of locational quotient and accessibility are also considered.

REFERENCES

1. B.N. Asthana and S.S. Srivastava, *Applied Statistics of India*, Chaitanya Publishing House, Allahabad, 1962, pp. 184-85.
2. M.F. Karennavar, "Industrial Landscape of Bhadrawati town", 21st International Geographical Congress, 1968.
3. H. Gallacher, "Employment Vulnerability in the East Midlands", *The East Midland Geographer*, Vol. 5, Part 4, 1971.
4. S.C. Chatterji, Symposium on Industrialization in Developing Countries, Section-II, 1968, p. 5.
5. H. Gallacher, "Employment Vulnerability in the East Midlands", *The East Midland Geographer*, Vol. 5, Part 4, 1971, p. 209.
6. P.R.P. Coelho, "Illusory Regional Income Differentials", *The Annnals of Regional Science*, Vol. VIII, No. 2, 1974, p. 88.
7. M.I. Logan, "The Australian Scene Changes in the Distribution of New South Wales Manufacturing, 1951-61", *The Australian Geographer*, Vol. IX, No. 2, 1963, p. 123.
8. H. Gallacher, "Employment Vulnerability in the East Midlands", *The East Midland Geographer*, 1971, Vol. 5, Part, 8, pp. 204-215.
9. R.B. Singh, "Growth of Employment and Export Base and Change in the Local Employment Multiplier of Delhi Metropolitan Area", *The National Geographical Journal of India*, Vol. XIV, Part 4, 1968, pp. 243-245.
10. Tomkeen Kidwai, *Measurement of Economic Base of Metropolitan Areas*, p. 2.
11. J.P. Gibbs, *Urban Research Methods*, Von Nostrand East-West Press, New Delhi, 1966, p. 337.
12. P.J. Mcdermott, "Market Linkages and Spatial Monopoly in New Zealand Manufacturing", *New Zealand Geographer*, Vol. 30, No. 1, April, 1974, pp. 1-17(8).
13. M.J. Taylor and P.J. Wood, "West Midlands Metal Industries", *Transactions of the Council of British Geographers*, Vol. 59, 1973, p.136.
14. R.B. Mandal, "Manufacturing Region of North Bihar", *The National Geographical Journal of India*, Vol. XVI, Part 1, 1971, p. 50.
15. M. Mazhari, "Lesson on Summer Institute in Geography", Gauhati University, 1973. (Manuscript Received).
16. J. Brinke, "Position of Engineering Industry of Central Bohemian Region", Sbornik, Universite Karlova, Praha 1970, p. 52.
17. G.I. Gladkevich and A.T. Khrushchev, "Principles of an Economic Evaluation of Mineral Deposits for Purposes of Geographical Prediction", *Soviet Geography*, Vol. XV, No. 1, 1974, pp. 12-18.

NINETEEN

R. B. Mandal

TRANSPORT GEOGRAPHY

ALTHOUGH transport geography has a very wide scope as it includes roads, railways, airways and inland transport, in the field of quantitative geography each one of them has a very terse mathematical approach mainly by the engineers and it is not of any significance in regional analysis according to the need of geographers. Keeping in view these aspects some approaches of regional analysis have been presented here.

Dominant Type of Transportation

The three means of transport prevalent in the region, i.e., roads, railways and cart tracks are taken into consideration and the dominant type has been derived on the basis of the following formula for each region[1]:

$$D = \frac{LT \times 100 \ \& \ 10{,}000}{A}$$

where D=Dominant type of transport,

LT=Length of roads or railways or cart tracks,

100=Area in square kilometres,

10,000=Population,

A=Area of the physiographic region.

The highest values per 100 square kilometres of area and per 10,000 persons give us a fairly correct idea about the dominant type in each region. The result would have been otherwise if only one value, i.e., either area or population was considered for the simple reason that each region varies in size and population.

TABLE 19.1 : **Dominant Types of Transportation**

Region	Total Length of Roads in kms	Length of Roads per 100 sq. kms.	Length of Roads per 10,000 persons	Total Length of Railway in kms	Length of Rlys. 100 sq. kms.	Length of Rlys. 10,000 persons	Total Length of Cart Tracks in kms	Length of Cart Tracks per 100 sq. kms.	Length of Cart Tracks per 10,000 Persons
Hilly Crescent	550	9.16	19.9	48	0.83	1.8	300	5.0	10.8
Riverain Tract	839	10.5	10.05	185	2.30	3.3	818	10.0	14.5

Source : After H.M. Saxena, *Geography of Transport and Market Centres* p. 83.

Features of Highway Network

Definitions of traffic flows and capacity cost are presented in a list below :

X_i = It is presumed that traffic may be measured for the *i*th route in an amount X_i. The measurement might be the per day flow of tons of goods, passengers or passenger cars. The routes are those recognized in the highway network for planning, traffic control, and investment purposes, $i=1$ to n, which are the n routes in the system.

K_i=The highway development problem is to select capacity increments to be allocated to different places in the highway system during the development capacity inputs are made on the *ith* route in the amount K_i.

Y_i = 1 for an even traffic stream (at optimum speed) and is less than time 1 when traffic varies from time to time.

Traffic is influenced by the capacity of the *ith* route when the equation :

$$K_i = \frac{X_i}{Y_i}$$

holds.

P_i = Associated with each possible improvement in the cost of that input identified by P_i, the dollar cost of a capacity increment on the *ith* route. Total capacity costs are,[2]

$$\sum i P_i K_i$$

Degree of Connectivity

Method: 1

With the help of graph theory Prithar has suggested the degree of network for which the number of places is known can be expressed by the ratio of maximum number of routes to the observed number of routes, i.e.

$$\text{Degree of connectivity} = \frac{L^*}{L}$$

Prithar[3] has also suggested the formula for minimum degree of connectivity as $\frac{L^*}{m-l}$ and maximum degree

of connectivity as $\frac{L^*}{\frac{m(m-1)}{2}} = 1.$

where L^* is the maximum number of routes that could exist under an idealised and theoretical situation and L is the observed number of routes.

L^* is obtained by the formula

$$L^* = \frac{m(m-1)}{2}$$

where 'm' is the number of places (nodes) in the network.

$$= \frac{13(13-1)}{2} = 78 \text{ (here } m=13).$$

Therefore, actual degree of connectivity

$$= \frac{L^*}{L} = \frac{78}{38} = 2.05$$

(where L = observed number of routes which is 38).

Again the minimum connectivity among 13 nodes in a theoretical situation is given by :

$$\frac{L^*}{m-1} = \frac{78}{13-1} = \frac{78}{12} = 6.5$$

The matrix corresponding to the direct connections has been prepared. The matrix value 'one' is given if direct link exists and 'zero' if there is no direct link.[4]

TABLE 19.2 : **Direct Connections between the Transport Nodes**

Sl. No.	District	1	2	3	4	5	Total Connectivity
1.	Gauhati	0	1	1	1	1	4
2.	Dhubri	1	0	0	0	0	1
3.	Nowgong	1	0	0	0	0	1
4.	Tezpur	1	0	0	0	1	2
5.	N. Lakhim-pur	1	0	0	1	0	2
Total Connectivity		4	1	1	2	2	10

Source : After Dharitri Deka.

Method : 2

The degree of connectivity is the relative position of places on the network as measured with the 'associated number' of a place on the network. Associated number is the maximum number of edges from a given vertex to each of the other vertices. The central place on a network is one whose associated number is minimum. The maximum associated number, on the other hand, measures the diameter of the network. The degree of connectivity of the whole network is expressed as percentage of the maximum connectivity which is defined as [5]:

$$M_{ax}e = V\frac{(V-1)}{2} = 1$$

The degree of connectivity is obtained by dividing the maximum connectivity by the observed number of edges which may then be expressed in percentage for comparison purposes. The degree of connectivity is obtained by other ratio measures such as (*a*) (*B*) and (*r*) as well. The (*a*) index is an adjusted cyclomatic number which is one of the fundamental indices of graph theory. It is expressed as $d=e-v+P$ where *e* is the number of edges, *v* is the number of vertices and *P* is the number of non-connected subgraphs. The (*a*) index is obtained by the following formula :

$$d = \frac{e-v+P}{\frac{Iv(v-1)}{2} - V - 1}$$

It thus expresses the ratio between the observed number of circuits and the maximum number of circuits in a network. The index expressed as $\frac{e}{v}$ is the simplest measure describing connectivity. Disconnected or branching tree-like networks have values <1, a network having only one circuit has$=1$ and complicated networks have higher values. *r* index is independent of the number of vertices in a network. The values obtained between 0 and 1 express the various degrees of connectivity. It is calculated from the following formula :

$$r = \frac{e}{\frac{v\,(v-1)}{2}} \text{ for non planar,}$$

$$\text{and } r = \frac{e}{3\,(v-2)} \text{ for planar graphs.}$$

Similarly graph theoretic measures are helpful in assessing the accessibility of a vertex *i* to all other vertices on the network by adding up all the distances *d*.

Some of the ratios expressing the relations between the whole network and the individual routes give further insights into the structure of the network. *L* index expresses the ratio of the total route length of the network and the diameter length.

Value greater than 1 are suggestive of complicated networks. 0 values obtained by dividing the total traffic volume by the number of vertices reveals the average volume of freight per vertex. Iota *v* measure expressed as $i=\frac{M}{W}$ (where *M* is the total mileage and *W* observed number of vertices weighted by their functions) throws light on length and function of the net.

The graph theory in the evaluation of the structure of transport network analyses the degree of connectivity, extent, dispersion, and relative accessibility of points located on the network.

Air and Surface Movement Costs

In order to compare user costs for the various sites the research team first had to predict the amounts of traffic that would flow to the different airports from the various parts of the country. To do this they used the following model[6] :

$$T_{ij} = \frac{O_i \, D_j \, F_{ij}}{\sum\limits_{j} D_j \, F_{ij}}$$

where T_{ij}=The number of trips between zone i and airport j,

O_i=The total number of trips generated in zone i (i. e., the various part of England),

D_j=The total number of trips attracted to airport j.

and F_{ij}=The cost or friction of distance between i and j.

The model allocates passengers from regions i to airports j in direct proportion to the attractiveness of airports D_j, and in inverse proportion to distance from i to j. Using this model and an accessibility measure derived from it, the commission calculated the volume of traffic generated and its total costs. It reasoned that the more efficient the configuration of airports, the greater would be the total amount of traffic generated by it, and this additional traffic constituted a benefit that could be assigned to the more efficient system.

Gravity models have been used with considerable success to allocate given amounts of traffic among alternative destinations, but their ability to estimate the total amount of traffic generated by a network configuration is completely unproven. The above model could be called a 'loosely attraction constrained' model. O_i, D_j and F_{ij} are inputs to the model and must be provided by the programmer: T_{ij}, the number of trips between i and j, is the output. A recursive solution procedure is used which permits O_i, D_j and F_{ij} to vary within certain limits but the total amount of traffic generated by the system is O_i, D_j is a function of the population, income, and accessibility to airports, of the various zones of origin. 'Accessibility in calculated by

$$\text{Accessibility of zone } i = \frac{\sum_j (D_i \, F_i l)}{\sum_j D_j}$$

The calculation of the amount of traffic generated depends heavily both on the assumed capacities of the various airports, and on the assumption that the functional relationship between accessibility and traffic will remain constant while the volume of traffic will increase ten fold. The first assumption is arbitrary and the second may be seriously questioned.

Variations in Distance Friction

In course of devising a method to predict interregional road freight flows in Great Britain a gravity model of total commodity flows between 78 regions was calibrated. It was thought that the distance friction parameter obtained when flows between all origin/destination pairs of regions were used in the calibration might be concealing considerable regional variations which should be investigated. A similar model was calibrated taking each origin in turn to reveal departures from the overall distance friction coefficient. The differences which came to light suggested that it would be better to work on the basis of individual commodities.

The Gravity Model

Suppose we have a method of estimating the tonnages of all goods originating in a region (O_i) and of goods terminating in a region (D_j), and that we know the distance between[7] these regions (D_{ij}), we wish to find a way to estimate the flows between pairs of regions (T_{ij}).

Wilson (1968) has provided a theoretical justification for using a gravity model of the following form :

$$\overline{T}_{ij} = K.\ O_i,\ D_j - d_{ij} - \beta \qquad \text{...(1)}$$

where $\overline{T}_{ij}$ = an estimated flow between i and j,

β = a parameter to be estimated empirically,

K = a constant of proportionality that ensures that

$$\Sigma_i \, \Sigma_j \, T_{ij} = \Sigma_j \quad D_j = \Sigma_i \, O_i$$

However this expression of the model does not ensure that

$$-\Sigma_i \overline{T}_{ij}=D_j \qquad \ldots(2)$$

$$\text{and } j\ \overline{T}_{ij}=D_j \qquad \ldots(3)$$

Equations 2 and 3 are the same as the constraints of the linear programming transportation problem (Hadley, 1962) that the rows and columns of the predicted matrix of flows should sum to the amount available in associated origins and required in associated destinations.

These constraints can be satisfied in the gravity model by replacing K by two factors A_i and B_j :

$$T_{ij}=A_i.\ \beta_j.\ O_i.\ D_j.\ d_{ij}-\beta \qquad \ldots(4)$$

This expression for T_{ij} is then substituted into each of the constraints in turn :

$$\Sigma_i\ A_i.\ \beta_j.\ O_i.\ D_j.\ d_{ij}\ \beta = D_j$$

which can be rearranged to give :

$$\beta_j = \frac{1}{\Sigma_i\ A_i,\ O_i\ d_{ij}-\beta} \qquad \ldots(5)$$

similarly substitution into 3 gives :

$$A_i = \frac{1}{\Sigma_j\ \beta_j.\ D_j-d_{ij}-\beta} \qquad \ldots(6)$$

5 and 6 can be solved iteratively to find suitable values for A_i and β_j.

Calibration of the Gravity Model

If one has an origin/destination matrix of actual flows for some time period, an initial estimate of β can be obtained using least squares. O_i and D_j are given by the row and column sums of the matrix and we can measure d_{ij}. If, from 1, we regress

$log\left(\frac{T_{ij}}{O_i-D_j}\right)$ on $log\ d_{ij}$, then

from the estimating equation

$$log\left(\frac{\overline{T}_{ij}}{O_i-D_j}\right)=a-b\ log\ d_{ij} \qquad \ldots(7)$$

antilog a gives the constant K, and b gives β.

This volume of β can be applied to 5 and 6 and the A_{is} and B_{js} can be obtained by an iterative procedure. A_{is} can be set equal to one and a set of β_{js} calculated from 5. These can be substituted into 6 and the procedure repeated till the values converge to an equilibrium. The values of A_i and β_j are substituted into 4 along with β to produce an estimated matrix of flows. This matrix is compared with the actual pattern-comparing individual flows, frequency distributions of length of haul and the mean length of haul for each matrix. β may be varied systematically about the least squares value until the analyst judges the best fit on all criteria is obtained.

TABLE 19.3 : **Variations in Distance Friction Coefficients by Zone of Origin**

(Least Squares Estimates)

Centroid	β	Centroid	β
1. Morpeth	—2.0	4. Newcastle	—1.7
2. Tynemouth	—1.6	5. Sunderland	—2.7
3. Gateshead	—1.8	6. Durham	—2.2

Mean Speed Index

Let d_{ij} be the straight line distance in kilometres between two cities, i and j 'and t_{ij} be the number of hours to go from i to j and return. Then the mean speed (MS) to travel from city i to all the others and return is :

$$M_{si} = \sum_{j=1}^{n} d_{ij} / \sum_{j=1}^{n} t_{ij} \quad \text{with } j = i$$

where n is the number of cities (65). This index does not depend upon the topological position in the network and so estimates the impact on a given city of public investments.

Index of Investment Rentability of a Transportation Surface

Such a measure is still unstable because cities grow. For each time period the population in each of the 65 cities is a percentage of the total population of those centres. This index can be written mathematically for each i and each period j as[8] ;

$$1_{ij}=(P_{ij}/\sum_{i}P_{ij}) : dx^2_{ij}+dy^2_{ij})\frac{1}{2}$$

with P_{ij} the population, and dx and dy the partial derivatives of the fitted surface.

This index can be altered by changes in the quality of network or by movement of inhabitants from one city to another.

Accessibility and Technological Change

For each of these villages, an accessibility value was calculated[9] :

$$A=d_{ir}^{-1}+\frac{1}{.2b_r^2}$$

where d_{ir} is the distance in miles by the shortest foot route to the nearest primary or secondary road; b_r is the mean number of regularly scheduled passenger buses per daylight hour in one direction on that road, and .2 in an arbitrarily established weighting coefficient.

TABLE 19.4 : **Accessibility Scores of Study Villages**

Accessibility Variable	A	Barabanki B	C
dir	1.5	2.5	7.0
b_r	2.50	0.42	2.50
A	0.99	0.53	0.46

Testing the Hypothesis

If rate of technological modernization is a function of accessibility, as hypothesized, one would expect less accessible villages to be less modernized and to have adopted fewer of the technological innovations espoused by the Community Development Programme. Because of the crude nature of data, the small size of the sample, and the qualitative nature of the hypothesis, this expectation was tested by an ordinal measure of association.

Consider a set of K trials, each consisting of a subset of n elementary events, the observations of which may be ordinally scaled. Three statistics may be defined—D_k, S_k and T_k—which measure monotonicity of ordering. For an observed ordering of sequential observations, it is possible to measure the number of departures (inversions) from a natural (or hypothesized monotonic) ordering. Let this number of departure be denoted by D. Where Y is a trial, the K trial D_k is given by :

$$D_k = \sum_{y=1}^{P} D_y$$

D_k, as a random variable, treats each of the permutations of order as equally likely. Since the exact sampling distribution of D_k is relatively simply found for small samples, and a normal approximation may be used for larger samples, an observed D_k may be used for a significance test.

S_k may be defined as :

$$S_k = m_{ax}(D_k) - 2D_k$$

and T_k, a k trial adaptation of Kendall's T is :

$$T_k = \frac{S_k}{m_{ax}\,(D_k)}$$

T_k gives values analogous to those of a rank correlation coefficient.

Since in this study, the number of inversions is of primary interest in comparing the results of experiments with k and n held constant, D_k will be used as the test statistic. Statistically significant values of D_k with k-4 and n-3 at conventional confidence levels are :

a	Significant D_k	Equivalent T_k
.05	0	1.00
	1	0.83
	2	0.67
.01	0	1.00

Six modernisation indices were constructed for aspects of agricultural modernization, each involving Thurstone scaling of the relative contributions of different items and practices to increasing agricultural productivity. Even interpreting all the scores as inversions, four indices show ordinal relationships which correspond significantly with the accessibility ordering.

TABLE 19.5 : **Technological Development of Villages (Primary Indices)**

Modernization Category	Barabanki *A*	*B*	*C*
Irrigation	2.5	1.0	0.5
Fertilizer	2.5	0.5	1.5
Cultivating Equipments	1.5	0.0	0.0
Improved seeds	3.8	1.5	0.0
Aggregate	9.3	3.0	0.0

Trip Generation Model

$T=a+b_1 X_1+6_2 X_2+b_3 X_3+6_4 X_4+......X$ etc.,

T=Trip per person (or vehicle) per day,

a=Intercept on taxis,

b_1, b_2, b_3 etc.=Multiple regression coefficients of trips per person (or vehicle of different factors),

X_1, X_2, X_3 etc.=Are different factors considered.

Graph Theory as Tool for Network Studies

A network can be defined as a set of geographic locations interconnected in a system by a number of (Kansk, 1963) routes. The network is characterized by the following statements :

1. A network has a finite number of places,
2. Each route joins two different places,
3. At the most only one route may join a pair of places,
4. No distinction is made between initial and the terminal points of routes; in other words routes are two-way.

Accessibility : (A_i)

It indicates spatial relation between a given vertex and the remaining vertices in the network; in other words it measures geometrical centrality of centres in the network. Mathematically it can be represented as :

$$A_i = \sum_{j=1}^{n} D_{ij}$$

where D is the distance between i and j, i. e. for the ith vertex this may be calculated by the above formula.

The Use of Matrices

The information given in a graph can be represented in terms of matrices. Consider the following matrix (corresponding to Fig. 19.1).[10]

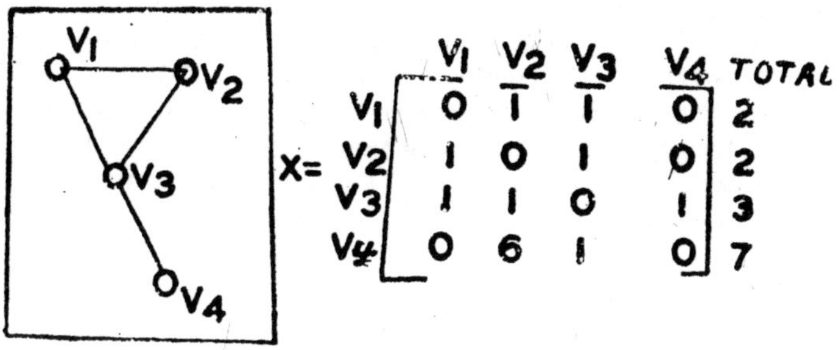

FIG. 19.1 : **Terms of Matrices**

Where a place is directly connected by an edge with another place, the value i is given to both the cells (i.e. v_{ij} and v_{ij}), and when they are not connected directly by an edge, both cells take values of zero. Obviously, therefore, the resultant matrix is system.

The row totals will indicate the connectiveness of the vertex. Thus in the above matrix V_3 is the most connected vertex. These matrices can also be powered. The powered matrix indicates the number of ways one can travel between given places in two three ....n steps. A matrix powered to n represents the number of n step connections available to different vertices.

REFERENCES

1. H.M. Saxena, *Geography of Transport and Market Centres : A Case Study of Hadaoti Plateau*, S. Chand and Co. Ltd., New Delhi, p. 83.
2. W.L. Garrison, B.J.L. Berry, D.F. Marble, J.D. Nystuen and R.L. Morrill, "Studies of Highway Development and Geographic Change", University of Washington Press, Seattle, 1959, p. 29.
3. S.I. Hullur, and B.N. Sinha, "Accessibility of Roads in Mysore State", *The National Geographical Journal of India*, Vol. XVII, Part, 2-3, 1971, p. 82.
4. P. Deka, "Passenger Traffic in North-East India : A Diagnostic Survey", *Journal of the North-East India Geographical Society*, Vol. V, No. 1-2, 1973, p. 76.
5. Jagdish Singh, "Research Methodology in Spatial Functional Interaction", *The Deccan Geographer*, Vol. XII, No. 2, 1974, pp. 122-123.
6. J. Adams, "Westminister: The Fourth London Airport", *Area*, No. 2, 1970, pp. 3-4.
7. Patrick O. Sullivan, "Variations in Distance Friction in Great Britain", *Area*, No. 2, 1970, pp. 36-37.
8. Bernard Merchand, "Deformation of a Transportation Surface", *Annals of the Association of American Geographers*, Vol. 63, No. 4, 1973, p. 515.
9. T.J. Wilbanks, "Accessibility and Technological Change," *Annals of the Association of American Geographers*, Vol. 62, No. 3, 1972, p. 431.
10. H. Ramachandran "Analysis of Network Geometry : An Evaluation of Connectivity Indices", *Geographical Review of India*, Vol. 37, No, 1, 1975, pp. 81-86.

APPENDICES

APPENDIX I

RANDOM SAMPLING NUMBERS

1	2	3	4	5	6	7	8	9
20 17	42 28	23 17	59 66	38 61	02 10	86 10	51 55	92 52
74 49	04 49	03 04	10 33	53 70	11 54	48 63	94 60	94 49
94 70	49 31	38 67	23 42	29 65	40 88	78 71	37 18	48 64
22 15	78 15	69 84	32 52	32 54	15 12	54 02	01 37	38 37
93 29	12 18	27 30	30 55	91 87	50 57	58 51	49 36	12 53
45 04	77 97	36 14	99 45	52 95	69 85	03 83	51 87	85 56
44 91	99 49	89 39	94 60	48 49	06 77	64 72	59 26	08 51
16 23	91 02	16 96	47 59	89 65	27 84	30 92	63 37	26 24
04 50	65 04	65 65	82 42	70 51	55 04	61 47	88 83	99 34
32 70	17 72	03 61	66 26	24 71	22 77	88 33	17 78	08 92
03 64	59 07	42 95	81 39	06 41	22 81	92 34	51 90	39 08
62 49	00 90	67 86	83 48	31 83	19 07	67 68	49 03	27 47
61 00	95 86	98 36	14 03	48 88	51 07	33 40	06 86	33 76
89 03	90 49	28 74	21 04	09 96	60 45	22 03	52 80	01 79
01 72	33 85	52 40	60 07	06 71	89 27	14 29	55 24	85 79
27 56	49 79	34 34	32 32	60 53	91 17	33 26	44 70	92 14
49 05	74 48	10 55	35 25	24 28	20 22	35 66	66 34	26 35
49 74	37 25	97 26	33 94	42 23	01 28	59 58	92 69	03 66
20 26	22 43	88 08	19 85	08 12	47 65	65 63	56 07	97 85
48 87	77 96	43 39	76 93	08 79	22 18	54 55	93 75	97 26
08 72	87 46	75 73	00 11	27 07	05 20	30 85	22 21	04 67
95 97	98 62	17 27	31 42	64 71	46 22	32 75	19 32	20 99
37 99	57 31	70 40	46 55	46 12	24 32	36 74	69 20	72 10
05 79	58 37	85 33	75 18	88 71	23 44	54 28	00 48	96 23
55 85	63 42	00 79	91 22	29 01	41 49	51 50	36 65	26 11
67 28	96 25	68 36	24 72	03 85	49 24	05 69	64 86	08 19
85 86	94 78	32 59	51 82	86 43	73 84	45 60	89 57	06 87
40 10	60 09	05 88	78 44	63 13	58 25	37 11	18 47	75 62
94 55	89 48	90 80	77 80	26 89	87 44	23 74	66 20	20 19
11 63	77 77	23 20	33 62	62 19	29 03	94 15	56 37	14 09
64 00	26 04	54 55	38 57	94 62	68 40	26 04	24 25	03 61
50 94	13 23	78 41	60 58	10 60	88 46	30 21	45 98	70 96
66 98	37 96	44 13	45 05	34 59	75 85	48 97	27 19	17 85
66 91	42 83	60 77	90 91	60 90	79 62	57 66	72 28	08 70
33 58	12 18	02 07	19 40	21 29	39 45	90 42	58 84	85 43

Appendix I (Contd.)

1	2	3	4	5	6	7	8	9
52 49	70 16	72 40	73 05	50 90	02 04	98 24	05 30	27 25
74 98	93 99	78 30	79 47	96 62	45 58	40 37	80 76	84 41
50 26	54 30	01 88	69 57	54 45	69 88	23 21	05 69	93 44
49 46	61 89	33 79	96 84	28 34	19 35	28 73	39 59	56 34
19 64	13 44	78 39	73 88	62 03	36 00	25 96	86 76	67 90
64 17	47 67	87 59	81 40	72 61	14 00	28 28	55 86	23 38
18 43	97 37	68 97	56 56	57 95	01 88	11 89	48 07	42 07
65 58	60 87	51 09	95 61	15 53	66 81	66 88	44 75	37 01
79 90	31 00	91 14	85 65	31 75	43 15	45 93	64 78	34 53
07 23	00 15	59 05	16 09	94 42	20 40	63 76	65 67	34 11
90 98	14 24	01 51	96 46	30 32	33 19	00 14	19 28	40 51
53 82	62 02	21 82	34 13	41 03	12 85	65 30	00 97	56 30
98 17	26 15	04 50	76 25	20 33	54 84	39 31	23 33	59 64
08 91	12 44	82 40	30 62	45 50	64 54	65 17	89 25	59 44
37 21	46 77	84 87	67 39	85 54	97 37	33 41	11 74	90 50

APPENDIX II

VALUES OF MOD. t WITH A PROBABILITY OF BEING EXCEEDED IN RANDOM SAMPLING

ν=number of degrees of freedom

d. f.		Level of Significance			
	0·50	0·10	0·05	0·02	0·01
1	1·000	6·31	12·71	31·82	63·66
2	0·816	2·92	4·30	6·97	9·92
3	0·765	2·35	3·18	4·54	5·34
4	0·741	2·13	2·78	3·75	4·60
5	0·727	2·02	2·57	3·37	4·03
6	0·718	1·94	2·45	3·14	3·71
7	0·711	1·90	2·36	3·00	3·50
8	0·706	1·86	2·31	2·90	3·36
9	0·703	1·83	2·26	2·82	3·25
10	0·700	1·81	2·23	2·76	3·17
11	0·697	1·80	2·20	2·72	3·11
12	0·695	1·78	2·18	2·68	3·06
13	0·694	1·77	2·16	3·65	3·01
14	0·692	1·76	2·14	2·62	2·98
15	0·691	1·75	2·13	2·60	2·95
16	0·690	1·75	2·12	2·58	2·92
17	0·689	1·74	8·11	2·57	2·90
18	0·688	1·73	2·10	2·58	2·88
19	0·688	1·73	2·09	2·54	2·86
20	0·687	1·73	2·09	2·53	2·85
21	0·686	1·72	2·08	2·52	2·83
22	0·686	1·72	2·07	2·51	2·82
23	0·685	1·71	2·07	2·50	2·81
24	0·685	1·71	2·06	2·49	2·80
25	0·684	1·71	2·06	2·40	2·79
26	0·784	1.71	2·06	2·48	2·78
27	0·684	1·70	2·05	2·47	2·77
28	0·683	1·70	2·05	2·47	2·76
29	0·683	1·70	2·04	2·46	2·76
30	0·683	1·70	2·04	2·46	2·75
35	0·682	1·69	2·03	2·44	2·72
40	0·681	1·68	2·02	2·42	2·71
45	0·680	1·68	2·02	2·41	2·69
50	0·679	1·68	2·01	2·40	2·68
60	0·678	1·67	2·00	2·39	2·66
00	0·674	1·64	1·96	2·33	2·58

APPENDIX III

SIGNIFICANCE POINTS OF THE VARIANCE—RATIO F

5 per cent points

ν_1 / ν_2	1	2	3	4	5	6	8	12	24	∞
1	2	3	4	5	6	7	8	9	10	11
1	161·40	199·5	215·7	224·6	230·2	234·0	238·9	243·9	249·0	254·3
2	18·51	19·00	19·16	19·25	19·30	19·33	19·37	19·41	19·45	19·50
3	10·13	9·55	9·28	9·12	9·01	8·94	8·84	8·74	8·64	8·53
4	7·70	6·94	6·59	6·39	6·26	6·16	6·04	5·91	5·77	5·63
5	6·61	5·79	5·41	5·19	5·05	4·95	4·82	4·68	4·53	4·36
6	5·99	5·14	4·76	4·53	4·39	4·28	4·15	4·00	3·84	3·67
7	5·59	4·74	4·35	4·12	3·97	3·87	3·73	3·57	3·41	3·23
8	5·32	4·46	4·07	3·84	3·69	3·58	3·44	3·28	3·12	2·93
9	5·12	4·26	3·86	3·63	3·48	3·37	3·23	3·07	2·90	2·71
10	4·96	4·10	3·71	3·48	3·33	3·22	3·07	2·91	2·74	2·54
11	4·84	3·98	3·59	3·36	3·20	3·09	2·95	2·79	2·61	2·40
12	4·75	3·88	3·49	3·26	3·11	3·00	2·85	2·69	2·50	2·30
13	4·67	3·80	3·41	3·18	3·02	2·92	2·77	2·60	2·42	2·21
14	4·60	3·74	3·34	3·11	2·96	2·85	2·70	2·53	2·35	2·13
15	4·54	3·68	3·29	3·06	2·9[illegible]	2·79	2·64	2·48	2·29	2·07

1	2	3	4	5	6	7	8	9	10	11
16	4·49	3·63	3·24	3·01	2·85	2·74	2·59	2·42	2·24	2·01
17	4·45	3·59	3·20	2·96	2·81	2·70	2·55	2·38	2·19	1·96
18	4·41	3·55	3·16	2·93	2·77	2·66	2·51	2·34	2·15	1·92
19	4·38	3·52	3·13	2·90	2·74	2·63	2·48	2·31	2·21	1·88
20	4·35	3·49	3·10	2·87	2·71	2·60	2·45	2·28	2·08	1·84
21	4·32	3·47	3·07	2·84	2·68	2·57	2·42	2·25	2·05	1·81
22	4·30	3·44	3·05	2·82	2·66	2·55	2·40	2·23	2·03	1·78
23	4·28	3·42	3·03	2·80	2·64	2·53	2·38	2·20	2·00	1·76
24	4·26	4·40	3·01	2·78	2·62	2·51	2·36	2·18	1·91	1·73
25	4·24	3·38	2·99	2·76	2·60	2·49	2·34	2·16	1·96	1·71
26	4·22	3·37	2·98	2·74	2·59	2·47	2·32	2·15	1·95	1·69
27	4·21	3·35	2·96	2·73	2·57	2·46	2·30	2·13	1·93	1·67
28	4·20	3·34	2·95	2·71	2·56	2·44	2·29	2·12	1·91	1·65
29	4·18	3·33	2·93	2·70	2·54	2·43	2·28	2·10	1·90	1·64
30	4·17	3·32	2·92	2·69	2·53	2·42	2·27	2·09	1·89	1·62
40	4·08	3·23	2·84	2·61	2·45	2·34	2·18	2·00	1·79	1·51
60	4·00	3·15	2·76	2·52	2·37	2·25	2·10	1·92	1·70	1·89
120	3·92	3·07	2·68	2·45	2·29	2·17	2·02	1·83	1·61	1·25
∞	3·84	2·99	2·60	2·37	2·21	2·09	1·94	1·75	1·52	1·00

Appendix III (Contd.)

1% Points of F

v_1 / v_1		2	3	4	5	6	7	8	9	∞
1	2	3	4	5	6	7	8	9	10	11
1	40·52	4999·5	54·03	56·25	57·64	58·59	59·82	61·06	62·35	63·66
2	98·50	99·00	99·17	99·25	99·30	99·33	99·37	99·42	99·46	99·50
3	34·12	30·82	29·46	28·71	28·24	27·91	27·49	27·05	26·60	26·13
4	21·20	18·00	16·69	15·98	15·52	15·21	14·80	14·37	13·93	13·45
5	16·86	13·27	12·06	11·39	10·97	10·67	10·29	9·89	9·47	9·02
6	13·75	10·92	9·76	9·15	8·75	8·47	8·10	7·72	7·31	6·88
7	12·25	9·55	8·48	7·85	7·45	7·19	6·84	6·47	6·07	5·65
8	11·26	8·65	7·59	7·01	6·66	6·37	6·03	5·67	5·28	4·86
9	10·56	8·02	6·99	6·42	6·04	5·80	5·47	5·11	4·73	4·31
10	10·04	7·56	6·55	5·99	5·66	5·39	5·06	4·71	4·33	3·91
11	9·65	7·21	6·22	5·87	5·32	5·07	4·74	4·40	4·02	3·60
12	9·33	6·93	5·95	5·41	5·06	4·82	4·50	4·16	3·78	3·36
13	9·07	6·70	5·74	5·21	4·86	4·62	4·30	3·96	3·59	3·17
14	8·86	6·51	5·56	5·04	4·69	4·46	4·14	3·80	3·43	3·00
15	8·68	6·36	5·42	4·89	4·56	4·32	4·00	3·67	3·29	2·87
16	8·53	6·23	5·29	4·77	4·44	4·20	3·89	3·55	3·18	2·75
17	8·40	6·11	5·18	4·67	4·34	4·10	3·79	3·46	3·08	2·65

Appendix III (Contd.)

1	2	3	4	5	6	7	8	9	10	11
18	8·29	6·01	5·09	4·58	4·25	4·01	3·71	3·37	3·00	2·59
19	8·18	5·93	5·01	4·50	4·17	3·94	3·63	3·30	3·92	2·49
20	8·10	5·85	4·94	4·43	4·10	3·87	3·56	3·23	2·86	2·42
21	8·02	5·78	4·87	4·37	4·04	3·81	3·51	3·17	2·80	2·36
22	7·95	5·72	4·82	4·31	3·99	3·76	3·45	3·12	2·75	2·31
23	7·88	5·66	4·76	4·26	3·94	3·71	3·41	3·07	2·70	2·26
24	7·82	5·61	4·72	4·22	3·90	3·67	3·36	3·03	2·66	2·21
25	7·77	5·57	1·68	4·18	3·85	3·63	3·32	2·99	2·62	2·17
26	7·72	5·53	4·64	4·14	3·82	3·59	3·29	2·96	2·58	2·10
27	7·68	5·49	4·60	4·11	3·78	3·56	3·26	2·93	2·45	2·13
28	7·64	5·45	4·57	4·07	3·75	3·53	3·23	2·90	2·52	2·06
29	7·60	5·42	4·54	4·04	3·73	3·50	3·20	2·87	2·49	2·03
30	7·56	5·39	4·51	4·02	3·70	3·47	3·17	2·84	2·47	2·01
40	7·31	5·18	4·31	3·83	3·51	3·29	2·99	2·66	2·29	1·80
60	7·08	4·98	4·13	3·65	3·34	3·12	2·82	2·50	2·12	1·60
120	6·85	4·79	3·95	3·48	3·17	2·96	2·66	2·34	1·95	1·38
∞	6·64	4·60	3·78	3·32	3·02	2·89	2·51	2·18	1·97	1·00

APPENDIX IV

VALUES OF x^2 WITH PROBABILITY *P* OF BEING EXCEEDED IN RANDOM SAMPLING

ν=number of degrees of freedom

ν \ P	0·20	0·10	0·05	·02	0·01
1	1·64	2·71	3·84	5·41	6·63
2	3·32	4·61	5·99	7·82	9·21
3	4·64	6·25	7·81	9·84	11·34
4	5·90	7·78	9·49	11·67	13·28
5	7·29	9·24	11·07	13·39	15·09
6	8·56	10·64	12·59	15·03	16·81
7	9·80	12·02	14·07	16·62	18·48
8	11·03	13·36	15·51	18·17	20·09
9	12·24	14·68	16·92	19·68	21·67
10	13·44	15·99	18·31	21·16	23·21
11	14·63	17·28	19·68	22·62	24·73
12	15·81	18·55	21·03	24·05	26·22
13	16·98	19·81	22·36	25·47	27·69
14	18·15	21·06	23·68	26·87	29·14
15	19·31	22·31	25·00	28·26	30·58
16	20·46	23·54	26·30	29·63	32·00
17	21·62	24·77	27·59	30·99	33·41
18	22·76	25·99	28·87	32·35	34·81
19	23·90	27·20	30·14	33·69	36·19
20	25·04	28·41	31·41	35·02	37·57
21	26·17	29·62	32·67	36·34	38·93
22	27·30	30·81	33·92	37·66	40·29
23	28·43	32·01	35·17	38·97	41·64
24	29·55	33·20	36·42	40·27	42·98
25	30·68	34·38	37·65	41·57	44·31
26	31·80	35·56	38·88	41·86	45·64
27	32·91	36·74	40·11	44·14	46·96
28	34·03	37·92	41·34	45·42	48·28
29	35·14	39·09	42·56	46·69	49·59
30	36·25	40·26	43·77	47·96	50·89

INDEX